Fixed Point Theory in Probabilistic Metric Spaces

Mathematics and Its Applications

Volume 536

Fixed Point Theory in Probabilistic Metric Spaces

by

Olga Hadžić and Endre Pap

Institute of Mathematics,
University of Novi Sad, Yugoslavia

KLUWER ACADEMIC PUBLISHERS
DORDRECHT / BOSTON / LONDON

A C.I.P. Catalogue record for this book is available from the Library of Congress.

ISBN 978-90-481-5875-1

Published by Kluwer Academic Publishers,
P.O. Box 17, 3300 AA Dordrecht, The Netherlands.

Sold and distributed in North, Central and South America
by Kluwer Academic Publishers,
101 Philip Drive, Norwell, MA 02061, U.S.A.

In all other countries, sold and distributed
by Kluwer Academic Publishers,
P.O. Box 322, 3300 AH Dordrecht, The Netherlands.

Printed on acid-free paper

Printed in the Netherlands

Contents

Introduction

Fixed point theory in probabilistic metric spaces can be considered as a part of Probabilistic Analysis, which is a very dynamic area of mathematical research. A primary aim of this monograph is to stimulate interest among scientists and students in fixed point theory in probabilistic metric spaces, [268].

The main text is self-contained for the reader with a modest knowledge of the metric fixed point theory [52].

The introduction of the general concept of statistical metric spaces is owed to K. Menger, who dealt with probabilistic geometry. The new theory of fundamental probabilistic structures was developed later on by many authors.

The main influence upon the development of the theory of probabilistic metric spaces is owed to B. Schweizer and A. Sklar and their coworkers. The excellent monograph [268] enables the reader to obtain much information about the importance of this field of mathematics to mathematicians as well as to applied scientists.

Several themes run through the present book. The first is the theory of triangular norms (t-norms), which is closely related to the fixed point theory in probabilistic metric spaces. The recent development of the theory of t-norms, presented in [164], has had a strong influence upon the fixed point theory in probabilistic metric spaces. In Chapter 1 some basic properties of t-norms are presented and several special classes of t-norms are investigated. The class of q-convergent t-norms is introduced and the results obtained are applied in the later chapters to the fixed point theory.

Chapter 2 is an overview of some basic definitions and examples from the theory of probabilistic metric spaces that are of importance for the next chapters. An example of a Menger space connected with decomposable measures [217] is included.

Chapters 3, 4, and 5 deal with some single-valued and multi-valued probabilistic versions of Banach contraction principle. In [272] the notion of a probabilistic q-contraction $f : S \to S$ was introduced and it was proved that if $(S, \mathcal{F}, \min)$ is a complete Menger space the set of all fixed points of the mapping f, Fix(f), is non-empty. Without some growth conditions on the mapping $\mathcal{F} : S \times S \to \Delta^+$ (the set of distribution functions) the problem of the existence of a unique fixed point of a probabilistic q-contraction $f : S \to S$ for complete Menger spaces $(S, \mathcal{F}, T)$ with a continuous t-norm T is completely solved. A necessary and sufficient condition that the set Fix(f) is non-empty, in this case, is that T is the so called t-norm of H-type.

In the paper [310] Tardiff imposed some growth conditions on the mapping $\mathcal{F}$:

$S \times S \to \Delta^+$, which enabled the class of t-norms T to be enlarged for which the set Fix(f) is non-empty, where $f : S \to S$ and $(S, \mathcal{F}, T)$ is a Menger space. Namely, Tardiff's fixed point theorem holds if $T \geq T_{\mathbf{L}}$, where $T_{\mathbf{L}}$ is defined by $T_{\mathbf{L}}(x,y) = \max(x+y-1, 0)$. Using some results about infinitary operations from the theory of t-norms, under some stronger growth conditions on $\mathcal{F}$, we obtain some probabilistic versions of Banach contraction principles where T satisfies $T_{\mathbf{L}} \geq T$, or even is incomparable with $T_{\mathbf{L}}$.

A special class of probabilistic metric spaces is that of random normed spaces [273] $(S, \mathcal{F}, T)$, where $T \geq T_{\mathbf{L}}$. It is known that a random normed space $(S, \mathcal{F}, T)$, where T is a continuous t-norm, belongs to the class of topological vector spaces (with the (ε, λ)-topology), which are, in general, not necessarily locally convex spaces.

A very important function space $S(\Omega, \mathcal{A}, P)$ of all classes of equivalence of measurable mappings $X : \Omega \to M$, where $(\Omega, \mathcal{A}, P)$ is a probability measure space and $(M, \|\cdot\|)$ is a separable normed space, is a random normed space $(S(\Omega, \mathcal{A}, P), \mathcal{F}, T_{\mathbf{L}})$, where $\mathcal{F}$ is defined by $\mathcal{F}_{X,Y}(u) = P(\|X - Y\| < u) \quad (X, Y \in S, u \geq 0)$.

From the very extensive literature [96] on fixed point theory in not necessarily locally convex topological vector spaces we pick in Chapter 6 only some basic results and apply them to the fixed point theory in random normed spaces. An important notion of an admissible subset of a topological vector space, introduced by V. Klee [160], has the fundamental influence for the fixed point theory in topological vector spaces.

It is still an open question whether every nonempty, compact, and convex subset of a topological vector space is admissible? The positive answer to this question would solve the problem of whether Tychonoff's fixed point theorem holds in a general topological vector space. Namely, it is known that Tychonoff's fixed point theorem is valid if $f : K \to K$ is a continuous mapping and K a nonempty, compact, convex and admissible subset of a topological vector space.

Many important function spaces are admissible non-locally convex topological vector spaces and $(S(\Omega, \mathcal{A}, P), \mathcal{F}, T_{\mathbf{L}})$ with the (ε, λ)-topology, where $(M, \|\cdot\|)$ is $(\mathbb{R}^n, \|\cdot\|_{\mathbb{R}^n})$, belongs to this class of topological vector spaces [174]. Chapter 6 contains some applications of the fixed point theory in random normed spaces to the space $(S(\Omega, \mathcal{A}, P), \mathcal{F}, T_{\mathbf{L}})$.

The authors of many results presented in the book are participants in the Seminar on Analysis at the University of Novi Sad and the Seminar on Probabilistic Analysis at the West University in Timisoara. Some of their results are also included in the book [40].

Much has been left unsaid—KKM theory [286] in probabilistic metric spaces with some additional structures, variational inequalities, minimization problems, and some recent results from the fixed point theory in not necessarily locally convex topological spaces.

It is our hope that the material presented in the book will be enough to lead scientists and students into the field and to stimulate them to investigate further this challenging field.

Fixed point theory in probabilistic metric spaces is still in its formative stage. We feel that within the not too distant future we shall witness the development of a unified theory of random equations that will be based, partly, on concepts and tools of the fixed point theory in probabilistic metric spaces.

The authors thank all colleagues who read and commented upon the manuscript. Our special thanks go to R. Mesiar, A. Takači, and V. Radu for their help with the proof reading of some parts of the book and very useful suggestions. We also want to thank the Institute of Mathematics in Novi Sad for our working conditions. We thank for the support of the National Project in the Fields of Basic Research "Mathematical models of nonlinearity, uncertainty and decision" supported by Ministry of Science, Technology and Development of Serbia.

The authors wish to express their gratitude to the Kluwer Publishing House for accepting this book in their series, especially to Dr Liesbeth Mol, Ms Angelique Hempel, and Ms Patricia deVries.

Novi Sad, May 2001

The Authors

Chapter 1

Triangular norms

Triangular norms first appeared in the framework of probabilistic metric spaces in the work K. Menger [186] (see [164, 268]). It turns also out that this is a crucial operation in several fields, e.g., in fuzzy sets, fuzzy logics (see [164]) and their applications, but also, among other fields, in the theory of generalized measures [164, 168, 201, 217, 321] and in nonlinear differential and difference equations [217].

In this chapter we recall some basic results from the theory of triangular norms, see [164], giving also some new results closely related to the theory of the fixed point in probabilistic metric spaces. Section 1.1 is devoted to basic definitions and some important examples of triangular norms and conorms. In section 1.2 we give some basic properties of triangular norms. The ordinal sum construction of t-norms is given in section 1.3 and the basic section 1.4 is devoted to the representation of continuous t-norms. Especially important classes of t-norms with left-continuous diagonals in the point $(1,1)$ and t-norms of H-type are given in section 1.5 and section 1.6, respectively. Section 1.7 is devoted to the practical conditions for comparison of t-norms with respect to the pointwise order, as well as the domination relation between t-norms, especially important for the product of probabilistic metric spaces. Countable infinitary composition of t-norms with respect to special classes of t-norms is investigated in section 1.8.

1.1 Triangular norms and conorms

Definition 1.1 *A triangular norm (t-norm for short) is a binary operation on the unit interval* $[0,1]$*, i.e., a function* $T : [0,1]^2 \to [0,1]$ *such that for all* $x, y, z \in [0,1]$

the following four axioms are satisfied:

(T1) $T(x,y) = T(y,x)$ (commutativity);

(T2) $T(x,T(y,z)) = T(T(x,y),z)$ (associativity);

(T3) $T(x,y) \leq T(x,z)$ *whenever* $y \leq z$ (monotonicity);

(T4) $T(x,1) = x$ (boundary condition).

The commutativity (T1), the monotonicity (T3), and the boundary condition (T4) imply that for each t-norm T and for each $x \in [0,1]$ the following boundary conditions are also satisfied:

$$\begin{aligned} T(x,1) &= T(1,x) = x, \\ T(x,0) &= T(0,x) = 0, \end{aligned}$$

and therefore all t-norms coincide on the boundary of the unit square $[0,1]^2$.

The monotonicity of a t-norm T in its second component (T3) is, together with the commutativity (T1), equivalent to the (joint) monotonicity in both components, i.e., to

$$T(x_1,y_1) \leq T(x_2,y_2) \quad \text{whenever } x_1 \leq x_2 \text{ and } y_1 \leq y_2. \tag{1.1}$$

Definition 1.2 *If T is a t-norm, then its dual t-conorm* $\mathbf{S} : [0,1]^2 \to [0,1]$ *is given by*

$$\mathbf{S}(x,y) = 1 - T(1-x, 1-y).$$

It is obvious that a t-conorm is a commutative, associative, and monotone operation on $[0,1]$ with unit element 0.

Example 1.3 The following are the four basic t-norms together with their dual t-conorms:

(i) *Minimum* $T_{\mathbf{M}}$ and *maximum* $\mathbf{S}_{\mathbf{M}}$ given by

$$\begin{aligned} T_{\mathbf{M}}(x,y) &= \min(x,y), \\ \mathbf{S}_{\mathbf{M}}(x,y) &= \max(x,y); \end{aligned}$$

(ii) *Product* $T_{\mathbf{P}}$ and *probabilistic sum* $\mathbf{S}_{\mathbf{P}}$ given by

$$\begin{aligned} T_{\mathbf{P}}(x,y) &= x \cdot y, \\ \mathbf{S}_{\mathbf{P}}(x,y) &= x + y - x \cdot y; \end{aligned}$$

(iii) *Lukasiewicz t-norm* $T_{\mathbf{L}}$ and *Lukasiewicz t-conorm* $\mathbf{S_L}$ given by

$$T_{\mathbf{L}}(x,y) = \max(x+y-1,0),$$
$$\mathbf{S_L}(x,y) = \min(x+y,1);$$

(iv) *Weakest t-norm (drastic product)* $T_{\mathbf{D}}$ and *strongest t-conorm* $\mathbf{S_D}$ given by

$$T_{\mathbf{D}}(x,y) = \begin{cases} \min(x,y) & \text{if } \max(x,y)=1, \\ 0 & \text{otherwise,} \end{cases} \tag{1.2}$$

$$\mathbf{S_D}(x,y) = \begin{cases} \max(x,y) & \text{if } \min(x,y)=0, \\ 1 & \text{otherwise.} \end{cases} \tag{1.3}$$

Example 1.4 (i) The family $(T_\lambda^{\mathbf{F}})_{\lambda\in[0,+\infty]}$ of *Frank t-norms* is given by

$$T_\lambda^{\mathbf{F}}(x,y) = \begin{cases} T_{\mathbf{M}}(x,y) & \text{if } \lambda=0, \\ T_{\mathbf{P}}(x,y) & \text{if } \lambda=1, \\ T_{\mathbf{L}}(x,y) & \text{if } \lambda=+\infty, \\ \log_\lambda\left(1+\dfrac{(\lambda^x-1)(\lambda^y-1)}{\lambda-1}\right) & \text{otherwise.} \end{cases}$$

The family $(\mathbf{S}_\lambda^{\mathbf{F}})_{\lambda\in[0,+\infty]}$ of *Frank t-conorms* is given by

$$\mathbf{S}_\lambda^{\mathbf{F}}(x,y) = \begin{cases} \mathbf{S_M}(x,y) & \text{if } \lambda=0, \\ \mathbf{S_P}(x,y) & \text{if } \lambda=1, \\ \mathbf{S_L}(x,y) & \text{if } \lambda=+\infty, \\ 1-\log_\lambda\left(1+\dfrac{(\lambda^{1-x}-1)(\lambda^{1-y}-1)}{\lambda-1}\right) & \text{otherwise.} \end{cases} \tag{1.4}$$

(ii) The family $(T_\lambda^{\mathbf{Y}})_{\lambda\in[0,+\infty]}$ of *Yager t-norms* is given by

$$T_\lambda^{\mathbf{Y}}(x,y) = \begin{cases} T_{\mathbf{D}}(x,y) & \text{if } \lambda=0, \\ T_{\mathbf{M}}(x,y) & \text{if } \lambda=+\infty, \\ \max\left(0,1-((1-x)^\lambda+(1-y)^\lambda)^{\frac{1}{\lambda}}\right) & \text{otherwise.} \end{cases} \tag{1.5}$$

The family $(\mathbf{S}_\lambda^{\mathbf{Y}})_{\lambda\in[0,+\infty]}$ of *Yager t-conorms* is given by

$$\mathbf{S}_\lambda^{\mathbf{Y}}(x,y) = \begin{cases} \mathbf{S_D}(x,y) & \text{if } \lambda = 0, \\ \mathbf{S_M}(x,y) & \text{if } \lambda = +\infty, \\ \min\left(1, (x^\lambda + y^\lambda)^{\frac{1}{\lambda}}\right) & \text{otherwise.} \end{cases} \tag{1.6}$$

(iii) The family $(T_\lambda^{\mathbf{SW}})_{\lambda\in[-1,+\infty]}$ of *Sugeno–Weber t-norms* is given by

$$T_\lambda^{\mathbf{SW}}(x,y) = \begin{cases} T_\mathbf{D}(x,y) & \text{if } \lambda = -1, \\ T_\mathbf{P}(x,y) & \text{if } \lambda = \infty, \\ \max\left(0, \dfrac{x+y-1+\lambda xy}{1+\lambda}\right) & \text{otherwise.} \end{cases} \tag{1.7}$$

The family $(\mathbf{S}_\lambda^{\mathbf{SW}})_{\lambda\in[-1,+\infty]}$ of *Sugeno–Weber t-conorms* is given by

$$\mathbf{S}_\lambda^{\mathbf{SW}}(x,y) = \begin{cases} \mathbf{S_D}(x,y) & \text{if } \lambda = -1, \\ \mathbf{S_M}(x,y) & \text{if } \lambda = \infty, \\ \min(1, x+y+\lambda xy) & \text{otherwise.} \end{cases} \tag{1.8}$$

(iv) Another interesting t-norm is the *nilpotent minimum* $T^{\mathbf{nM}}$ given by

$$T^{\mathbf{nM}}(x,y) = \begin{cases} \min(x,y) & \text{if } x+y>1, \\ 0 & \text{otherwise.} \end{cases} \tag{1.9}$$

If, for two t-norms T_1 and T_2, the inequality $T_1(x,y) \leq T_2(x,y)$ holds for all $(x,y) \in [0,1]^2$, then we say that T_1 is *weaker* than T_2 or, equivalently, that T_2 is *stronger* than T_1.

We shall write $T_1 < T_2$ whenever $T_1 \leq T_2$ and $T_1 \neq T_2$, i.e., if $T_1 \leq T_2$, but $T_1(x_0,y_0) < T_2(x_0,y_0)$ holds for some $(x_0,y_0) \in [0,1]^2$.

As a consequence of (1.1) we obtain for each $(x,y) \in [0,1]^2$

$$T(x,y) \leq T(x,1) = x,$$
$$T(x,y) \leq T(1,y) = y.$$

Since, for all $(x,y) \in (0,1)^2$ trivially $T(x,y) \geq 0 = T_{\mathbf{D}}(x,y)$, we get for an arbitrary t-norm T

$$T_{\mathbf{D}} \leq T \leq T_{\mathbf{M}},$$

i.e., $T_{\mathbf{D}}$ is weaker and $T_{\mathbf{M}}$ is stronger than any other t-norm.

As it is easy to see that $T_{\mathbf{L}} < T_{\mathbf{P}}$, we obtain the following ordering of the four basic t-norms

$$T_{\mathbf{D}} < T_{\mathbf{L}} < T_{\mathbf{P}} < T_{\mathbf{M}}.$$

If T is a t-norm, then an element $x \in [0,1]$ with $T(x,x) = x$ is called an *idempotent element* of T. It is immediate that 0 and 1 are idempotent elements (the so called *trivial idempotent elements*) for each t-norm T. The set of idempotent elements is equal to $[0,1]$ in the case of the minimum $T_{\mathbf{M}}$, and $\{0\} \cup (0.5,1]$ in the case of the nilpotent minimum $T^{\mathbf{nM}}$; all the other t-norms mentioned so far only have trivial idempotents.

An interesting question is whether a t-norm is determined uniquely by its values on the diagonal or some other subset of the unit square. The boundary t-norms $T_{\mathbf{D}}$ and $T_{\mathbf{M}}$ are completely determined by their values on the diagonal $\{(x,x) \mid x \in (0,1)\}$ of the (open) unit square.

Proposition 1.5 *(i) The minimum $T_{\mathbf{M}}$ is the only t-norm satisfying $T(x,x) = x$ for all $x \in (0,1)$ (i.e., each $x \in [0,1]$ is an idempotent element).*

(ii) The weakest t-norm $T_{\mathbf{D}}$ is the only t-norm satisfying $T(x,x) = 0$ for all $x \in (0,1)$.

Proof. If for a t-norm T we assume $T(x,x) = x$ for each $x \in (0,1)$, then whenever $y \leq x < 1$ the monotonicity (T3) implies

$$y = T(y,y) \leq T(x,y) \leq \min(x,y) = y.$$

Together with (T1) and the boundary conditions this gives exactly $T = T_{\mathbf{M}}$.

If we assume $T(x,x) = 0$ for all $x \in (0,1)$, then we obtain for each $y \in [0,x)$

$$0 \leq T(x,y) \leq T(x,x) = 0,$$

hence yielding $T = T_{\mathbf{D}}$. □

Observe that the relationship of continuous t-norms and their diagonals is completely characterized in [188].

1.2 Properties of t-norms

It is well known that a real function of two variables with domain $[0,1]^2$ can be continuous in each variable without being continuous on $[0,1]^2$. But triangular norms (and conorms, of course) are exceptions to this.

Proposition 1.6 *A t-norm T is continuous if and only if it is continuous in its first component, i.e., if for each $y \in [0,1]$ the one-place function*

$$T(\cdot, y) : [0,1] \to [0,1], \quad x \mapsto T(x,y),$$

is continuous.

Proof. If a function from $[0,1]^2$ into $[0,1]$ is continuous, then it is obviously continuous in each component.

Conversely, if T is continuous in the first component, fix $(x_0, y_0) \in [0,1]^2$, $\varepsilon > 0$ and let $(x_n)_{n\in\mathbb{N}}$ and $(y_n)_{n\in\mathbb{N}}$ be sequences in $[0,1]$ converging to x_0 and y_0, respectively. From this we can construct four monotone sequences $(a_n)_{n\in\mathbb{N}}, (b_n)_{n\in\mathbb{N}}, (c_n)_{n\in\mathbb{N}}$ and $(d_n)_{n\in\mathbb{N}}$ such that for all $n \in \mathbb{N}$ we have

$$\begin{array}{lll} a_n \leq x_n \leq b_n & \text{and} & (a_n)_{n\in\mathbb{N}} \nearrow x_0, \quad (b_n)_{n\in\mathbb{N}} \searrow x_0, \\ c_n \leq y_n \leq d_n & \text{and} & (c_n)_{n\in\mathbb{N}} \nearrow y_0, \quad (d_n)_{n\in\mathbb{N}} \searrow y_0. \end{array}$$

The continuity in the first component and the commutativity of T imply the continuity of the function $T(x_0, \cdot)$, which means that there exists an $N \in \mathbb{N}$ such that, as a consequence of the monotonicity of T, for all $n \geq N$

$$T(x_0, y_0) - \varepsilon < T(x_0, c_N) \leq T(x_0, y_n) \leq T(x_0, d_N) < T(x_0, y_0) + \varepsilon.$$

Since also the two functions $T(\cdot, c_N)$ and $T(\cdot, d_N)$ are continuous there is a number $M \in \mathbb{N}$ such that for all $m \geq M$ and $n \geq N$ (again taking into account the monotonicity of T) we obtain

$$\begin{aligned} T(x_0, c_N) - \varepsilon &< T(a_M, c_N) \\ &\leq T(x_m, y_n) \\ &\leq T(b_M, d_N) \\ &< T(x_0, d_N) + \varepsilon. \end{aligned}$$

Putting $K = \max(M, N)$, then for all $k \geq K$ we have

$$T(x_0, y_0) - 2\varepsilon < T(x_k, y_k) < T(x_0, y_0) + 2\varepsilon,$$

proving that $(T(x_k, y_k))_{k\in\mathbb{N}}$ converges to $T(x_0, y_0)$, i.e., that the t-norm T is continuous in (x_0, y_0). □

In an analogous way we can prove the following result.

Proposition 1.7 *A t-norm T is left-continuous (right-continuous) if and only if it is left continuous (right continuous) in its first component, i.e., if for each $y \in [0,1]$ and for each sequence $(x_n)_{n\in\mathbb{N}} \in [0,1]^{\mathbb{N}}$ we have, respectively,*

$$\sup_{n\in\mathbb{N}} T(x_n, y) = T\left(\sup_{n\in\mathbb{N}} x_n, y\right),$$

$$\inf_{n\in\mathbb{N}} T(x_n, y) = T\left(\inf_{n\in\mathbb{N}} x_n, y\right).$$

Definition 1.8 *(i) A t-norm T is said to be strictly monotone if it is strictly increasing on* $(0,1]^2$ *as a function from* $[0,1]^2$ *into* $[0,1]$ *or, equivalently, if (taking into account the commutativity (T1) and the boundary condition (T4))*

$$T(x,y) < T(x,z) \quad \text{whenever } x \in (0,1) \text{ and } y < z. \tag{1.10}$$

(ii) A t-norm T is called strict if it is continuous and strictly monotone.

Among the four basic t-norms presented in Example 1.3, only the product $T_{\mathbf{P}}$ is a strict t-norm. The minimum $T_{\mathbf{M}}$ and the Łukasiewicz t-norm $T_{\mathbf{L}}$ are continuous but not strictly monotone.

The t-norm T defined by

$$T^{\Delta}(x,y) = \begin{cases} \frac{xy}{2} & \text{if } \max(x,y) < 1, \\ xy & \text{otherwise,} \end{cases} \tag{1.11}$$

is strictly monotone but not continuous.

If T is a t-norm, $x \in [0,1]$ and $n \in \mathbb{N} \cup \{0\}$ then we shall write

$$x_T^{(n)} = \begin{cases} 1 & \text{if } n = 0, \\ T\left(x_T^{(n-1)}, x\right) & \text{otherwise.} \end{cases}$$

If T is a strictly monotone t-norm then for each $x \in (0,1)$ we have $0 < T(x,x) < x$ and, as a consequence, the sequence $(x_T^{(n)})_{n\in\mathbb{N}}$ is strictly decreasing.

Proposition 1.9 *A t-norm T is strictly monotone if and only if the cancellation law holds, i.e., if* $T(x,y) = T(x,z)$ *and* $x > 0$ *imply* $y = z$.

Proof. Obviously, the strict monotonicity (1.10) of T implies the validity of the cancellation law. Conversely, the strict monotonicity follows from the cancellation law together with the monotonicity (T3). □

Definition 1.10 *A t-norm T is called Archimedean if for all* $(x,y) \in (0,1)^2$ *there is an integer* $n \in \mathbb{N}$ *such that*

$$x_T^{(n)} < y.$$

Note that if a left-continuous t-norm T is Archimedean then it is necessarily continuous, see [170].

The product $T_{\mathbf{P}}$, the Łukasiewicz t-norm $T_{\mathbf{L}}$ (which is not strictly monotone and, hence, not strict), and the weakest t-norm $T_{\mathbf{D}}$ are all Archimedean (observe that the latter is not continuous), the minimum $T_{\mathbf{M}}$, however, is not an Archimedean t-norm.

Proposition 1.11 *A t-norm T is Archimedean if and only if for each $x \in (0,1)$ we have*

$$\lim_{n\to\infty} x_T^{(n)} = 0. \tag{1.12}$$

Proof. If T is Archimedean then for each $x \in (0,1)$ and each $\varepsilon > 0$ we have $0 \leq x_T^{(n_0)} < \varepsilon$ for some $n_0 \in \mathbb{N}$ and, because of the monotonicity (T3), we even have $0 \leq x_T^{(n)} < \varepsilon$ for all natural numbers $n \geq n_0$, thus implying (1.12). Conversely, suppose that for a t-norm T and for an arbitrary $x \in (0,1)$ we have $\lim_{n\to\infty} x_T^{(n)} = 0$. Then for each $y \in (0,1)$ there exists a natural number n such that $x_T^{(n)} < y$, i.e., T is Archimedean. □

At least for right continuous t-norms it is possible to characterize the Archimedean property (1.12) by their diagonal mapping:

Theorem 1.12 *(i) If T is an Archimedean t-norm then for each $x \in (0,1)$ we have*

$$T(x,x) < x. \tag{1.13}$$

(ii) If T is a right continuous t-norm and if for each $x \in (0,1)$ the property (1.13) is satisfied, then T is Archimedean.

Proof. The assumption that for some $x \in (0,1)$ we have $T(x,x) = x$ would imply $x_T^{(n)} = x$ for all $n \in \mathbb{N}$, contradicting that T is Archimedean, thus proving (i).
In order to show (ii), suppose that (1.13) holds for all $x \in (0,1)$, and fix an arbitrary element $u \in (0,1)$. By the monotonicity (T3) of T the sequence

$$\left(u_T^{(2^k)}\right)_{k\in\mathbb{N}}$$

is non-increasing and, consequently, has some limit $u_0 \leq u < 1$. The right continuity of T implies

$$T(u_0,u_0) = \lim_{k\to\infty} T\left(u_T^{(2^k)}, u_T^{(2^k)}\right) = \lim_{k\to\infty} \left(u_T^{(2^k)}\right)_T^{(2)} = \lim_{k\to\infty} u_T^{(2^{k+1})} = u_0,$$

which, since (1.13) holds for all $x \in (0,1)$, only leaves the possibility $u_0 = 0$. Then, because of Proposition 1.11 T is Archimedean. □

An immediate consequence of Theorem 1.12(ii) is that each strict t-norm T is Archimedean. However, there are strictly monotone t-norms which are not Archimedean (see [164, 287]). Such t-norms are necessarily non-continuous. Note that there are left-continuous t-norms of such type with dense set of discontinuity points, [287]. Note also that a t-norm with non-trivial idempotent elements cannot be Archimedean.

The assumption of right continuity in Theorem 1.12(ii) is necessary. The following example shows that, for t-norms which are not right continuous, (1.13) for all $x \in (0,1)$ does not necessarily imply the Archimedean property (1.12) or the strict monotonicity (1.10).

Example 1.13 ([164, 268]) *The t-norm T given by*

$$T(x,y) = \begin{cases} 0 & \text{if } (x,y) \in [0,0.5]^2, \\ 2(x-0.5)(y-0.5)+0.5 & \text{if } (x,y) \in (0.5,1]^2, \\ \min(x,y) & \text{otherwise.} \end{cases}$$

is obviously not right continuous. It only has trivial idempotent elements (i.e., (1.13) holds for all $x \in (0,1)$), but it is neither Archimedean nor strictly monotone.

Definition 1.14 *(i) A t-norm T is called nilpotent if it is continuous and if each element $x \in (0,1)$ is nilpotent, i.e., if there exists some $n \in \mathbb{N}$ such that $x_T^{(n)} = 0$.*

(ii) An element $x \in (0,1)$ is called a zero divisor of T if there exists some $y \in (0,1)$ such that $T(x,y) = 0$.

Obviously, each nilpotent element of a t-norm T is a zero divisor of T. For the t-norm $T^{\mathbf{nM}}$ each $x \in (0,1)$ is a zero divisor but each $x \in (0.5,1]$ is an idempotent element, and therefore it cannot be a nilpotent element of $T^{\mathbf{nM}}$.

However, if a t-norm T has a zero divisor then it also has nilpotent elements: if $T(a,b) = 0$ for some $a > 0$ and $b > 0$, then for $c = \min(a,b) > 0$ we get $T(c,c) = 0$.

A t-norm T has no zero divisors if and only if for each $x \in (0,1]$ we have $T(x,x) > 0$.

It turns out that strict and nilpotent t-norms can be completely characterized using Theorem 1.12(ii).

Theorem 1.15 *Let T be a continuous Archimedean t-norm. Then the following are equivalent:*

(i) T is nilpotent.

(ii) There exists some nilpotent element of T.

(iii) There exists some zero divisor of T.

(iv) T is not strict.

Proof. It is trivial that (i) implies (ii) and that (ii) implies (iv). We already know that (ii) and (iii) are equivalent. In order to show that (iv) implies (i), assume that T is continuous and Archimedean but not strict. Then there are numbers $u, v, w \in [0,1]$with $u > 0$ and $v < w$ such that $T(u,v) = T(u,w)$. Since we also have $T(v,w) \leq v < w = T(1,w)$, the continuity of T implies that there is a $z \in [v,1)$ such that $v = T(z,w) = T(w,z)$. Then we have

$$T(u,w) = T(u,v) = T(u,T(w,z)) = T(T(u,w),z)$$

and, by induction, for each $n \in \mathbb{N}$

$$T(u,w) = T(T(u,w), z_T^{(n)}).$$

Using the continuity and Proposition 1.11 we get

$$\begin{aligned} T(u,w) &= \lim_{n\to\infty} T(T(u,w), z_T^{(n)}) \\ &= T(T(u,w), \lim_{n\to\infty} z_T^{(n)}) \\ &= T(T(u,w),0) \\ &= 0. \end{aligned}$$

Hence, u and v are zero divisors of T. Then $b = \min(u,w) \in (0,1)$ is a nilpotent element of T. Now, for an arbitrary $a \in (0,1)$ the Archimedean property ensures that $a_T^{(n_0)} < b$ for some $n_0 \in \mathbb{N}$ and, consequently,

$$0 \leq a_T^{(2n_0)} \leq b_T^{(2)} = 0,$$

showing that T is nilpotent. □

1.3 Ordinal sums

A method of construction a new t-norm from a system of given t-norms is based on algebraic result from the semigroup theory, see [68, 164, 178, 197].

Theorem 1.16 *Let $(T_k)_{k\in K}$ be a family of t-norms and let $((\alpha_k, \beta_k))_{k\in K}$ be a family of pairwise disjoint open subintervals of the unit interval $[0,1]$ (i.e., K is an at most countable index set). Consider the linear transformations $\{\varphi_k : [\alpha_k, \beta_k] \to [0,1]\}_{k\in K}$ given by*

$$\varphi_k(u) = \frac{u - \alpha_k}{\beta_k - \alpha_k}. \tag{1.14}$$

Then the function $T : [0,1]^2 \to [0,1]$ defined by

$$T(x,y) = \begin{cases} \varphi_k^{-1}(T_k(\varphi_k(x), \varphi_k(y))) & \text{if } (x,y) \in (\alpha_k, \beta_k)^2, \\ \min(x,y) & \text{otherwise,} \end{cases} \tag{1.15}$$

is a triangular norm.

Proof. The function T obviously is commutative and fulfills the boundary conditions.

To check the monotonicity of T, we have to show the monotonicity of the functions $T(x,\cdot):[0,1]\to[0,1]$ for all $x\in[0,1]$.

If $x\notin\bigcup_{k\in K}(\alpha_k,\beta_k)$ then for each $y\in[0,1]$ we have $T(x,y)=\min(x,y)$, therefore $T(x,\cdot)$ is monotone.

If $x\in(\alpha_{k_0},\beta_{k_0})$ for some $k_0\in K$ then we have

$$T(x,y)=\begin{cases} y & \text{if } y\in[0,\alpha_{k_0}],\\ \varphi_k^{-1}(T_k(\varphi_k(x),\varphi_k(y))) & \text{if } y\in(\alpha_{k_0},\beta_{k_0}),\\ x & \text{if } y\in[\beta_{k_0},1].\end{cases}$$

Now, because for all $y\in(\alpha_{k_0},\beta_{k_0})$ we have

$$\alpha_{k_0}=\varphi_{k_0}^{-1}(0)\le T(x,y)\le\varphi_{k_0}^{-1}(\min(\varphi_{k_0}(x),\varphi_{k_0}(y)))\le x,$$

and the monotonicity of T_{k_0} implies the monotonicity of $T(x,\cdot)$, thus concluding the proof of the monotonicity of T.

Note that the boundary conditions of T_k imply $T(x,y)=\varphi_k^{-1}(T_k(\varphi_k(x),\varphi_k(y)))$ not only for $(x,y)\in(\alpha_k,\beta_k)^2$, but also for all $(x,y)\in[\alpha_k,\beta_k]^2$.

To show the associativity of T, we distinguish the following four possible cases for the values $x,y,z\in[0,1]$.

Case I: $x,y,z\in[\alpha_{k_0},\beta_{k_0}]$ for some $k_0\in K$. Then the associativity of T_{k_0} implies $T(x,T(y,z))=T(T(x,y),z)$.

Case II: $x,z\in[\alpha_{k_0},\beta_{k_0}]$ and $y\notin[\alpha_{k_0},\beta_{k_0}]$ for some $k_0\in K$. Then we have

$$\begin{aligned} T(x,T(y,z)) &= T(x,\min(y,z))\\ &= \min(T(x,y),T(x,z))\\ &= \min(\min(x,y),T(x,z))\\ &= \min(y,T(x,z))\\ &= \min(\min(y,z),T(x,z))\\ &= \min(T(y,z),T(x,z))\\ &= T(\min(x,y),z)\\ &= T(T(x,y),z), \end{aligned}$$

where the equalities follow from the definition and the monotonicity of T.

Case III: ($x, y \in [\alpha_{k_0}, \beta_{k_0}]$ and $z \notin [\alpha_{k_0}, \beta_{k_0}]$) or ($y, z \in [\alpha_{k_0}, \beta_{k_0}]$ and $x \notin [\alpha_{k_0}, \beta_{k_0}]$) for some $k_0 \in K$. Analogously as in Case II: $T(x, T(y, z)) = T(T(x, y), z)$.

Case IV: In all the other cases we obtain

$$T(x, T(y, z)) = \min(x, y, z) = T(T(x, y), z).$$

This shows that T is associative and, consequently, a t-norm. □

Definition 1.17 *Let $(T_k)_{k\in K}$ be a family of t-norms and let $((\alpha_k, \beta_k))_{k\in K}$ be a family of pairwise disjoint open subintervals of the unit interval $[0, 1]$. The t-norm T defined by (1.15) is called the ordinal sum of the summands $(\langle(\alpha_k, \beta_k), T_k\rangle)_{k\in K}$, and we shall write*

$$T = (\langle(\alpha_k, \beta_k), T_k\rangle)_{k\in K}.$$

Example 1.18 (i) An empty ordinal sum of t-norms, i.e., an ordinal sum of t-norms with index set $\varnothing$, yields the minimum $T_{\mathbf{M}}$:

$$T_{\mathbf{M}} = (\varnothing) = (\langle(\alpha_k, \beta_k), T_k\rangle)_{k\in\varnothing}.$$

(ii) Each t-norm T can be read as a trivial ordinal sum with one summand $\langle(0, 1), T\rangle$:

$$T = (\langle(0, 1), T\rangle).$$

(iii) The ordinal sum T of the two summands $\langle(\frac{1}{4}, \frac{1}{2}), T_{\mathbf{P}}\rangle$ and $\langle(\frac{2}{3}, \frac{3}{4}), T_{\mathbf{L}}\rangle$, i.e., $T = (\langle(\frac{1}{4}, \frac{1}{2}), T_{\mathbf{P}}\rangle, \langle(\frac{2}{3}, \frac{3}{4}), T_{\mathbf{L}}\rangle)$, is given by

$$T(x, y) = \begin{cases} \frac{1}{4}(1 + (4x - 1)(4y - 1)) & \text{if } (x, y) \in \left(\frac{1}{4}, \frac{1}{2}\right)^2, \\ \frac{2}{3} + \max(0, x + y - \frac{17}{12}) & \text{if } (x, y) \in \left(\frac{2}{3}, \frac{3}{4}\right)^2, \\ \min(x, y) & \text{otherwise.} \end{cases}$$

(iv) An ordinal sum of t-norms may have infinitely many summands. For instance, $T = (\langle(\frac{1}{2^n}, \frac{1}{2^{n-1}}), T_{\mathbf{P}}\rangle)_{n\in\mathbb{N}}$ means that

$$T(x, y) = \begin{cases} \dfrac{1}{2^n}(1 + (2^n x - 1)(2^n y - 1)) & \text{if } (x, y) \in \left(\dfrac{1}{2^n}, \dfrac{1}{2^{n-1}}\right)^2, \\ \min(x, y) & \text{otherwise.} \end{cases}$$

Corollary 1.19 *Let* $(\mathbf{S}_k)_{k\in K}$ *be a family of t-conorms and* $((\alpha_k,\beta_k))_{k\in K}$ *be a family of pairwise disjoint open subintervals of* $[0,1]$. *Then the function* $\mathbf{S}:[0,1]^2\to[0,1]$ *defined by*

$$\mathbf{S}(x,y)=\begin{cases}\alpha_k+(\beta_k-\alpha_k)\cdot S_k\left(\dfrac{x-\alpha_k}{\beta_k-\alpha_k},\dfrac{y-\alpha_k}{\beta_k-\alpha_k}\right) & \text{if } (x,y)\in(\alpha_k,\beta_k)^2,\\ \max(x,y) & \text{otherwise,}\end{cases} \tag{1.16}$$

is a t-conorm which is called the ordinal sum of the summands $\langle(\alpha_k,\beta_k),\mathbf{S}_k\rangle$, $k\in K$, *and we shall write*

$$\mathbf{S}=(\langle(\alpha_k,\beta_k),\mathbf{S}_k\rangle)_{k\in K}.$$

All results for ordinal sums of t-norms remain valid for t-conorms with the obvious changes where necessary.

In particular, concerning the duality of ordinal sums we get the following: If $T=(\langle(\alpha_k,\beta_k),T_k\rangle)_{k\in K}$ is an ordinal sum of t-norms, then the dual t-conorm S given by (1.16) can be be written as an ordinal sum of t-conorms:

$$\mathbf{S}=(\langle(1-\beta_k,1-\alpha_k),\mathbf{S}_k\rangle)_{k\in K},$$

where the t-conorm $\mathbf{S}_k$ is the dual of the t-norm T_k. Note, however, that the t-norm T_k and the t-conorm $\mathbf{S}_k$ in general act on different intervals.

1.4 Representation of continuous t-norms

1.4.1 Pseudo-inverse

The following extension of the notion of inverse function will occur in the representation theorem 1.23. First we shall consider a special case when the function is a strictly monotone bijection. Then we introduce a notion which enables us to consider extension of the inverse function also on the arguments out of the range of the function g, but still to have the behavior of the inverse function on the range of the function g.

Definition 1.20 *Let* $g:[a,b]\to[c,d]$ *be a strictly monotone bijection of* $[a,b]$ *onto* $[c,d]$, *where* $[a,b]$ *and* $[c,d]$ *are closed subintervals of the extended real line* $[-\infty,+\infty]$. *A monotone continuous mapping* $g^{(-1)}:[-\infty,+\infty]\to[a,b]$, *such that for all* $y\in[c,d]$ *it is* $g^{(-1)}(y)=g^{-1}(y)$, *if* $y>d$ *then* $g^{(-1)}(y)=g^{-1}(d)$ *and if* $y<c$ *then* $g^{(-1)}(y)=g^{-1}(c)$, *is called a pseudo-inverse of* g.

For example, if we take $g(x) = e^{x-1}$ on $[0,1]$ then

$$g^{(-1)}(x) = \begin{cases} 0 & \text{if } x < e^{-1} \\ \ln x + 1 & \text{if } x \in [e^{-1}, 1] \\ 1 & \text{if } x > 1. \end{cases}$$

We have in general that $g^{(-1)} \circ g(x) = x$ for $x \in [a,b]$. But we can consider also expression of the form $g^{(-1)}(g(x) + g(y))$ for any $x, y \in [a,b]$, although $g(x) + g(y)$ can be out of the range of the function g. On the other hand, we can consider also the function $g \circ g^{(-1)} : [-\infty, \infty] \to [c,d]$, for which we have

$$g \circ g^{(-1)}(y) = \begin{cases} c & \text{if } y < c \\ y & \text{if } y \in [c,d] \\ d & \text{if } y > d. \end{cases}$$

We extend now the notion of the pseudo-inverse also for the general (not necessarily strictly) monotone function and which can be also not surjective. This extension will be very useful for the construction of new t-norms given in Theorem 1.22.

Definition 1.21 *Let $g : [a,b] \to [c,d]$ be a monotone function from $[a,b]$ to $[c,d]$, where $[a,b]$ and $[c,d]$ are closed subintervals of the extended real line $[-\infty, +\infty]$. A mapping $g^{(-1)} : [c,d] \to [a,b]$ defined by*

$$g^{(-1)}(y) = \sup\{x \in [a,b] \mid (g(x) - y)(g(b) - g(a)) < 0\} \tag{1.17}$$

is called a pseudo-inverse of g.

We remark that the Definition 1.20 is covered by Definition 1.21 taking for a bijection g in Definition 1.21 Range $(g) \subset [c,d] = [-\infty, +\infty]$. In this general case for a strictly monotone function $g : [a,b] \to [c,d]$ we still have $g^{(-1)} \circ g(x) = x$ for $x \in [a,b]$ and $g \circ g^{(-1)}(y) = y$ for $y \in$ Range (g). For any monotone function g we always have $g^{(-1)} \circ g(x) \le x$ for $x \in [a,b]$.

For a non-decreasing (non-constant) function g (1.17) reduces on

$$g^{(-1)}(y) = \sup\{x \in [a,b] \mid g(x) < y\}.$$

For a non-increasing (non-constant) function g (1.17) reduces on

$$g^{(-1)}(y) = \sup\{x \in [a,b] \mid g(x) > y\}.$$

There is a simple algorithm for drawing the graph of the pseudo-inverse $g^{(-1)}$ of the given function. First draw vertical line segments at discontinuities of g. Then reflect the graph of the function g with respect to the line $y = x$. Finally, remove any vertical line segments from the reflected graph except for their lowest points.

Using now the notion of pseudo-inverse from Definition 1.21 we can formulate a theorem which gives a method for the construction of new t-norms starting from a given t-norm, [164], Theorem 3.6 (see also [149, 163, 268])).

Theorem 1.22 *Let $g : [0,1] \to [0,1]$ be a non-decreasing function and T a t-norm such that*

$$T(g(x), g(y)) \in \text{Range}\ (g) \cup [0, g(0+)],$$

for all $x, y \in [0,1)$ and for all $x, y \in [0,1]$, $T(g(x), g(y)) \in \text{Range}\ (g)$,

$$g \circ g^{(-1)}(T(g(x), g(y)) = T(g(x), g(y)).$$

Then the function $T_g : [0,1]^2 \to [0,1]$ defined by

$$T_g(x,y) = \begin{cases} g^{(-1)}(T(g(x), g(y)) & \textit{if } x, y \in [0,1) \\ \min(x,y) & \textit{otherwise} \end{cases} \tag{1.18}$$

is a t-norm.

For a continuous function g and any t-norm T the above conditions are always satisfied and (1.18) always give a t-norm. For many interesting examples see [164], section 3.1.

1.4.2 Additive generators

An important subclass of continuous t-norms has nice representations in terms of single-argument functions.

Theorem 1.23 *A function $T : [0,1]^2 \to [0,1]$ is a continuous Archimedean triangular norm if and only if there exists a continuous, strictly decreasing function $\mathbf{t} : [0,1] \to [0,+\infty]$ with $\mathbf{t}(1) = 0$ such that for all $x, y \in [0,1]$*

$$T(x,y) = \mathbf{t}^{-1}(\min(\mathbf{t}(x) + \mathbf{t}(y), \mathbf{t}(0))) = \mathbf{t}^{(-1)}(\mathbf{t}(x) + \mathbf{t}(y)). \tag{1.19}$$

The function $\mathbf{t}$ is then called an additive generator of T; it is uniquely determined by T up to a positive multiplicative constant.

Proof. Assume first that $\mathbf{t} : [0,1] \to [0,+\infty]$ is a continuous, strictly decreasing function with $\mathbf{t}(1) = 0$ and that T is constructed by (1.19). The commutativity

(T1) and the monotonicity (T3) of T are obvious. Also, the boundary condition (T4) holds, since for all $x \in [0,1]$

$$T(x,1) = \mathbf{t}^{-1}(\min(\mathbf{t}(x) + \mathbf{t}(1), \mathbf{t}(0))) = \mathbf{t}^{-1}(\min(\mathbf{t}(x), \mathbf{t}(0))) = x.$$

Concerning the associativity (T2), for all $x, y, z \in [0,1]$ we obtain

$$\begin{aligned} T(T(x,y),z) &= \mathbf{t}^{-1}(\min(\mathbf{t}(T(x,y)) + \mathbf{t}(z), \mathbf{t}(0))) \\ &= \mathbf{t}^{-1}(\min(\min(\mathbf{t}(x) + \mathbf{t}(y), \mathbf{t}(0)) + \mathbf{t}(z), \mathbf{t}(0))) \\ &= \mathbf{t}^{-1}(\min(\mathbf{t}(x) + \mathbf{t}(y) + \mathbf{t}(z), \mathbf{t}(0)) \\ &= \mathbf{t}^{-1}(\min(\mathbf{t}(x) + \min(\mathbf{t}(y) + \mathbf{t}(z), \mathbf{t}(0)), \mathbf{t}(0))) \\ &= \mathbf{t}^{-1}(\min(\mathbf{t}(x) + \mathbf{t}(T(y,z)), \mathbf{t}(0))) \\ &= T(x, T(y,z)), \end{aligned}$$

showing that T is indeed a t-norm.

To prove the converse let T be a continuous Archimedean t-norm. To simplify notations, in this proof we shall write simply $x^{(n)}$ instead of $x_T^{(n)}$ for $x \in [0,1]$ and $n \in \mathbb{N}$. If, for some $x \in [0,1]$ and some $n \in \mathbb{N} \cup \{0\}$, we have $x^{(n)} = x^{(n+1)}$ then by induction we obtain

$$x^{(n)} = x^{(2n)} = \left(x^{(n)}\right)^{(2)}$$

and, since T is continuous Archimedean, $x^{(n)} \in \{0,1\}$. This means that we have $x^{(n)} > x^{(n+1)}$ whenever $x^{(n)} \in (0,1)$.

Define now for $x \in [0,1]$ and $m, n \in \mathbb{N}$

$$\begin{aligned} x^{(\frac{1}{n})} &= \sup\{y \in [0,1] \mid y^{(n)} = x\}, \\ x^{(\frac{m}{n})} &= \left(x^{(\frac{1}{n})}\right)^{(m)}. \end{aligned}$$

Since T is Archimedean, we have for all $x \in (0,1]$

$$\lim_{n\to\infty} x^{(\frac{1}{n})} = 1. \tag{1.20}$$

Note that the expression $x^{(\frac{m}{n})}$ is well defined because of $x^{(\frac{m}{n})} = x^{(\frac{km}{kn})}$ for all $k \in \mathbb{N}$. Moreover, we have for all $x \in [0,1]$ and $m, n, p, q \in \mathbb{N}$

$$\begin{aligned} x^{(\frac{m}{n}+\frac{p}{q})} &= x^{(\frac{mq+np}{nq})} \\ &= \left(x^{(\frac{1}{nq})}\right)^{(mq+np)} \\ &= T\left(\left(x^{(\frac{1}{nq})}\right)^{(mq)}, \left(x^{(\frac{1}{nq})}\right)^{(np)}\right) \\ &= T\left(x^{(\frac{m}{n})}, x^{(\frac{p}{q})}\right). \end{aligned}$$

Let c be an arbitrary but fixed element from $(0,1)$. Denote $\mathbb{Q}^+ = \mathbb{Q} \cap [0,+\infty)$. We define the function $\xi : \mathbb{Q}^+ \to [0,1]$ by $\xi(r) = c^{(r)}$. Since T is continuous and since (1.20) holds for all $x \in (0,1]$, ξ is a continuous function. As a consequence of

$$\xi(r+s) = c^{(r+s)} = T(c^{(r)}, c^{(s)}) \leq c^{(r)} = \xi(r),$$

ξ is also non-increasing. The function ξ is even strictly decreasing on the preimage of $(0,1]$ since for all $\frac{m}{n}, \frac{p}{q} \in \mathbb{Q}^+$ with $\xi(\frac{m}{n}) > 0$ we obtain

$$\xi\left(\frac{m}{n} + \frac{p}{q}\right) \leq \xi\left(\frac{mq+1}{nq}\right) = \left(c^{(\frac{1}{nq})}\right)^{(mq+1)} < \left(c^{(\frac{1}{nq})}\right)^{(mq)} = \xi\left(\frac{m}{n}\right).$$

The monotonicity and continuity of ξ on $\mathbb{Q}^+$ allows us to extend it uniquely to a function $\overline{\xi} : [0,+\infty] \to [0,1]$ via

$$\overline{\xi}(x) = \inf\left\{\xi(r) \mid r \in \mathbb{Q}^+, r \leq x\right\}.$$

Then $\overline{\xi}$ is continuous and non-increasing, and we have

$$\overline{\xi}(x+y) = T(\overline{\xi}(x), \overline{\xi}(y)).$$

Moreover, $\overline{\xi}$ is strictly decreasing on the preimage of $(0,1]$.

Define the function $\mathbf{t} : [0,1] \to [0,+\infty]$ by

$$\mathbf{t}(x) = \sup\left\{y \in [0,+\infty] \mid \overline{\xi}(y) > x\right\},$$

where, as usual, $\sup \emptyset = 0$. Then $\mathbf{t}$ is continuous, strictly decreasing, and we have $\mathbf{t}(1) = 0$. Observe that $\overline{\xi}(x) = 0$ if and only if $x \geq \mathbf{t}(0)$, and that for all $x \in [0,+\infty]$ with $\overline{\xi}(x) > 0$ we obtain $\overline{\xi}(x) = \mathbf{t}^{-1}(x)$.

Putting everything together, for all $(x,y) \in [0,1]^2$ we have

$$\begin{aligned} T(x,y) &= T(\overline{\xi}(\mathbf{t}(x)), \overline{\xi}(\mathbf{t}(y))) \\ &= \overline{\xi}(\mathbf{t}(x) + \mathbf{t}(y)) \\ &= \mathbf{t}^{-1}(\min(\mathbf{t}(x) + \mathbf{t}(y), \mathbf{t}(0))), \end{aligned}$$

showing that (1.19) holds for all $x, y \in [0,1]$.

Finally, assume that that $f, g : [0,1] \to [0,+\infty]$ are continuous, strictly decreasing functions with $f(1) = g(1) = 0$ such that for all $x, y \in [0,1]$

$$f^{-1}(\min(f(x) + f(y), f(0))) = g^{-1}(\min(g(x) + g(y), g(0))).$$

Writing $u = g(x)$ and $v = g(y)$, we obtain for all $u, v \in [0, g(0)]$

$$g \circ f^{-1}(\min(f \circ g^{-1}(u) + f \circ g^{-1}(v), f(0))) = \min(u + v, g(0).$$

Defining the continuous, strictly increasing function $k : [0, g(0)] \to [0, +\infty]$ by $k = f \circ g^{-1}$, this implies that for all $u, v \in [0, g(0)]$

$$\min(k(u) + k(v), f(0)) = k(\min(u + v, g(0))),$$

which, as a consequence of the continuity of k, implies that for all $u, v \in [0, g(0)]$ with $u + v \in [0, g(0)]$

$$k(u) + k(v) = k(u + v).$$

This is a Cauchy functional equation, of which the only continuous, strictly increasing solutions (see [1]) satisfy $k(u) = c \cdot u$ for all $u \in [0, g(0)]$ and for some positive number $c \in \mathbb{R}$. As a consequence we obtain $f = c \cdot g$, showing that additive generators of continuous Archimedean t-norms are unique up to a positive multiplicative constant. □

We have already seen in Theorem 1.15 that a continuous Archimedean t-norm is either strict or nilpotent, a distinction which can also be made with the help of their additive generators.

Proposition 1.24 *Let T be a continuous Archimedean t-norm.*

(i) T is nilpotent if and only if for each additive generator $\mathbf{t}$ of T we have $\mathbf{t}(0) < +\infty$.

(ii) T is strict if and only if for each additive generator $\mathbf{t}$ of T we have $\mathbf{t}(0) = +\infty$.

Proof. If $\mathbf{t}$ is an additive generator of T with $\mathbf{t}(0) < +\infty$ then the element $x = \mathbf{t}^{-1}\left(\frac{f(0)}{2}\right) \in (0, 1)$ is nilpotent because of

$$x_T^{(2)} = \mathbf{t}^{-1}(\min(2\mathbf{t}(x), \mathbf{t}(0))) = 0,$$

and therefore T is a nilpotent t-norm.
If $\mathbf{t}$ is an additive generator of T with $\mathbf{t}(0) = +\infty$ then for all $x, y \in (0, 1]$ we have $\mathbf{t}(x) + \mathbf{t}(y) < \mathbf{t}(0)$, i.e.,

$$T(x, y) = \mathbf{t}^{-1}(\mathbf{t}(x) + \mathbf{t}(y)) > 0,$$

showing that T has no zero divisors. As a consequence of Theorem 1.15, the proof is complete. □

Example 1.25 (i) A family of additive generators $(\mathbf{t}_\lambda^{\mathbf{F}} : [0, 1] \to [0, +\infty])_{\lambda \in (0, +\infty]}$ for the family $(T_\lambda^{\mathbf{F}})_{\lambda \in (0, +\infty]}$ of Frank t-norms is given by

$$\mathbf{t}_\lambda^{\mathbf{F}}(x) = \begin{cases} -\log x & \text{if } \lambda = 1, \\ 1 - x & \text{if } \lambda = +\infty, \\ -\log \dfrac{\lambda^x - 1}{\lambda - 1} & \text{otherwise.} \end{cases}$$

(ii) A family of additive generators $(\mathbf{t}_\lambda^{\mathbf{Y}} : [0,1] \to [0,+\infty])_{\lambda\in(0,+\infty)}$ for the family $(T_\lambda^{\mathbf{Y}})_{\lambda\in(0,+\infty)}$ of Yager t-norms is given by

$$\mathbf{t}_\lambda^{\mathbf{Y}}(x) = (1-x)^\lambda.$$

(iii) For the family of Sugeno–Weber's t-conorms $(\mathbf{S}_\lambda^{\mathbf{SW}})_{\lambda\in(-1,\infty)}$, the corresponding additive generators are given by

$$\mathbf{s}_\lambda^{\mathbf{SW}}(x) = \begin{cases} x & \text{if } \lambda = 0, \\ \dfrac{\log(1+\lambda x)}{\log(1+\lambda)} & \text{if } \lambda \in (-1,\infty)\setminus\{0\}. \end{cases}$$

1.4.3 Multiplicative generators

The representation of continuous Archimedean t-norms given in Theorem 1.23 is based on the addition on the interval $[0,+\infty]$. There is a completely analogous representation based on the multiplication on $[0,1]$.

If $\mathbf{t} : [0,1] \to [0,\infty]$ is an additive generator of the continuous t-norm T, then we define strictly increasing function $\theta : [0,1] \to [0,1]$ by $\theta(x) = e^{-\mathbf{t}(x)}$, which is a multiplicative generator of T (duality between additive and multiplicative generators, see [164]). The following representation theorem holds, see [164, 268].

Theorem 1.26 *A function $T : [0,1]^2 \to [0,1]$ is a continuous Archimedean triangular norm if and only if there exists a continuous, strictly increasing function $\theta : [0,1] \to [0,1]$ such that $\theta(1) = 1$ (called multiplicative generator), and such that for all $x, y \in [0,1]$*

$$T(x,y) = \theta^{-1}(\max(\theta(x)\cdot\theta(y),\theta(0))) = \theta^{(-1)}(\theta(x)\theta(y)).$$

Moreover, T is a strict t-norm if and only if every continuous multiplicative generator θ of T satisfies $\theta(0) = 0$, and T is a nilpotent t-norm if and only if each continuous multiplicative generator θ of T satisfies $\theta(0) > 0$.

Proof. If $\mathbf{t} : [0,1] \to [0,+\infty]$ is an additive generator of a continuous Archimedean t-norm T then $\theta : [0,1] \to [0,1]$ given by $\theta(x) = e^{-\mathbf{t}(x)}$ obviously is a multiplicative generator of T. □

The argument in the proof above can be reversed: if $\theta : [0,1] \to [0,1]$ is a multiplicative generator of a continuous Archimedean t-norm T then $\mathbf{t} : [0,1] \to [0,+\infty]$ given by $\mathbf{t}(x) = -\log(\theta(x))$ is an additive generator of T.

As known, a multiplicative generator θ is uniquely determined by T up to a positive constant exponent.

Example 1.27 *For strict $T_{\mathbf{P}}$ a multiplicative generator is given by $\theta(x) = x$. For nilpotent $T_{\mathbf{L}}$ a multiplicative generator is given by $\theta(x) = e^{x-1}$.*

If T is a t-norm with a multiplicative generator θ then the function $\xi : [0,1] \to [0,1]$ given by $\xi(x) = \theta(1-x)$ is a multiplicative generator of the dual t-conorm $\mathbf{S}$.

We shall need the following families of t-norms with the corresponding multiplicative generators, see [164].

Example 1.28 (i) *The Dombi family* of t-norms $(T_\lambda^{\mathbf{D}})_{\lambda\in[0,\infty]}$ is defined by

$$T_\lambda^{\mathbf{D}}(x,y) = \begin{cases} T_{\mathbf{D}}(x,y) & \text{if } \lambda = 0, \\ T_{\mathbf{M}}(x,y) & \text{if } \lambda = \infty, \\ \left(1 + \left(\left(\frac{1-x}{x}\right)^\lambda + \left(\frac{1-y}{y}\right)^\lambda\right)^{1/\lambda}\right)^{-1} & \text{if } \lambda \in (0,\infty). \end{cases}$$

$T_\lambda^{\mathbf{D}}$ is Archimedean if and only if $\lambda \in [0,\infty)$. A family of multiplicative generators $(\theta_\lambda^{\mathbf{D}})_{\lambda\in(0,\infty)}$ is given by

$$\theta_\lambda^{\mathbf{D}}(x) = e^{-\left(\frac{1-x}{x}\right)^\lambda}.$$

$T_\lambda^{\mathbf{D}}$ is strict if and only if $\lambda \in (0,\infty)$.

(ii) *The Schweizer–Sklar family* of t-norms $(T_\lambda^{\mathbf{SS}})_{\lambda\in[-\infty,\infty]}$ is defined by

$$T_\lambda^{\mathbf{SS}}(x,y) = \begin{cases} T_{\mathbf{M}}(x,y) & \text{if } \lambda = -\infty, \\ (x^\lambda + y^\lambda - 1)^{1/\lambda} & \text{if } \lambda \in (-\infty,0), \\ T_{\mathbf{P}}(x,y) & \text{if } \lambda = 0, \\ (\max(x^\lambda + y^\lambda - 1, 0))^{1/\lambda} & \text{if } \lambda \in (0,\infty), \\ T_{\mathbf{D}}(x,y) & \text{if } \lambda = \infty. \end{cases}$$

$T_\lambda^{\mathbf{SS}}$ is continuous Archimedean if and only if $\lambda \in (-\infty,\infty)$. A family of multiplicative generators $(\theta_\lambda^{\mathbf{SS}})_{\lambda\in(-\infty,\infty)}$ is given by

$$\theta_\lambda^{\mathbf{SS}}(x) = \begin{cases} x & \text{if } \lambda = 0, \\ e^{(x^\lambda-1)/\lambda} & \text{if } \lambda \in (-\infty,0)\cup(0,\infty). \end{cases}$$

$T_\lambda^{\mathbf{SS}}$ is strict if and only if $\lambda \in (-\infty,0]$.

(iii) *The Aczél–Alsina family* of t-norms $(T_\lambda^{\mathbf{AA}})_{\lambda\in[0,\infty]}$ is defined by

$$T_\lambda^{\mathbf{AA}}(x,y) = \begin{cases} T_{\mathbf{D}}(x,y) & \text{if } \lambda = 0, \\ T_{\mathbf{M}}(x,y) & \text{if } \lambda = \infty, \\ e^{-(|\log x|^\lambda + |\log y|^\lambda)^{1/\lambda}} & \text{if } \lambda \in (0,\infty). \end{cases}$$

$T_\lambda^{\mathbf{AA}}$ is Archimedean if and only if $\lambda \in [0,\infty)$. A family of multiplicative generators $(\theta_\lambda^{\mathbf{AA}})_{\lambda\in(0,\infty)}$ is given by

$$\theta_\lambda^{\mathbf{AA}}(x) = e^{-(-\log x)^\lambda}.$$

$T_\lambda^{\mathbf{AA}}$ is strict if and only if $\lambda \in (0,\infty)$.

1.4.4 Isomorphism of continuous Archimedean t-norms with either $T_{\mathbf{P}}$ or $T_{\mathbf{L}}$

Proposition 1.29 *Let T be a continuous Archimedean t-norm and $\mathbf{t} : [0,1] \to [0,+\infty]$ an additive generator of T. If T^* is a t-norm which is isomorphic to T, i.e., if there is a strictly increasing bijection $\varphi : [0,1] \to [0,1]$ such that for all $x, y \in [0,1]$*

$$T^*(x,y) = \varphi^{-1}(T(\varphi(x),\varphi(y))),$$

then T^* is also a continuous Archimedean t-conorm, and the function $h : [0,1] \to [0,+\infty]$ defined by $h = \mathbf{t} \circ \varphi$ is an additive generator of T^*.

Proof. Taking into account $h(0) = \mathbf{t}(\varphi(0)) = \mathbf{t}(0)$, we obtain

$$\begin{aligned} T^*(x,y) &= \varphi^{-1} \circ \mathbf{t}^{-1}(\min(\mathbf{t}\circ\varphi(x) + \mathbf{t}\circ\varphi(y), \mathbf{t}(0))) \\ &= h^{-1}(\min(h(x)+h(y), h(0))) \end{aligned}$$

for all $x, y \in [0,1]$. □

It is straightforward that each isomorphism $\varphi : [0,1] \to [0,1]$ preserves the other algebraic and analytical properties of t-norms. In particular, each t-norm which is isomorphic to a strict or nilpotent t-norm, is strict or nilpotent, respectively.

Proposition 1.30 *(i) Any two strict t-norms are isomorphic.*

(ii) Any two nilpotent t-norms are isomorphic.

Proof. If T_1 and T_2 are two strict or two nilpotent t-norms with additive generators $\mathbf{t}_1$ and $\mathbf{t}_2$, respectively, such that $\mathbf{t}_1(0) = \mathbf{t}_2(0)$ (observe that this condition is always satisfied in the case of additive generators of strict t-norms), then in both cases the function $\varphi : [0,1] \to [0,1]$ defined by $\varphi = \mathbf{t}_1^{-1} \circ \mathbf{t}_2$ is an isomorphism between T_2 and T_1. □

An immediate consequence of Propositions 1.29 and 1.30 is that the product $T_{\mathbf{P}}$ and the Łukasiewicz t-norm $T_{\mathbf{L}}$ are not only prototypical examples of strict and nilpotent t-norms, respectively, but that each continuous Archimedean t-norm is isomorphic either to $T_{\mathbf{P}}$ or to $T_{\mathbf{L}}$.

Theorem 1.31 *(i) A function $T : [0,1]^2 \to [0,1]$ is a strict t-norm if and only if it is isomorphic to the product $T_{\mathbf{P}}$.*

(ii) A function $T : [0,1]^2 \to [0,1]$ is a nilpotent t-norm if and only if it is isomorphic to the Łukasiewicz t-norm $T_{\mathbf{L}}$.

Note that the representation of continuous Archimedean t-norms by means of multiplicative generators can be derived directly from more general results for I-semigroups (see [164, 197, 215]).

1.4.5 General continuous t-norms

The following representation of continuous t-norms by means of ordinal sums follows also from results of [197] in the context of I-semigroups.

Theorem 1.32 *A function $T : [0,1]^2 \to [0,1]$ is a continuous t-norm if and only if T is an ordinal sum of continuous Archimedean t-norms.*

Proof. Obviously, each ordinal sum of continuous t-norms is a continuous t-norm. Conversely, if T is a continuous t-norm we first show that the set I_T of all idempotent elements of T is a closed subset of $[0,1]$. Indeed, if $(x_n)_{n\in\mathbb{N}}$ is a sequence of idempotent elements of T which converges to some $x \in [0,1]$, then the continuity of T implies

$$x = \lim_{n\to\infty} x_n = \lim_{n\to\infty} T(x_n, x_n) = T(x, x).$$

Therefore x is also an idempotent element of T, and I_T is closed.

In the case $I_T = [0,1]$ we have $T = T_{\mathbf{M}}$, i.e., an empty ordinal sum, as a consequence of Proposition 1.5.

If I_T is a proper subset of $[0,1]$, then there is an at most countable, non-empty index set K and a family of pairwise disjoint open subintervals $((\alpha_k, \beta_k))_{k\in K}$ of $[0,1]$ such that

$$[0,1] \setminus I_T = \bigcup_{k\in K} (\alpha_k, \beta_k).$$

Choose and fix an element $k \in K$, and observe that both α_k and β_k are idempotent elements of T, but that there is no idempotent element of T in (α_k, β_k). The monotonicity of T implies that for all $(x, y) \in [\alpha_k, \beta_k]^2$ we have

$$\alpha_k = T(\alpha_k, \alpha_k) \leq T(x, y) \leq T(\beta_k, \beta_k) = \beta_k,$$

i.e., the range of $T|_{[\alpha_k,\beta_k]^2}$ equals $[\alpha_k, \beta_k]$ owing to the continuity of T. Further, for each $x \in [\alpha_k, 1]$ we have

$$\alpha_k \geq T(\alpha_k, x) \geq T(\alpha_k, \alpha_k) = \alpha_k,$$

i.e, α_k is an annihilator of $T|_{[\alpha_k,1]^2}$. Because of $\{T(x, \beta_k) \mid x \in [0,1]\} = [0, \beta_k]$, for each $y \in [0, \beta_k]$ there is an $x \in [0,1]$ with $y = T(x, \beta_k)$, and the associativity of$T$ yields

$$T(y, \beta_k) = T(T(x, \beta_k), \beta_k) = T(x, T(\beta_k, \beta_k)) = T(x, \beta_k) = y.$$

This means that β_k is a neutral element of $T|_{[0,\beta_k]^2}$.

All these, together with the continuity of T, imply that the function $T_k : [0,1]^2 \to [0,1]$ defined by

$$T_k(x, y) = \varphi_k(T(\varphi_k^{-1}(x), \varphi_k^{-1}(y))),$$

where $\varphi_k : [\alpha_k, \beta_k] \to [0,1]$ is the strictly increasing bijection given in (1.14), is a continuous t-norm. Since there is no idempotent element of T in (α_k, β_k) we have $T_k(x, x) < x$ for all $x \in (0,1)$ and, as a consequence of Theorem 1.12, T_k is a continuous Archimedean t-norm.

If $(x, y) \in (\alpha_k, \beta_k)^2$ then obviously

$$T(x, y) = \varphi_k^{-1}(T_k(\varphi_k(x), \varphi_k(y))).$$

Finally, if $(x, y) \notin (\alpha_k, \beta_k)^2$ for all $k \in K$ and, say, $x \leq y$, then there is some idempotent element $c \in I_T$ with $x \leq c \leq y$. Recall that c is a neutral element of $T|_{[0,c]^2}$ and an annihilator of $T|_{[c,1]^2}$. Then

$$\begin{aligned} T(x, y) &= T(T(x, c), y) \\ &= T(x, T(c, y)) \\ &= T(x, c) \\ &= x = \min(x, y), \end{aligned}$$

showing that $T = (\langle(\alpha_k, \beta_k), T_k\rangle)_{k \in K}$. □

Approximation of t-norms is specified in the following definition.

Definition 1.33 *Let T_1 and T_2 be t-norms and $\varepsilon \in (0,1)$. We say that T_1 and T_2 are ε-close to each other if for every $(x, y) \in [0,1]^2$*

$$|T_1(x, y) - T_2(x, y)| \leq \varepsilon.$$

We have by Proposition 8.3. from [164].

Proposition 1.34 *If* $(T_n)_{n\in\mathbb{N}}$ *is a sequence of t-norms (not necessarily continuous) which converges pointwise to a continuous t-norm* T, *then for every* $\varepsilon > 0$ *there exists* $n_0(\varepsilon) \in \mathbb{N}$ *such that* T *and* T_n *are* ε*-close for every* $n \geq n_0(\varepsilon)$.

The following result is related to the approximation of an arbitrary continuous t-norms, see [164], Theorem 8.12.

Theorem 1.35 *Let* T *be a continuous t-norm. For every* $\varepsilon > 0$ *there exists a strict t-norm* T_ε *such that* T *and* T_ε *are* ε*-close.*

1.5 Triangular norms with left-continuous diagonals in the point (1, 1)

We recall that a t-norm T is left-continuous if for every sequence $(x_n)_{n\in\mathbb{N}}$ from $[0,1]$ and every $y \in [0,1]$ we have

$$\sup_{n\in\mathbb{N}} T(x_n, y) = T(\sup_{n\in\mathbb{N}} x_n, y).$$

Special attention will be taken on the continuity in the point $(1,1)$. It is obvious that left-continuity in the point $(1,1)$ implies

$$\sup_{x\in[0,1)} T(x,x) = 1. \tag{1.21}$$

The opposite is also true (using $T(x,y) \geq \min(T(x,x), T(y,y))$). The property (1.21) means that the diagonal section $\delta_T : [0,1] \to [0,1]$ of a t-norm T, given by $\delta_T(x) = T(x,x)$, is left-continuous at the point 1. We remark that generally even the continuity of the whole diagonal $\delta_T(x) = T(x,x), x \in [0,1]$, of an Archimedean t-norm T does not imply the continuity of the t-norm T (see [164], Appendix B, Krause t-norm). The t-norm T defined by

$$T^{\Delta}(x,y) = \begin{cases} \dfrac{xy}{2} & \text{if } x, y \in [0,1), \\ \min(x,y) & \text{otherwise} \end{cases}$$

is strictly monotone continuous for $(x,y) \in [0,1)$ but it does not satisfy the condition (1.21). Hence it is not continuous in the point $(1,1)$.

The diagonal section $\delta_T : [0,1] \to [0,1]$ of a t-norm T characterizes some algebraic properties of the t-norm T, e.g., the Archimedean property and idempotent elements are fixed points of δ_T. The only continuous t-norm which is uniquely determined by its function δ_T is the minimum $T_{\mathbf{M}}$. For some classes of t-norms it is possible to obtain their determination with some additional values of t-norms on some subsets of the unit square, see [164].

We now introduce a general notion of an additive generator even for non-continuous t-norms.

Definition 1.36 *An additive generator* $\mathbf{t} : [0,1] \to [0,\infty]$ *(if it exists) of a t-norm* T *is a strictly decreasing function such that:*

(i) $\mathbf{t}$ *is right-continuous at* 0 *with* $\mathbf{t}(1) = 0$;

(ii) for $x, y \in [0,1]$ *we have* $\mathbf{t}(x) + \mathbf{t}(y) \in \text{Range}(\mathbf{t}) \cup [\mathbf{t}(0+), \infty]$;

(iii) $T(x,y) = \mathbf{t}^{(-1)}(\mathbf{t}(x) + \mathbf{t}(y)), x, y \in [0,1]$.

For example, the non-continuous t-norm $T_{\mathbf{D}}$ has as additive generator

$$\mathbf{t}(x) = \begin{cases} 2 - x & \text{if } x \in [0,1), \\ 0 & \text{if } x = 1, \end{cases}$$

and the non-continuous t-norm $T^{\triangle}$ (given above) has an additive generator

$$\mathbf{t}(x) = \begin{cases} -\log \dfrac{x}{2} & \text{if } x \in [0,1), \\ 0 & \text{if } x = 1. \end{cases}$$

But,for a t-norm T having an additive generator is a strong property, since adding the continuity only at the point $(1,1)$ (and then (1.21)) also forces the complete continuity of T.

Theorem 1.37 *If* T *is a left-continuous t-norm in the point* $(1,1)$ *with an additive generator then* T *is continuous.*

Proof. We shall show that for a left-continuous t-norm in the point $(1,1)$ T with an additive generator $\mathbf{t}$ this additive generator is continuous. First we shall prove that $\mathbf{t}$ is left-continuous in 1. Suppose the contrary, i.e., $\mathbf{t}(1-0) > 0$. Therefore

$$T(1-0, 1-0) = \mathbf{t}^{(-1)}(2\mathbf{t}(1-0)) < 1,$$

a contradiction with the left-continuity of T in the point $(1,1)$. Now we shall prove that $\mathbf{t}$ is continuous on the interval $(0,1]$. Suppose the contrary. Then there is some interval $(c,d) \subset (0, \mathbf{t}(0))$ such that $(c,d) \cap \text{Range}(\mathbf{t}) = \varnothing$. Therefore for each $n \in \mathbb{N}$ we have also $(c/n, d/n) \cap \text{Range}(\mathbf{t}) = \varnothing$. If n_0 is the smallest integer greater than

$c/(d-c)$ then we have $c/n < d/(n+1)$ for all $n \geq n_0$, i.e., the intervals $(c/n, d/n)$ and $(c/(n+1), d/(n+1))$ are overlapping. Hence

$$\bigcup_{n=n_0}^{\infty} (c/n, d/n) = (0, d/n_0).$$

Therefore $\mathbf{t}(1-0) \geq d/n_0$, a contradiction with the left-continuity of $\mathbf{t}$ in 1. □

Observe that there are alternative approaches to additive generators of t-norm relaxing the property (ii) in Definition 1.36, see [318, 319], which allows to generate t-norms with non-trivial idempotent elements as well as t-norms left-continuous in the point $(1,1)$. An example of such additive generator is the function $\mathbf{t} : [0,1] \to [0,\infty]$ defined by

$$\mathbf{t}(x) = \begin{cases} 3 - x & \text{if } x \in [0, 1/2), \\ 1 - x & \text{if } x \in [1/2, 1], \end{cases}$$

where the generated t-norm T has idempotent element $1/2$ and is left-continuous in the point $(1,1)$. However, also in this case the left-continuity of T (or, equivalently, the left-continuity of its diagonal) forces the continuity of T.

1.6 Triangular norms of H-type

Definition 1.38 *A t-norm T is of H-type if the family $(x_T^{(n)})_{n\in\mathbb{N}}$ is equicontinuous at the point $x = 1$, where $x_T^{(n)}$ is defined by*

$$x_T^{(1)} = x, \qquad x_T^{(n)} = T(x_T^{(n-1)}, x), \quad n \geq 2,\ x \in [0,1].$$

There is a nontrivial example of a t-norm T such that $(x_T^{(n)})_{n\in\mathbb{N}}$ is an equicontinuous family at the point $x = 1$.

Example 1.39 Let $\bar{T}$ be a continuous t-norm and let for every $m \in \mathbb{N} \cup \{0\}$:

$$I_m = [1 - 2^{-m}, 1 - 2^{-m-1}].$$

If

$$T(x,y) = 1 - 2^{-m} + 2^{-m-1}\bar{T}(2^{m+1}(x - 1 + 2^{-m}),\ 2^{m+1}(y - 1 + 2^{-m}))$$

for $(x,y) \in I_m \times I_m$ and $T(x,y) = \min(x,y)$ for $(x,y) \notin \bigcup_{m\in\mathbb{N}\cup\{0\}} I_m \times I_m$ then the family $(x_T^{(n)})_{n\in\mathbb{N}}$ is equicontinuous at the point $x = 1$, i.e., T is a t-norm of H-type.

Proposition 1.40 *If a continuous t-norm T is Archimedean than it can not be a t-norm of H-type.*

Proof. The reason is that for every $x, y \in (0,1)$ there exists $n \in \mathbb{N}$ such that $x_T^{(n)} < y$, and therefore the family $(x_T^{(n)})_{n\in\mathbb{N}}$ can not be equicontinuous at the point $x = 1$. Hence if in the ordinal sum representation $T = (<(\alpha_k,\beta_k),T_k>)_{k\in K}$ (see Theorem 1.32) there is some $k^* \in K$ such that $\beta_{k^*} = 1$, then T_{k^*} is Archimedean on $(\alpha_{k^*},1)$ and $T = (<(\alpha_k,\beta_k),T_k>)_{k\in K}$ can not be of H-type. □

The following proposition was proved in [224].

Proposition 1.41 *A continuous t-norm T is of H-type if and only if $T = (<(\alpha_k,\beta_k),T_k>)_{k\in K}$ and $\sup\beta_k < 1$ or $\sup\alpha_k = 1$.*

Proof. By Theorem 1.32 T is continuous if and only if $T = (<(\alpha_k,\beta_k),T_k>)_{k\in K}$ with Archimedean continuous summands T_k, $(k \in K)$.

Suppose that T is of H-type. Then $\beta_k < 1$ for every $k \in \mathbb{N}$. Since the system $((\alpha_k,\beta_k))_{k\in K}$ is disjoint we have that $\sup\beta_k < 1$ or $\sup\beta_k = \sup\alpha_k = 1$. The case $\sup\beta_k = 1$ and $\sup\alpha_k < 1$ is impossible, since in this case $\sup\beta_k = \max\beta_k = \beta_{k^*} = 1$, for some $k^* \in K$, which leads to a contradiction.

Suppose that $\sup\beta_k < 1$ or $\sup\alpha_k = 1$ and prove that T is of H-type. If $\sup\beta_k < 1$ then $x_T^{(n)} = x$, for any $x > \sup\beta_k$, and therefore T is of H-type. Let $\sup\alpha_k = 1$. Then for any $\varepsilon > 0$ there is a $k \in K$ such that $1-\varepsilon < \alpha_k$ and so for $x \geq \alpha_k$, $x_T^{(n)} \geq \alpha_k > 1-\varepsilon$. This means that T is of H-type. □

Remark 1.42 If $T = (<(\alpha_k,\beta_k),T_k>)_{k\in K}$ and $\sup\beta_k < 1$ or $\sup\alpha_k = 1$, then T is of H-type for any summands T_k (not only for continuous and Archimedean summands T_k, $k \in K$, see [224]). Hence , if

$$T = \left(<(1-2^{-k}, 1-2^{-k-1}),\ \bar{T}> \right)_{k\in\mathbb{N}\cup\{0\}}$$

we have $\sup\alpha_k = \sup(1-2^{-k}) = 1$ (cf. Example 1.39).

For an arbitrary t-norm of H-type we have by [250] the following characterization.

Theorem 1.43 *Let T be a t-norm. Then (i) and (ii) hold, where:*

(i) Suppose that there exists a strictly increasing sequence $(b_n)_{n\in\mathbb{N}}$ from the interval $[0,1)$ such that $\lim_{n\to\infty} b_n = 1$ and $T(b_n,b_n) = b_n$. Then T is of H-type.

(ii) If T is continuous and of H-type, then there exists a sequence $(b_n)_{n\in\mathbb{N}}$ as in (i).

Proof. (i) Let $\lambda \in (0,1)$ be given. We shall prove that there exists $\eta \in (0,1)$ such that $x > 1-\eta$ implies that $x_T^{(n)} > 1-\lambda$ for every $n \in \mathbb{N}$. Since $\lim_{n\to\infty} b_n = 1$, there exists $k = k(\lambda) \in \mathbb{N}$ such that $b_k > 1-\lambda$. Let $\eta = 1-b_k$. If $x > 1-\eta = b_k$ then $T(x,x) \geq T(b_k,b_k) = b_k > 1-\lambda$ which means that T is of H-type.

(ii) Suppose now that T is of H-type. Let k be a fixed natural number, $\varepsilon_k = 1/k$ and let η_k be such that

$$x > 1 - \eta_k \Rightarrow x_T^{(n)} > 1 - \varepsilon_k, \text{ for every } n \in \mathbb{N}.$$

The sequence $(x_T^{(n)})_{n\in\mathbb{N}}$ $(x > 1 - \eta_k)$ is non-increasing and therefore there exists $b_k = \lim_{n\to\infty} x_T^{(n)}, k \in \mathbb{N}$. Since $x_T^{(2n+1)} = T(x_T^{(n)}, x_T^{(n)})$, using the continuity of T we obtain that

$$\lim_{n\to\infty} x_T^{(2n+1)} = b_k = T(\lim_{n\to\infty} x_T^{(n)}, \lim_{n\to\infty} x_T^{(n)}) = T(b_k, b_k)$$

for every $k \in \mathbb{N}$. □

From the proof of the above theorem we see that the condition of continuity of whole sequence $(x_T^{(n)})_{n\in\mathbb{N}}$ can be replaced by the condition that the function $\delta_T(x) = T(x,x)\,(x \in [0,1])$ is right-continuous on an interval $[b,1)$ for $b < 1$.

Theorem 1.44 *Let T be a t-norm such that the function $\delta_T(x) = T(x,x)\,(x \in [0,1])$ is right-continuous on an interval $[b,1)$ for $b < 1$. Then T is a t-norm of H-type if and only if there exists a sequence $(b_n)_{n\in\mathbb{N}}$ from the interval $(0,1)$ of idempotents of T such that $\lim_{n\to\infty} b_n = 1$.*

In particular, for continuous t-norms we obtain the following characterization.

Theorem 1.45 *Let T be a continuous t-norm. Then the following are equivalent:*

a) T is not of H-type.

b) There exist $a_T \in [0,1)$ and a continuous strictly increasing and surjective mapping $\varphi_{a_T} : [a_T, 1] \to [0,1]$ such that

$$T(x,y) = \varphi_{a_T}^{-1}(\varphi_{a_T}(x) \star \varphi_{a_T}(y)), \text{ for every } x, y \geq a_T,$$

where the operation $\star$ is either $T_{\mathbf{P}}$ or $T_{\mathbf{L}}$.

Proof. a) implies b). Suppose T is not of H-type. Then there exists $b < 1$ such that $T(x,x) < x$, for every $x \in (b,1)$. If $a_T = \lim_{n\to\infty} b_T^{(n)}$, from $b_T^{(2n+1)} = T(b_T^{(n)}, b_T^{(n)})$, $n \in \mathbb{N}$, using the continuity of T, we obtain that $a_T = T(a_T, a_T)$. We shall prove that a_T is unique idempotent element of T on the interval $[a_T, 1]$. Suppose that $c \geq a_T$, $T(c,c) = c$. Since $c \leq b$ we obtain that $c_T^{(n)} \leq b_T^{(n)}$ and therefore $a_T \leq c = c_T^{(n)} \leq b_T^{(n)}$. Hence, $a_T \leq \lim_{n\to\infty} b_T^{(n)} = a_T$, which implies that $c = a_T$, i.e., c is not an interior idempotent element of T in $[a_T, 1]$. Then b) follows by Theorem 1.31.

b) implies a). By b) we have that for $x > a_T$ with $T(x,x) = x$ always $x = T(x,x) = \varphi_{a_T}^{-1}(\varphi_{a_T}(x) \star \varphi_{a_T}(x))$, and therefore $\varphi_{a_T}(x) = \varphi_{a_T}(x) \star \varphi_{a_T}(x)$. Since $\star$ is either $T_{\mathbf{P}}$ or $T_{\mathbf{L}}$ we obtain that it have to be $x = 1$ and a) follows by Theorem 1.44.
□

Definition 1.46 *Let T be a t-norm. We say that $T \in \mathcal{H}$ if and only if there exists a non-decreasing sequence $(b_n)_{n\in\mathbb{N}}$ from $(0,1)$ such that $\lim_{n\to\infty} b_n = 1$ the following implication holds:*

For every $n \in \mathbb{N}$

$$1 \geq x > b_n,\ 1 \geq y > b_n \Rightarrow T(x,y) > b_n.$$

We remark that if a t-norm T is of H-type and summands in the representing ordinal sum form are strict then $T \in \mathcal{H}$.

1.7 Comparison of t-norms

The main purpose of this section is to give some practical conditions assuring $T_1 \leq T_2$ for some continuous t-norms T_1, T_2. It is evident that for their duals it should then be $\mathbf{S}_1 \geq \mathbf{S}_2$. In what follows we restrict ourselves to continuous t-norms and t-conorms.

We are now interested in the question whether, given two t-norms T_1 and T_2, T_1 is weaker than T_2 or, equivalently, T_2 is stronger than T_1 (in symbols $T_1 \leq T_2$), i.e., $T_1(x,y) \leq T_2(x,y)$ for all points (x,y) in the unit square. We have already shown that

$$T_{\mathbf{D}} \leq T_{\mathbf{L}} \leq T_{\mathbf{P}} \leq T_{\mathbf{M}}.$$

Recall also that the monotonicity and boundary conditions imply $T_{\mathbf{D}} \leq T \leq T_{\mathbf{M}}$ for each t-norm T. However, $\leq$ is not a total order on the set of all t-norms: take, e.g., the product t-norm $T_{\mathbf{P}}$ and the Yager t-norm $T_2^{\mathbf{Y}}$, e.g., see their diagonal sections.

We shall give some practical criterions for the comparison of t-norms.

1.7.1 Comparison of continuous Archimedean t-norms

Let T_1, T_2 be two continuous Archimedean t-norms with additive generators f_1 and f_2, respectively. The full information about T_i is contained in $\mathbf{t}_i$ and, as a consequence, it should be possible to decide whether T_1 is weaker than T_2 only by means of $\mathbf{t}_1$ and $\mathbf{t}_2$. The first step into this direction was done by [265, 268], who proved that if both T_1 and T_2 are strict, then $T_1 \leq T_2$ if and only if the composite $h = \mathbf{t}_1 \circ \mathbf{t}_2^{-1}$ is a subadditive function, i.e., if for all $s, t \geq 0$

$$h(s+t) \leq h(s) + h(t).$$

We already know that for all continuous Archimedean t-norms T_1, T_2 we have $T_1 = T_2$ if and only if $\mathbf{t}_2 = c\mathbf{t}_1$ for some positive constant c, i.e., the composite $h = \mathbf{t}_1 \circ \mathbf{t}_2^{-1}$ is a linear function on $[0, f_2(0))$. The following result is owed to [268]:

Theorem 1.47 *Let T_1, T_2 be two continuous Archimedean t-norms with additive generators $\mathbf{t}_1, \mathbf{t}_2$, respectively. Let $h = \mathbf{t}_1 \circ \mathbf{t}_2^{-1}$ be the composite function defined on $[0, \mathbf{t}_2(0)]$. Then T_1 is weaker than T_2 if and only if h is subadditive.*

Note that in any case $h(0) = 0$. An important class of such subadditive functions is the set of all concave, continuous and strictly increasing functions h defined on a closed interval $[0, M]$ with $M \in (0, +\infty]$, fulfilling $h(0) = 0$. Note that the subadditivity of h on $[0, M]$ does not imply its concavity: the continuous and strictly increasing function $h : [0, 4] \to [0, 3]$ given by

$$h(x) = \begin{cases} x & \text{if } 0 \le x \le 1, \\ \frac{1+x}{2} & \text{if } 1 < x \le 3, \\ x - 1 & \text{if } 3 < x \le 4. \end{cases}$$

is subadditive but non-concave.

Using Theorem 1.47 we obtain the following sufficient condition for $T_1 \le T_2$:

Corollary 1.48 *Let T_1, T_2 be two continuous Archimedean t-norms with additive generators $\mathbf{t}_1$ and $\mathbf{t}_2$, respectively. If $h = \mathbf{t}_1 \circ \mathbf{t}_2^{-1}$ is a concave function on $[0, \mathbf{t}_2(0)]$ then we have $T_1 \le T_2$.*

If the composite function $h = \mathbf{t}_1 \circ \mathbf{t}_2^{-1}$ is differentiable on $(0, \mathbf{t}_2(0))$ then its concavity is a consequence of the non-increasingness of its first derivative. Note that in such a case we obtain for each $x \in (0, \mathbf{t}_2(0))$

$$h'(x) = \frac{\mathbf{t}_1'(\mathbf{t}_2^{-1}(x))}{\mathbf{t}_2'(\mathbf{t}_2^{-1}(x))} = \frac{\mathbf{t}_1'(u)}{\mathbf{t}_2'(u)},$$

where $u = \mathbf{t}_2^{-1}(x) \in (0, 1)$. Then the non-increasingness of $\mathbf{t}_2^{-1}$ reverses the monotonicity, hence h' is non-increasing (in x) if and only if $\frac{\mathbf{t}_2'}{\mathbf{t}_2'}$ is non-decreasing (in u). We therefore have shown the following result:

Corollary 1.49 *Let T_1, T_2 be two continuous Archimedean t-norms with differentiable additive generators $\mathbf{t}_1$ and $\mathbf{t}_2$, respectively. If $g = \frac{\mathbf{t}_1'}{\mathbf{t}_2'}$ is a non-decreasing function on $(0, 1)$ then we have $T_1 \le T_2$.*

The duality between t-norms and t-conorms yields the following result for the comparison of two continuous Archimedean t-conorms:

Corollary 1.50 *Let $\mathbf{S}_1, \mathbf{S}_2$ be two continuous Archimedean t-conorms with differentiable additive generators $\mathbf{s}_1$ and $\mathbf{s}_2$, respectively. If $g = \frac{\mathbf{s}_1'}{\mathbf{s}_2'}$ is a non-increasing function on $(0, 1)$, then we have $\mathbf{S}_1 \ge \mathbf{S}_2$.*

Remark 1.51 We have by [162] another sufficient criterion for the comparability of t-norms: if the function

$$x \mapsto \frac{(\mathbf{t}_1 \circ \mathbf{t}_2^{-1})(x)}{x}$$

is decreasing then $T_1 \leq T_2$.

We now give two examples of families of t-norms which are monotone with respect to their parameter.

Example 1.52 Recall that Schweizer and Sklar t-norms $(T_\lambda^{\mathbf{SS}})_{\lambda\in[-\infty,+\infty]}$ are given by

$$T_\lambda^{\mathbf{SS}}(x,y) = \begin{cases} T_{\mathbf{M}}(x,y) & \text{if } \lambda = -\infty, \\ T_{\mathbf{P}}(x,y) & \text{if } \lambda = 0, \\ T_{\mathbf{D}}(x,y) & \text{if } \lambda = +\infty, \\ (\max(x^\lambda + y^\lambda - 1, 0))^{\frac{1}{\lambda}} & \text{otherwise,} \end{cases}$$

having, for $-\infty < \lambda < +\infty$, the additive generators

$$\mathbf{t}_\lambda(x) = \begin{cases} -\log x & \text{if } \lambda = 0, \\ \dfrac{1 - x^\lambda}{\lambda} & \text{if } \lambda \in (-\infty, 0) \cup (0, +\infty). \end{cases}$$

Obviously $T_{-\infty}^{\mathbf{SS}} \geq T_s^{\mathbf{SS}} \geq T_{+\infty}^{\mathbf{SS}}$. It is easy to see that for $-\infty < s < t < +\infty$ we always have

$$\frac{\mathbf{t}_t'(x)}{\mathbf{t}_s'(x)} = x^{t-s},$$

i.e., $\mathbf{t}_1'/\mathbf{t}_2'$ is a non-decreasing function on $(0,1)$. Hence Corollary 1.49 implies $T_t^{\mathbf{SS}} \leq T_s^{\mathbf{SS}}$.

Example 1.53 Many applications deal with the Frank family of t-norms $(T_\lambda^{\mathbf{F}})_{\lambda\in[0,+\infty]}$. Recall that

$$T_\lambda^{\mathbf{F}}(x,y) = \begin{cases} T_{\mathbf{M}}(x,y) & \text{if } \lambda = 0, \\ T_{\mathbf{P}}(x,y) & \text{if } \lambda = 1, \\ T_{\mathbf{L}}(x,y) & \text{if } \lambda = +\infty, \\ \log_\lambda\left(1 + \dfrac{(\lambda^x - 1)(\lambda^y - 1)}{\lambda - 1}\right) & \text{otherwise.} \end{cases}$$

Frank showed that this family is continuous with respect to the parameter λ. Note that trivially $T_0^{\mathbf{F}} = T_{\mathbf{M}} \geq T_\lambda^{\mathbf{F}}$ for all $\lambda \in (0, +\infty]$. For each $\lambda \in (0, +\infty)$, $T_\lambda^{\mathbf{F}}$ is a strict t-norm whose additive generator is given by

$$\mathbf{t}_\lambda(x) = \begin{cases} -\log x & \text{if } \lambda = 1, \\ \log \dfrac{\lambda - 1}{\lambda^x - 1} & \text{if } \lambda \neq 1. \end{cases}$$

$T_{+\infty}^{\mathbf{F}}$ is a nilpotent t-norm, its additive generator is given by $\mathbf{t}_{+\infty}(x) = 1 - x$. Then

$$\frac{\mathbf{t}'_{+\infty}(u)}{\mathbf{t}'_s(u)} = \frac{-1}{\dfrac{-s^u \log s}{s^u - 1}} = \frac{1}{\log s}(1 - s^{-u}),$$

i.e., for each $s \in (0, +\infty) \setminus \{1\}$, $\mathbf{t}'_{+\infty}/\mathbf{t}'_s$ is non-decreasing on $(0, 1)$. The same is true for $\mathbf{t}'_{+\infty}/\mathbf{t}'_1$ because of

$$\frac{\mathbf{t}'_{+\infty}(u)}{\mathbf{t}'_1(u)} = \frac{-1}{-\dfrac{1}{u}} = u,$$

hence implying $T_{+\infty}^{\mathbf{F}} \leq T_\lambda^{\mathbf{F}}$ for all $\lambda \in (0, +\infty)$.

Let us now prove that $T_{\lambda_2}^{\mathbf{F}} \leq T_{\lambda_1}^{\mathbf{F}}$ whenever $1 < \lambda_1 < \lambda_2 < +\infty$ (the case $0 < \lambda_1 < \lambda_2 < 1$ is completely analogous). Define $g : (0, 1) \to [0, +\infty)$ by

$$\begin{aligned} g(u) &= \frac{\mathbf{t}'_{\lambda_2}(u)}{\mathbf{t}'_{\lambda_1}(u)} \\ &= \frac{(\lambda_1^u - 1) \cdot \lambda_2^u \log \lambda_2}{(\lambda_2^u - 1) \lambda_1^u \log \lambda_1} \\ &= C \cdot \frac{1 - a^u}{1 - b^u}, \end{aligned}$$

where $C = \dfrac{\log \lambda_2}{\log \lambda_1} > 0$ and $0 < b = \dfrac{1}{\lambda_2} < \dfrac{1}{\lambda_1} = a < 1$. Then g is non-decreasing on $(0, 1)$ if and only if

$$(1 - b^u)(-a^u \log a) \geq (1 - a^u)(-b^u \log b),$$

i.e., if and only if

$$\frac{a^u \log a}{b^u \log b} \geq \frac{1 - a^u}{1 - b^u}. \tag{1.22}$$

Put $f(u) = 1 - a^u$ and $h(u) = 1 - b^u$. Then by the Cauchy Mean Value Theorem, for each $u \in (0,1)$ there exists an $r \in (0,u)$ such that

$$\begin{aligned} \frac{1-a^u}{1-b^u} &= \frac{f(u)-f(0)}{h(u)-h(0)} \\ &= \frac{f'(r)}{h'(r)} \\ &= \frac{a^r \log a}{b^r \log b}. \end{aligned}$$

This implies inequality (1.22). Therefore $\mathbf{t}'_{\lambda_2}/\mathbf{t}'_{\lambda_1}$ is non-decreasing, i.e., $T^{\mathbf{F}}_{\lambda_2} \le T^{\mathbf{F}}_{\lambda_1}$. The case $0 < \lambda_1 < 1 < \lambda_2 < +\infty$ is a consequence of the two previous cases and the continuity of this family with respect to the parameter.

We remark that $T^{\mathbf{SW}}_{\lambda_2} \le T^{\mathbf{SW}}_{\lambda_1}$ for $\lambda_1 \le \lambda_2$.

For the fixed point theory it will be important to study order with respect to the Łukasiewicz t-norm $T_{\mathbf{L}}$. Using the properties of the considered families of t-norms (see [164]) we can easily prove the following proposition.

Proposition 1.54 *The following statements hold true:*

(i) There exists a member of the family $(T^{\mathbf{D}}_\lambda)_{\lambda\in(0,\infty)}$ *which is incomparable with* $T_{\mathbf{L}}$.

(ii) All members of the family $(T^{\mathbf{SS}}_\lambda)_{\lambda\in(-\infty,0]}$ *are greater than* $T_{\mathbf{L}}$.

(iii) There exists a member of the family $(T^{\mathbf{AA}}_\lambda)_{\lambda\in(0,\infty)}$ *which is incomparable with* $T_{\mathbf{L}}$.

(iv) All members of the family $(T^{\mathbf{SW}}_\lambda)_{\lambda\in[-1,0]}$ *are weaker than* $T_{\mathbf{L}}$*, and for* $\lambda \in [0,\infty]$ *are stronger than* $T_{\mathbf{L}}$.

1.7.2 Comparison of continuous t-norms

Let T_1 and T_2 be two continuous t-norms which, according to Theorem 1.32, can be represented in a unique way as ordinal sums of continuous Archimedean t-norms, $(<(\alpha_k^{(1)},\beta_k^{(1)}),T_k^{(1)}>)_{k\in K_1}$ and $(<(\alpha_k^{(2)},\beta_k^{(2)}),T_k^{(2)}>)_{k\in K_2}$, respectively. If the t-norm T_1 has an idempotent element x which is contained in some interval $\left(\alpha_k^{(2)},\beta_k^{(2)}\right)$, then T_1 cannot be weaker than T_2: since $T_k^{(2)}$ is Archimedean, we obtain in this case the contradiction $T_2(x,x) < x = T_1(x,x)$.

Theorem 1.55 *Let* T_1 *and* T_2 *be two continuous t-norms which are represented as ordinal sums of* $\left(<(\alpha_k^{(1)},\beta_k^{(1)}),T_k^{(1)}>\right)_{k\in K_1}$ *and* $\left(<(\alpha_k^{(2)},\beta_k^{(2)}),T_k^{(2)}>\right)_{k\in K_2}$*, respectively, such that for each* $k_2 \in K_2$ *there is a (unique)* $k_1 \in K_1$ *with*

$$\left(\alpha_{k_2}^{(2)},\beta_{k_2}^{(2)}\right) \subseteq \left(\alpha_{k_1}^{(1)},\beta_{k_1}^{(1)}\right),$$

and let $\mathbf{t}_k^{(i)}$ *be an additive generator of the continuous Archimedean t-norm* $T_k^{(i)}$. *Then* $T_1 \leq T_2$ *if and only if for each* $k_2 \in K_2$ *and the corresponding* $k_1 \in K_1$,

$$h_{k_2} = \mathbf{t}_{k_1}^{(1)} \circ f_{k_1,k_2}^{-1} \circ (\mathbf{t}_{k_2}^{(2)})^{-1}$$

is subadditive on $[0, \mathbf{t}_{k_2}^{(2)}(0)]$ *up to points* u *and* v *in which*

$$h_{k_2}(u) + h_{k_2}(v) \geq \mathbf{t}_{k_1}^{(1)} \left(\frac{\alpha_{k_2}^{(2)} - \alpha_{k_1}^{(1)}}{\beta_{k_1}^{(1)} - \alpha_{k_1}^{(1)}} \right),$$

where the affine transformation $f_{k_1,k_2} : \mathbb{R} \to \mathbb{R}$ *is defined by*

$$f_{k_1,k_2}(u) = \frac{\alpha_{k_1}^{(1)} - \alpha_{k_2}^{(2)}}{\beta_{k_2}^{(2)} - \alpha_{k_2}^{(2)}} + \frac{\beta_{k_1}^{(1)} - \alpha_{k_1}^{(1)}}{\beta_{k_2}^{(2)} - \alpha_{k_2}^{(2)}} \cdot u.$$

For the proof and more details see [162, 164].

Analogous reasoning as in the previous section yields the following sufficient condition for the comparison of continuous t-norms:

Corollary 1.56 *Let* T_1 *and* T_2 *be two continuous t-norms as in Theorem 1.55. Keeping the notations of Theorem 1.55, suppose that for each* $k_2 \in K_2$ *the function* g_{k_2} *defined by*

$$g_{k_2}(u) = \frac{(\mathbf{t}_{k_1}^{(1)})'(f_{k_1,k_2}(u))}{(\mathbf{t}_{k_2}^{(2)}(u))'}$$

is non-decreasing on $(0, 1)$. *Then we have* $T_1 \leq T_2$.

Remark 1.57 We shall see that often in the theory of probabilistic metric spaces we shall deal with t-norms T stronger than $T_{\mathbf{L}}$. In this case T obviously satisfies (1.21), i.e., δ_T is continuous at 1. Restricting ourselves to continuous t-norms T we can apply Theorem 1.55. Recall that $\mathbf{t}(x) = 1 - x$ is an additive generator of $T_{\mathbf{L}}$. Then $T_{\mathbf{L}} \leq T$, where T is the ordinal sum of $(< (\alpha_k, \beta_k), T_k >)_{k \in K}$, if and only if for all $u, v \in [0, \mathbf{t}_k(0)]$ and for all $k \in K$ (where $\mathbf{t}_k$ is an additive generator of T_k) we have

$$\mathbf{t}_k^{-1}(u) + \mathbf{t}_k^{-1}(v) - \mathbf{t}_k^{-1}(u + v) \leq \frac{1 - \alpha_k}{\beta_k - \alpha_k}.$$

The latter condition is obviously fulfilled if the t-norm T_k is stronger than $T_{\mathbf{L}}$ or if $\beta_k \leq \dfrac{1 + \alpha_k}{2}$.

Example 1.58 It is easy to see that each t-norm T with an additive generator $\mathbf{t}$, where $\mathbf{t}(1 - x)$ is subadditive, is weaker than $T_{\mathbf{L}}$, i.e., $T \leq T_{\mathbf{L}}$. By Theorem 1.55, the converse assertion is true for continuous t-norms t, i.e., $T \leq T_{\mathbf{L}}$ if and only if T is Archimedean and $x \mapsto \mathbf{t}(1 - x)$ is subadditive. Note that the concavity of $\mathbf{t}$ is sufficient to ensure the subadditivity of $x \mapsto \mathbf{t}(1 - x)$.

Furthermore, each continuous t-norm weaker than $T_{\mathbf{L}}$ is nilpotent. This means that neither a strict t-norm nor a t-norm with a nontrivial idempotent element can be weaker than $T_{\mathbf{L}}$.

It is an important relation to the diagonals of t-norms, see [164], Corollary 7.18.

Theorem 1.59 *Let T_1 and T_2 be two continuous t-norms with $\delta_{T_1} = \delta_{T_2}$. Then either $T_1 = T_2$ or T_1 and T_2 are incomparable.*

1.7.3 Domination of t-norms

The domination relation in the family of t-norms arises in the construction of the Cartesian product of probabilistic metric spaces. We have by [268] (see also [164, 307]):

Definition 1.60 *A t-norm T_2 dominates a t-norm T_1, $T_2 \gg T_1$, if for each $a, b, c, d \in [0, 1]$,*

$$T_2(T_1(a, b), T_1(c, d)) \geq T_1(T_2(a, c), T_2(b, d)). \tag{1.23}$$

Example 1.61 It is easy to see that for any t-norm T it holds $T_{\mathbf{M}} \gg T$ and $T \gg T_{\mathbf{D}}$.

The following characterization of the domination relation will be very useful for further investigations.

Theorem 1.62 *A strict t-norm T_2 with a generator $\mathbf{t}_2$ dominates a strict t-norm T_1 with a generator $\mathbf{t}_1$ if and only if the function $h = \mathbf{t}_2 \circ \mathbf{t}_1^{-1}$ satisfies the following inequality*

$$h^{-1}(h(x+y) + h(z+w)) \leq h^{-1}(h(x) + h(z)) + h^{-1}(h(y) + h(w)) \tag{1.24}$$

for arbitrary non-negative real numbers x, y, z and w.

Proof. Since $\mathbf{t}_1^{-1} : [0, +\infty] \to [0, 1]$ it follows for arbitrary non-negative real numbers x, y, z and w that

$$a = \mathbf{t}_1^{-1}(x), \quad b = \mathbf{t}_1^{-1}(z), \quad c = \mathbf{t}^{-1}(y), \quad \text{and} \quad d = \mathbf{t}_1^{-1}(w)$$

belong to the interval $(0, 1]$.

For this values the inequality (1.23) obtains the following form

$$T_2(T_1(\mathbf{t}_1^{-1}(x), \mathbf{t}_1^{-1}(y)), T_1(\mathbf{t}_1^{-1}(z), \mathbf{t}_1^{-1}(w))) \geq T_2(T_1(\mathbf{t}_1^{-1}(x), \mathbf{t}_1^{-1}(z)), T_1(\mathbf{t}_1^{-1}(y), \mathbf{t}_1^{-1}(w))).$$

Using the representations of strict t-norms T_1 and T_2 by their generators $\mathbf{t}_1$ and $\mathbf{t}_2$, respectively, we obtain

$$\mathbf{t}_2^{-1}((\mathbf{t}_2 \circ \mathbf{t}_1^{-1})(x+y) + (\mathbf{t}_2 \circ \mathbf{t}_1^{-1})(z+w))$$

$\geq \mathbf{t}_1^{-1}((\mathbf{t}_1 \circ \mathbf{t}_2^{-1})((\mathbf{t}_2 \circ \mathbf{t}_1^{-1})(x) + (\mathbf{t}_2 \circ \mathbf{t}_1^{-1})(z)) + (\mathbf{t}_2 \circ \mathbf{t}_1^{-1})((\mathbf{t}_2 \circ \mathbf{t}_1^{-1})(y) + (\mathbf{t}_2 \circ \mathbf{t}_1^{-1})(w)).$

Since, by the representation of T_1 the function $\mathbf{t}_1$ is strictly decreasing applying it on both sides of the preceding inequality it reverses the inequality and we obtain the inequality (1.24).

Now, let a, b, c and d be arbitrary real numbers from the interval $[0,1]$. Then, since $\mathbf{t}_1 : [0,1] \to [0,+\infty]$, we obtain that $x = \mathbf{t}_1(a)$, $y = \mathbf{t}_1(c)$, $z = \mathbf{t}_1(b)$, and $w = \mathbf{t}_1(d)$ are non-negative real numbers. Putting these values in (1.24) we obtain

$$(\mathbf{t}_1 \circ \mathbf{t}_2^{-1})((\mathbf{t}_2 \circ \mathbf{t}_1^{-1})(\mathbf{t}_1(a) + \mathbf{t}_1(c)) + (\mathbf{t}_2 \circ \mathbf{t}_1^{-1}))(\mathbf{t}_1(b) + \mathbf{t}_1(d))$$

$\leq (\mathbf{t}_1 \circ \mathbf{t}_2^{-1})((\mathbf{t}_2 \circ \mathbf{t}_1^{-1})(\mathbf{t}_1(a)) + (\mathbf{t}_2 \circ \mathbf{t}_1^{-1})(\mathbf{t}_1(b))) + (\mathbf{t}_1 \circ \mathbf{t}_2^{-1})((\mathbf{t}_1(c) + (\mathbf{t}_2 \circ \mathbf{t}_1^{-1})(\mathbf{t}_1(d))).$

If we rewrite the last inequality using representation theorem we obtain (1.23), i.e., that $T_2 \gg T_1$. □

Proposition 1.63 *If a t-norm T_2 dominates a t-norm T_1, then T_1 is weaker than T_2. The opposite in general is not true even for strict t-norms.*

Proof. Let $T_2 \gg T_1$. Then taking in (1.23) $b = c = 1$ we obtain $T_1 \leq T_2$. The following example, taken from [307] shows that $T_1 \leq T_2$ holds but $T_2 \not\gg T_1$.

Let strict t-norms T_1 and T_2 have generators $\mathbf{t}_1$ and $\mathbf{t}_2$, respectively, given in the following way

$$\mathbf{t}_1(x) = \begin{cases} \dfrac{1}{2x} & \text{if } 0 < x \leq \frac{1}{2}, \\ \dfrac{1}{x} - 1 & \text{if } \frac{1}{2} < x \leq 1, \end{cases}$$

and $\mathbf{t}_2(x) = \dfrac{1}{x} - 1$ for $x \in (0,1]$. Then we have for $h = \mathbf{t}_2 \circ \mathbf{t}_1^{-1}$

$$h(x) = \begin{cases} x & \text{if } 0 \leq x \leq 1, \\ 2x - 1 & \text{if } x > 1, \end{cases}$$

and

$$h^{-1}(x) = \begin{cases} x & \text{if } 0 \leq x \leq 1, \\ \dfrac{x+1}{2} & \text{if } x > 1. \end{cases}$$

Then taking $x = z = w = 1$ and $y = \frac{1}{2}$ in (1.24) we obtain

$$h(h(1+1) + h(1/2+1)) = 3 > 3/2 + 9/8 = h^{-1}(h(1) + h(1)) + h^{-1}(h(1/2) + h(1)),$$

which means that (1.24) does not hold. Hence by Theorem 1.62 T_2 does not dominate T_1.

But T_1 is weaker then T_2, since h^{-1} is concave and $h^{-1}(0) = 0$ which implies subadditivity of h^{-1} and by Corollary 1.48 it is $T_1 \leq T_2$. □

The preceding Proposition 1.63 implies that there are t-norms T_1 and T_2 which are incomparable with respect to $\gg$. As we have seen, $T_2^{\mathbf{Y}}$ and $T_{\mathbf{P}}$ are incomparable with respect to the order relation $\geq$, which implies by Proposition 1.63, that $T_2^{\mathbf{Y}}$ and $T_{\mathbf{P}}$ are incomparable also which respect to $\gg$.

Even more, taking the example from the proof of Proposition 1.63, we have two t-norms which are incomparable with respect to $\gg$ ($T_1 \gg T_2$ would imply by Proposition 1.63 $T_1 \geq T_2$, which is not true) but for which $T_1 \leq T_2$ holds.

It is unknown whether the domination relation in general is an order relation or not on the family of all t-norms since the transitivity is an open problem up to now. But we have

Proposition 1.64 *The relation $\gg$ between t-norms is reflexive and anti-symmetric.*

Proof. Using the associativity and commutativity of a t-norm T we have

$$T(T(a,b),T(c,d)) = T(T(a,c),T(b,d)),$$

i.e., $\gg$ is reflexive.

The relation $\gg$ is anti-symmetric: Suppose that $T_1 \gg T_2$ and $T_2 \gg T_1$. Then by Proposition 1.63 we have $T_1 \geq T_2$ and $T_2 \geq T_1$, which implies $T_1 = T_2$. □

Example 1.65 By [278] we have for the family of Schweizer and Sklar that $T_{\lambda_1}^{\mathbf{SS}} \gg T_{\lambda_2}^{\mathbf{SS}}$ if and only if $\lambda_1 \leq \lambda_2$. This implies that $\gg$ is transitive and therefore an order relation on the family $(T_\lambda^{\mathbf{SS}})_{\lambda\in[-\infty,+\infty]}$.

Namely, since we have proved in Example 1.52 that $T_{\lambda_1}^{\mathbf{SS}} \geq T_{\lambda_2}^{\mathbf{SS}}$ if and only if $\lambda_1 \leq \lambda_2$, we obtain by Proposition 1.63 that $T_{\lambda_1}^{\mathbf{SS}} \gg T_{\lambda_2}^{\mathbf{SS}}$ implies $\lambda_1 \leq \lambda_2$. The proof of the opposite implication is rather long and can be found in [278].

To obtain some comparison criteria for strict t-norms with respect to the relation $\gg$ we have to examine first by Theorem 1.62 the properties of the functions h which satisfies the relation (1.24).

Proposition 1.66 *If a strictly increasing continuous function $h, h : [0,\infty) \to [0,\infty]$, with $h(0) = 0$, satisfies (1.24) then it is convex and super-additive. The opposite is not true, i.e., continuous increasing convex functions h with $h(0) = 0$ need not satisfy (1.24).*

Proof. For arbitrary but fixed non-negative real numbers a and b, such that $a \leq b$, we take

$$x = h^{-1}(a), \quad y = h^{-1}\left(\frac{a+b}{2}\right) - h^{-1}(a), \quad z = h^{-1}\left(\frac{b-a}{2}\right), \quad \text{and } w = 0.$$

Putting these values in (1.24) we obtain

$$\frac{h^{-1}(a)+h^{-1}(b)}{2} \leq h^{-1}\left(\frac{a+b}{2}\right).$$

Hence h^{-1} is concave function and therefore h is convex function. Hence by using $h(0) = 0$, h is subadditive. To prove that the converse is not true we take as the example

$$h(x) = \begin{cases} x, & \text{if } 0 \leq x \leq 1, \\ 2x-1, & \text{if } x \geq 1, \end{cases}$$

and we put $x = z = w = 1$ and $y = \frac{1}{2}$. Then we see that (1.24) is not satisfied. □

The next theorem gives sufficient conditions for a function h to satisfy (1.24).

Theorem 1.67 *Let $h : [0, +\infty] \to [0, +\infty]$ be continuous strictly increasing, everywhere differentiable convex function with $h(0) = 0$. If*

$$\log(h'(e^x))$$

is a convex function then h satisfies (1.24).

The preceding theorem is crucial in the following construction of a strict t-norm T_2 which will dominate a given strict t-norm T_1, see [309].

Theorem 1.68 *Let T_1 be a strict t-norm with an additive generator $\mathbf{t}_1$. If $H(x)$ is a continuous non-decreasing convex function such that*

$$\int_0^{+\infty} e^{H(\log u)}\, du = +\infty,$$

then the strict t-norm T_2 with additive generator

$$\mathbf{t}_2(x) = \int_0^{\mathbf{t}_1(x)} e^{H(\log u)}\, du$$

dominates T_1.

1.8 Countable extension of t-norms

Each t-norm T can be extended (by associativity) in a unique way to an n-ary operation taking for $(x_1, \ldots, x_n) \in [0,1]^n$, $n \in \mathbb{N}$, the values $T(x_1, \ldots, x_n)$ which is defined by

$$\mathop{\mathrm{T}}_{i=1}^{0} x_i = 1, \quad \mathop{\mathrm{T}}_{i=1}^{n} x_i = T(\mathop{\mathrm{T}}_{i=1}^{n-1} x_i, x_n) = T(x_1, \ldots, x_n).$$

Specially, we have $T_{\mathbf{L}}(x_1,\ldots,x_n) = \max\left(\sum_{i=1}^{n} x_i - (n-1), 0\right)$ and $T_{\mathbf{M}}(x_1,\ldots,x_n) = \min(x_1,\ldots,x_n)$.

We can extend T to a countable infinitary operation taking for any sequence $(x_n)_{n\in\mathbf{N}}$ from $[0,1]$ the values

$$\mathop{\mathsf{T}}_{i=1}^{\infty} x_i = \lim_{n\to\infty} \mathop{\mathsf{T}}_{i=1}^{n} x_i. \tag{1.25}$$

The limit on the right side of (1.25) exists since the sequence $(\mathop{\mathsf{T}}_{i=1}^{n} x_i)_{n\in\mathbf{N}}$ is non-increasing and bounded from below.

Remark 1.69 An alternative approach to the infinitary extension of t-norms can be found in [189].

In the fixed point theory it is of interest to investigate the classes of t-norms T and sequences $(x_n)_{n\in\mathbf{N}}$ from the interval $[0,1]$ such that $\lim_{n\to\infty} x_n = 1$, and

$$\lim_{n\to\infty} \mathop{\mathsf{T}}_{i=n}^{\infty} x_i = \lim_{n\to\infty} \mathop{\mathsf{T}}_{i=1}^{\infty} x_{n+i} = 1. \tag{1.26}$$

In the classical case $T = T_{\mathbf{P}}$ we have $(T_{\mathbf{P}})_{i=1}^{n} = \prod_{i=1}^{n} x_i$ and for every sequence $(x_n)_{n\in\mathbf{N}}$ from the interval $[0,1]$ with $\sum_{i=1}^{\infty}(1-x_n) < \infty$ it follows that

$$\lim_{n\to\infty} (T_{\mathbf{P}})_{i=n}^{\infty} = \lim_{n\to\infty} \prod_{i=n}^{\infty} x_i = 1.$$

Namely, it is well known that

$$\prod_{i=1}^{\infty} x_i > 0 \quad \Leftrightarrow \quad \lim_{n\to\infty} \prod_{i=n}^{\infty} x_i = 1 \quad \Leftrightarrow \quad \sum_{i=1}^{\infty}(1-x_i) < \infty.$$

The equivalence

$$\sum_{i=1}^{\infty}(1-x_i) < \infty \quad \Leftrightarrow \quad \lim_{n\to\infty} \mathop{\mathsf{T}}_{i=n}^{\infty} x_i = 1 \tag{1.27}$$

holds also for $T \geq T_{\mathbf{L}}$. Indeed

$$(T_{\mathbf{L}})_{i=1}^{n} x_i = \max\left(\sum_{i=1}^{n} x_i - (n-1), 0\right) = \max\left(\sum_{i=1}^{n}(x_i - 1) + 1, 0\right),$$

and therefore $\sum_{n=1}^{\infty}(1-x_n) < \infty$ holds if and only if

$$\lim_{n\to\infty}(T_{\mathbf{L}})_{i=n}^{\infty} x_i = \max\left(\lim_{n\to\infty}\sum_{i=n}^{\infty}(x_i - 1) + 1, 0\right) = 1.$$

For $T \geq T_{\mathbf{L}}$ we have $\top_{i=1}^{n} x_i \geq (T_{\mathbf{L}})_{i=1}^{n} x_i$ and therefore for such a t-norm T the implication

$$\sum_{i=1}^{\infty}(1-x_i) < \infty \quad \Rightarrow \quad \lim_{n\to\infty} \top_{i=n}^{\infty} x_i = 1$$

holds.

The condition $T \geq T_{\mathbf{L}}$ is fulfilled by the families:

1. $T_\lambda^{\mathbf{SS}}$ for $\lambda \in [-\infty, 1]$; 2. $T_\lambda^{\mathbf{SW}}$ for $\lambda \in [0, \infty]$; 3. $T_\lambda^{\mathbf{F}}$ for $\lambda \in [0, \infty]$; 4. $T_\lambda^{\mathbf{Y}}$ for $\lambda \in [1, \infty]$.

We shall give some sufficient conditions for (1.26).

Proposition 1.70 *Let $(x_n)_{n\in\mathbb{N}}$ be a sequence of numbers from $[0,1]$ such that $\lim_{n\to\infty} x_n = 1$ and t-norm T is of H-type. Then (1.26) holds.*

Proof. Since t-norm T is of H-type for every $\lambda \in (0,1)$ there exists $\delta(\lambda) \in (0,1)$ such that

$$x \geq \delta(\lambda) \quad \Rightarrow \quad \top_{i=1}^{p} x > 1 - \lambda$$

for every $p \in \mathbb{N}$. Since $\lim_{n\to\infty} x_n = 1$ there exits $n_0(\lambda) \in \mathbb{N}$ such that $x_n \geq \delta(\lambda)$ for every $n \geq n_0(\lambda)$. Hence

$$\begin{aligned} \top_{i=1}^{p} x_{n+i} &\geq \top_{i=1}^{p} \delta(\lambda) \\ &> 1 - \lambda, \end{aligned}$$

for every $n \geq n_0(\lambda)$ and every $p \in \mathbb{N}$. This means that (1.26) holds. □

For some families of t-norms we shall characterize the sequences $(x_n)_{n\in\mathbb{N}}$ from $(0,1]$, which tend to 1 and for which (1.26) holds.

Lemma 1.71 *Let T be a strict t-norm with an additive generator $\mathbf{t}$, and the corresponding multiplicative generator θ. Then we have*

$$\top_{i=1}^{\infty} x_i = \mathbf{t}^{-1}\left(\sum_{i=1}^{\infty} \mathbf{t}(x_i)\right)$$

or

$$\mathop{\mathsf{T}}_{i=1}^{\infty} x_i = \theta^{-1}\left(\prod_{i=1}^{\infty} \theta(x_i)\right).$$

The preceding lemma and the continuity of the generators of strict t-norms imply the following proposition.

Proposition 1.72 *Let T be a strict t-norm with an additive generator* $\mathbf{t}$, *and the corresponding multiplicative generator θ. For a sequence $(x_n)_{n\in\mathrm{N}}$ from the interval* $(0,1)$ *such that* $\lim_{n\to\infty} x_n = 1$ *the condition*

$$\lim_{n\to\infty} \sum_{i=n}^{\infty} \mathbf{t}(x_i) = 0.$$

or the condition

$$\lim_{n\to\infty} \prod_{i=n}^{\infty} \theta(x_i) = 1,$$

holds if and only if (1.26) is satisfied.

Example 1.73 Let $(T_\lambda^{\mathbf{D}})_{\lambda\in(0,\infty)}$ be the Dombi family of t-norms and $(x_n)_{n\in\mathrm{N}}$ be a sequence of elements from $(0,1]$ such that $\lim\limits_{n\to\infty} x_n = 1$. Then we have the following equivalence:

$$\sum_{i=1}^{\infty}\left(\frac{1-x_i}{x_i}\right)^{\lambda} < \infty \quad \Leftrightarrow \quad \lim_{n\to\infty} (T_\lambda^{\mathbf{D}})_{i=n}^{\infty} x_i = 1.$$

For a t-norm $T_\lambda^{\mathbf{D}}, \lambda \in (0,\infty)$, the multiplicative generator $\theta_\lambda^{\mathbf{D}}$ is given by

$$\theta_\lambda^{\mathbf{D}}(x) = e^{-\left(\frac{1-x}{x}\right)^{\lambda}}$$

and therefore with the property $\theta_\lambda^{\mathbf{D}}(1) = 1$. Hence

$$\begin{aligned}\prod_{i=n}^{\infty} \theta_\lambda^{\mathbf{D}}(x_i) &= \prod_{i=n}^{\infty} e^{-\left(\frac{1-x_i}{x_i}\right)^{\lambda}} \\ &= e^{-\sum_{i=n}^{\infty}\left(\frac{1-x_i}{x_i}\right)^{\lambda}},\end{aligned}$$

and therefore the above equivalence follows by Proposition 1.72. Since $\lim\limits_{n\to\infty} x_n = 1$, we have that

$$\left(\frac{1-x_n}{x_n}\right)^{\lambda} \sim (1-x_n)^{\lambda} \text{ as } n \to \infty.$$

Hence

$$\sum_{n=1}^{\infty}(1-x_n)^{\lambda} < \infty \quad \Leftrightarrow \quad \sum_{n=1}^{\infty}\left(\frac{1-x_n}{x_n}\right)^{\lambda} < \infty,$$

and

$$\sum_{n=1}^{\infty}(1-x_n)^{\lambda}<\infty \quad\Leftrightarrow\quad \lim_{n\to\infty}(T_{\lambda}^{\mathbf{D}})_{i=n}^{\infty}x_i=1.$$

Example 1.74 Let $(T_{\lambda}^{\mathbf{AA}})_{\lambda\in(0,\infty)}$ be the Aczél-Alsina family of t-norms given by

$$T_{\lambda}^{\mathbf{AA}}(x,y)=e^{-(|\log x|^{\lambda}+|\log y|^{\lambda})^{1/\lambda}}$$

and $(x_n)_{n\in\mathbb{N}}$ be a sequence of elements from $(0,1]$ such that $\lim\limits_{n\to\infty}x_n=1$. Then we have the following equivalence

$$\sum_{i=1}^{\infty}(1-x_i)^{\lambda}<\infty \quad\Leftrightarrow\quad \lim_{n\to\infty}(T_{\lambda}^{\mathbf{AA}})_{i=n}^{\infty}x_i=1.$$

For a t-norm $T_{\lambda}^{\mathbf{AA}}$, $\lambda\in(0,\infty)$, the multiplicative generator $\theta_{\lambda}^{\mathbf{AA}}$ is given by

$$\theta_{\lambda}^{\mathbf{AA}}(x)=e^{-(-\log x)^{\lambda}}$$

and therefore with the property $\theta_{\lambda}^{\mathbf{AA}}(1)=1$. Hence

$$\begin{aligned}\prod_{i=n}^{\infty}\theta_{\lambda}^{\mathbf{AA}}(x_i) &= \prod_{i=n}^{\infty}e^{-(-\log x_i)^{\lambda}}\\ &= e^{-\sum_{i=n}^{\infty}(-\log x_i)^{\lambda}}.\end{aligned}$$

Since $\lim\limits_{i\to\infty}x_i=1$ and $\log x_i\sim x_i-1$ as $i\to\infty$ by Proposition 1.72 the above equivalence follows.

For nilpotent t-norms $T_{\lambda}^{\mathbf{SW}}$, $\lambda\in(-1,0)$ we have the following proposition.

Proposition 1.75 *Let $(x_n)_{n\in\mathbb{N}}$ be a sequence from $(0,1)$ such that the series $\sum\limits_{n=1}^{\infty}(1-x_n)$ is convergent. Then for every $\lambda\in(-1,\infty]$*

$$\lim_{n\to\infty}(T_{\lambda}^{\mathbf{SW}})_{i=n}^{\infty}x_i=1.$$

Proof. An additive generator of $T_{\lambda}^{\mathbf{SW}}$ for $\lambda\in(-1,0)$ is given by

$$\mathbf{t}_{\lambda}^{\mathbf{SW}}(x)=-\log\left(\frac{1+\lambda x}{1+\lambda}\right)\cdot\frac{1}{\log(1+\lambda)}.$$

We shall prove that for some $n_1\in\mathbb{N}$ and every $p\in\mathbb{N}$

$$\prod_{i=1}^{p}\theta_{\lambda}^{\mathbf{SW}}(x_{n+i-1})=\exp\left(\sum_{i=1}^{p}\log\left(\frac{1+\lambda x_{n+i-1}}{1+\lambda}\right)\cdot\frac{1}{\log(1+\lambda)}\right)>e^{-1} \tag{1.28}$$

for every $n \geq n_1$ since in this case

$$(T_\lambda^{\mathbf{SW}})_{i=1}^p x_{n+i-1} = (\theta_\lambda^{\mathbf{SW}})^{-1}\left(\prod_{i=1}^p \theta_\lambda^{\mathbf{SW}}(x_{n+i-1})\right). \tag{1.29}$$

We have to prove that for some $n_1 \in \mathbb{N}$ and every $p \in \mathbb{N}$

$$-\frac{1}{\log(1+\lambda)}\sum_{i=0}^p \log\left(\frac{1+\lambda x_{n+i-1}}{1+\lambda}\right) < 1 \text{ for every } n > n_1, \tag{1.30}$$

since (1.30) implies (1.28). From $\lim_{n\to\infty}(1-x_n) = 0$ it follows that

$$\log\left(1+\frac{\lambda}{1+\lambda}(x_n-1)\right) \sim \frac{\lambda}{1+\lambda}(x_n-1)$$

and therefore the series

$$-\frac{1}{\log(1+\lambda)}\sum_{n=1}^\infty \log\left(1+\frac{\lambda}{1+\lambda}(x_n-1)\right)$$

is convergent. Hence it follows that there exists $n_1 \in \mathbb{N}$ such that (1.28) holds for every $n \geq n_1$ and every $p \in \mathbb{N}$, and this implies (1.29). □

The above proposition holds also for $\lambda \geq 0$ since in this case $T_\lambda^{\mathbf{SW}} \geq T_{\mathbf{L}}$.

It is of special interest for the fixed point theory in probabilistic metric spaces to investigate condition (1.26) for a special sequence $(1-q^n)_{n\in\mathbb{N}}$ for $q \in (0,1)$.

Definition 1.76 *We say that for some $q \in (0,1)$ a t-norm T is q-convergent if*

$$\lim_{n\to\infty} \mathop{\mathsf{T}}_{i=n}^{\infty}(1-q^i) = 1.$$

Since $\lim_{n\to\infty}(1-q^n) = 1$ and $\sum_{n=1}^\infty (1-(1-q^n))^s < \infty$ for every $s > 0$ it follows that all t-norms from the family

$$\mathcal{T}_0 = \bigcup_{\lambda\in(0,\infty)}\{T_\lambda^{\mathbf{D}}\} \bigcup \bigcup_{\lambda\in(0,\infty)}\{T_\lambda^{\mathbf{AA}}\} \bigcup \mathcal{T}^H \bigcup_{\lambda\in(-1,\infty]}\{T_\lambda^{\mathbf{SW}}\}$$

are q-convergent for every $q \in (0,1)$.

The following example shows that not every strict t-norm is q-convergent.

Example 1.77 Let T be the strict t-norm with an additive generator $\mathbf{t}(x) = -\frac{1}{\log(1-x)}$. In this case the series $\sum_{i=1}^\infty \mathbf{t}(1-q^i)$ for any $q \in (0,1)$ is not convergent since

$$\sum_{i=1}^\infty \mathbf{t}(1-q^i) = -\sum_{i=1}^\infty \frac{1}{\log(q^i)} = -\sum_{i=1}^\infty \frac{1}{i\log q}.$$

In the following two propositions we shall give sufficient conditions for a t-norm T to be q-convergent.

Proposition 1.78 *Let T and T_1 be strict t-norms and* $\mathbf{t}$ *and* $\mathbf{t}_1$ *their additive generators, respectively, and there exists $b \in (0,1)$ such that $\mathbf{t}(x) \leq \mathbf{t}_1(x)$ for every $x \in (b,1]$. If T_1 is q-convergent, then T is q-convergent.*

Proof. Since T_1 is q-convergent we have $\lim_{n\to\infty}(T_1)_{i=n}^{\infty}(1-q^i) = 1$. Therefore

$$\lim_{n\to\infty}\sum_{i=n}^{\infty}\mathbf{t}_1(1-q^i) = 0. \tag{1.31}$$

Since there exists $n_0 \in \mathbb{N}$ such that $1 - q^{n_0} \in (b,1]$ we have by the condition of the proposition that

$$\mathbf{t}(1-q^n) \leq \mathbf{t}_1(1-q^n) \text{ for every } n \geq n_0.$$

Therefore, by (1.31) $\lim_{n\to\infty}\sum_{i=n}^{\infty}\mathbf{t}(1-q^i) = 0$, i.e., T is q-convergent. □

Proposition 1.79 *Let T be a strict t-norm with a generator* $\mathbf{t}$ *which has a bounded derivative on an interval $(b,1)$ for some $b \in (0,1)$. Then T is q-convergent for every $q \in (0,1)$.*

Proof. By the Lagrange mean value theorem we have for every $x \in (b,1)$ that

$$\mathbf{t}(x) - \mathbf{t}(1) = \mathbf{t}(x) = \mathbf{t}'(\xi)(x-1)$$

for some $\xi \in (x,1)$, and therefore

$$\sum_{i=i_0}^{\infty}\mathbf{t}(1-q^i) \leq M\sum_{i=i_0}^{\infty}q^i,$$

where $M = \sup_{x\in(b,1)}|\mathbf{t}'(x)|$, and $1 - q^{i_0} \in (b,1)$. □

Proposition 1.80 *Let T be a t-norm and $\psi : (0,1] \to [0,\infty)$. If for some $\delta \in (0,1)$ and every $x \in [0,1]$, $y \in [1-\delta,1]$*

$$|T(x,y) - T(x,1)| \leq \psi(y) \tag{1.32}$$

then for every sequence $(x_n)_{n\in\mathbb{N}}$ from the interval $[0,1]$ such that $\lim_{n\to\infty}x_n = 1$ the following implication holds:

$$\sum_{n=1}^{\infty}\psi(x_n) < \infty \quad \Rightarrow \quad \lim_{n\to\infty}\left(\mathop{\mathrm{T}}_{i=n}^{\infty} x_i - x_n\right) = 0. \tag{1.33}$$

Proof. Let $n_0(\delta) \in \mathbb{N}$ be such that $x_n \in [1-\delta, 1]$ for every $n \geq n_0(\delta)$. Then (1.32) implies that for every $n \geq n_0(\delta)$

$$|T(x_n, x_{n+1}) - T(x_n, 1)| = |T(x_n, x_{n+1}) - x_n| \leq \psi(x_{n+1})$$

and

$$|T(T(x_n, x_{n+1}), x_{n+2}) - T(T(x_n, x_{n+1}), 1)| \leq \psi(x_{n+2}).$$

Similarly for every $m \in N$ and every $n \geq n_0(\delta)$

$$\left| \mathop{\mathsf{T}}_{i=n}^{n+m} x_i - T(\mathop{\mathsf{T}}_{i=n}^{n+m-1} x_i, 1) \right| \leq \psi(x_{n+m})$$

and therefore

$$\begin{aligned} \left| \mathop{\mathsf{T}}_{i=n}^{n+m} x_i - x_n \right| &\leq \left| \mathop{\mathsf{T}}_{i=n}^{n+m} x_i - \mathop{\mathsf{T}}_{i=n}^{n+m-1} x_i \right| + \cdots + \left| \mathop{\mathsf{T}}_{i=n}^{n+1} x_i - x_n \right| \\ &\leq \sum_{i=n+1}^{n+m} \psi(x_i). \end{aligned}$$

If $\sum\limits_{n=1}^{\infty} \psi(x_n) < \infty$ we obtain that for every $n > n_0(\delta)$

$$\left| \mathop{\mathsf{T}}_{i=n}^{\infty} x_i - x_n \right| \leq \sum_{i=n+1}^{\infty} \psi(x_i) \tag{1.34}$$

and (1.34) implies (1.33). □

Corollary 1.81 *Let T and ψ be as in Proposition 1.80. If for some $q \in (0,1)$, $\sum\limits_{n=1}^{\infty} \psi(1-q^n) < \infty$ then T is q-convergent.*

Proof. Since $\lim\limits_{n\to\infty} (1-q^n) = 1$ from (1.33) we obtain that

$$\lim_{n\to\infty} \mathop{\mathsf{T}}_{i=n}^{\infty} (1-q^n) = \lim_{n\to\infty} (1-q^n) = 1.$$

□

Example 1.82 Let $\alpha > 0$, $p > 1$ and $h_{\alpha,p} : (0,1] \times [0,1] \to [0,\infty)$ be defined in the following way:

$$h_{\alpha,p}(x,y) = \begin{cases} y - \dfrac{\alpha}{|\ln(1-x)|^p} & \text{if } (x,y) \in (0,1) \times [0,1], \\ y & \text{if } (x,y) \in \{1\} \times [0,1] \end{cases}$$

In this case the function $h_{\alpha,p}$ is equal zero on the curve which connects the points $(1,0)$ and $(1-e^{-\alpha^{1/p}},1)$, where $1-e^{-\alpha^{1/p}}<1$.

Let T be a t-norm such that

$$T(x,y) \geq h_{\alpha,p}(x,y) \quad \text{for every} \quad (x,y) \in [1-\delta,1] \times [0,1].$$

Then for every $(x,y) \in [0,1] \times [1-\delta,1)$

$$\begin{aligned} |T(x,y)-T(x,1)| &= |T(y,x)-T(1,x)| \\ &= |T(y,x)-h_{\alpha,p}(1,x)| \\ &\leq |h_{\alpha,p}(x,y)-h_{\alpha,p}(1,x)| \\ &= \left| x - \frac{\alpha}{|\ln(1-y)|^p} - x \right| \\ &\leq \frac{\alpha}{|\ln(1-y)|^p}, \end{aligned}$$

i.e., (1.32) holds for

$$\psi(y) = \begin{cases} \dfrac{\alpha}{|\ln(1-y)|^p} & \text{if } \ y \in [1-\delta,1), \\ 0 & \text{if } \ y = 1. \end{cases}$$

Since

$$\begin{aligned} \sum_{n=1}^{\infty} \psi(1-q^n) &= \sum_{n=1}^{\infty} \frac{\alpha}{|\ln(q^n)|^p} \\ &= \sum_{n=1}^{\infty} \frac{\alpha}{n^p |\ln(q)|^p} < \infty, \end{aligned}$$

T is q-convergent.

Chapter 2

Probabilistic metric spaces

In 1942 K. Menger introduced the notion of a statistical metric space as a natural generalization of the notion of a metric space (M, d) in which the distance $d(p,q)$ $(p, q \in M)$ between p and q is replaced by a distribution function $F_{p,q} \in \Delta^+$. $F_{p,q}(x)$ can be interpreted as the probability that the distance between p and q is less than x.

Section 2.1 is devoted to the basic operations related distribution functions: copulas and triangle functions. Probabilistic metric spaces and some generalizations are recalled in section 2.2. Important probabilistic metric spaces such as Menger, Wald, transformation-generated and spaces related to E-processes, and Markov chains are briefly presented in section 2.3. Topology on probabilistic metric spaces is considered in section 2.4. Special important Menger space, random normed space with their some basic properties are presented in section 2.5. Fuzzy metric spaces and their relation with probabilistic metric spaces through a topological space of F-type as a common generalization are considered in section 2.6. The important notion of functional analysis, the measure of non-compactness is generalized to probabilistic settings in section 2.7. Special non-additive measures, the so called decomposable measures, are used in section 2.8 for the construction of a Menger space.

2.1 Copulas and triangle functions

2.1.1 Copulas

M. Fréchet, W. Höffding, R. Féron, and G. Dall'Aglio have considered functions that links a multi-dimensional probability distribution function to its one-dimensional margins. The explicit definition is owed to A. Sklar. We recommend as sources for further information [43, 259, 207, 268].

Definition 2.1 *A (two-dimensional) copula is a function* $C : [0,1]^2 \to [0,1]$ *such that:*

1) $C(x,0) = C(0,x) = 0$ *and* $C(x,1) = C(1,x) = x$ *for any* $x \in [0,1]$ *;*

2) $C(x_2,y_2) + C(x_1,y_1) \geq C(x_1,y_2) + C(x_2,y_1)$ *whenever* $x_1 \leq x_2$ *and* $y_1 \leq y_2$*.*

If C is a copula, then C is non-decreasing in each place and continuous. $T_{\mathbf{M}}$ (Fréchet–Höffding upper bound), $T_{\mathbf{P}}$ and $T_{\mathbf{L}}$ (Fréchet–Höffding lower bound) are associative and commutative copulas. We remark that $\mathbf{S_M}$ does not satisfy 2) in Definition 2.1, and therefore it is not a copula. We have for any copula C that

$$T_{\mathbf{L}} \leq C \leq T_{\mathbf{M}}.$$

Example 2.2 The family of Cuadras–Augé copulas $(C_\lambda^{\mathbf{CA}})_{\lambda\in[0,1]}$ defined by $C_\lambda^{\mathbf{CA}} = {T_{\mathbf{M}}}^{\lambda} \cdot {T_{\mathbf{P}}}^{1-\lambda}$ is not associative for $\lambda \neq 0,1$, and therefore it is not a t-norm. We have $C_0^{\mathbf{CA}} = T_{\mathbf{P}}$ and $C_1^{\mathbf{CA}} = T_{\mathbf{M}}$.

Definition 2.3 *A distribution function (on* $[-\infty,\infty]$*) is a function* $F : [-\infty,\infty] \to [0,1]$ *which is left-continuous on* $\mathbb{R}$*, non-decreasing and* $F(-\infty) = 0,\ F(\infty) = 1$*.*

We denote by Δ the family of all distribution functions on $[-\infty,\infty]$.

The *Dirac distribution function,* $H_a : [-\infty,\infty] \to [0,1]$ is defined for $a \in [-\infty,\infty)$ by

$$H_a(u) = \begin{cases} 0 & \text{if } u \in [-\infty,a], \\ 1 & \text{if } u \in (a,\infty], \end{cases}$$

and for $a = \infty$ by

$$H_\infty(u) = \begin{cases} 0 & \text{if } u \in [-\infty,\infty), \\ 1 & \text{if } u = \infty. \end{cases}$$

The order on Δ is taken in the usual way (pointwise): for two distribution functions F and G we have $F \leq G$ if and only if $F(x) \leq G(x)$ for every $x \in [-\infty,\infty]$. Then $(\Delta,\leq)$ is a complete lattice with the maximal element $H_{-\infty}$ and the minimal element H_∞.

Let $(\Omega,\mathcal{A},P)$ be a probability measure space and $X : \Omega \to \mathbb{R}$ an $\mathcal{A}$-measurable function (random variable). Then the function $F_X : [-\infty,\infty] \to [0,1]$ defined by

$$F_X(x) = P(\{\omega \mid \omega \in \Omega, X(\omega) < x\}) \quad (= P(X < x))$$

is a distribution function of the random variable X. A *joint distribution function* is a function $J : [-\infty,\infty]^2 \to [0,1]$ which is left-continuous on $\mathbb{R}$ with $J(-\infty,\infty) = 1$ and such that for every $x,y,x',y' \in [-\infty,\infty]$ $J(x,-\infty) = J(-\infty,x) = 0$ and

$$J(x,y) + J(x',y') \geq J(x,y') + J(x',y).$$

For a joint distribution J the marginal distribution functions $F, G \in \Delta$ are defined by

$$F(x) = J(x, \infty), \quad G(x) = J(\infty, x).$$

We say for a random vector (X, Y) that the function $J_{XY} : [-\infty, \infty] \to [0, 1]$ defined by

$$J_{XY}(x, y) = P(X < x, Y < y)$$

is a joint distribution function of (X, Y), and its marginal distributions coincide with the distribution functions F_X and F_Y of X and Y, respectively.

The following key theorem represents by a copula the joint distribution of a random vector by corresponding marginal distributions.

Theorem 2.4 (Sklar's theorem) *If J is a two-dimensional distribution function with one-dimensional marginal distribution functions F and G then there exists a copula C such that for all $x, y \in \mathbb{R}$,*

$$J(x, y) = C(F(x), G(y)).$$

If F and G are continuous then C is unique; otherwise C is uniquely determined on (Range F) $\times$ (Range G).

It follows that if X and Y are real random variables with distribution functions F_X and F_Y and joint distribution function J_{XY} then there is a copula C_{XY} such that

$$J_{XY} = C_{XY}(F_X(x), F_Y(y)).$$

The random variables X, Y are independent if and only if it is possible to take $C = T_{\mathbf{P}}$. Sklar's theorem shows that much of the study of joint distribution functions can be reduced to the investigation of copulas.

Remark 2.5 Under a.s.strictly increasing transformations of X and Y the copula C_{XY} is invariant, although we can change the margins. Thus (for random variables with continuous distribution functions) the study of rank statistics may be characterized as the study of copulas and copula-invariant properties. For random variables with continuous distribution functions, the extreme copulas $T_{\mathbf{M}}$ and $T_{\mathbf{L}}$ are attained precisely when X is a.s. an increasing (respectively, decreasing) function of Y. Therefore copulas can be used to construct non-parametric measures of dependence.

Let $\star$ be the binary operation defined on the set of two-dimensional copulas for copulas C_1 and C_2 by

$$C_1 \star C_2(x, y) = \int_0^1 \frac{\partial C_1(x, t)}{\partial t} \cdot \frac{\partial C_2(t, y)}{\partial t} \, dt,$$

(these partial derivatives exist almost everywhere). Then $C_1 \star C_2$ is a copula, and the set of copulas is a non-commutative semigroup under the operation $\star$. The strong interpretation in the context of Markov processes is the following: If $(X_t)_{t\in I}$ is a real stochastic process with parameter set I and if C_{st} is the copula of X_s and X_t, then the transition probabilities of the process satisfy the Kolmogorov–Chapman equation if and only if $C_{st} = C_{su} \star C_{ut}$ for all $s, t, u \in I$ with $s < u < t$, see [46].

Remark 2.6 The concept of a copula can be extended to n dimensions. An n-copula is an n-dimensional distribution function whose support is in the unit n-cube and whose one-dimensional margins are uniform, see [268]. If J is an n-dimensional distribution function with one-dimensional margins $F_1, \ldots, F_n$, then there is an n-copula C such that

$$J(x_1, \ldots, x_n) = C(F_1(x_1), \ldots, F_n(x_n))$$

for all $(x_1, \ldots, x_n) \in \mathbb{R}^n$. Moreover, for any n-copula:

$$T_{\mathbf{L}}(x_1, \ldots, x_n) \leq C(x_1, \ldots, x_n) \leq T_{\mathbf{M}}(x_1, \ldots, x_n).$$

The upper function $T_{\mathbf{M}}$ is an n-copula for any $n \in \mathbb{N}$, the lower function $T_{\mathbf{L}}$ is not an n-copula for any $n > 2$. The main problem in the theory of copulas is to determine which sets of copulas (of possible different dimensions) can appear as margins of a single higher-dimensional copula.

2.1.2 Triangle functions

Definition 2.7 *A distance distribution function $F : [-\infty, \infty] \to [0, 1]$ is a distribution function with support contained in $[0, \infty]$. The family of all distance distribution functions will be denoted by Δ^+. We denote $\mathcal{D}^+ = \{F \mid F \in \Delta^+, \lim_{x\to\infty} F(x) = 1\}$.*

Since any function from Δ^+ is equal zero on $[-\infty, 0]$ we can consider the set Δ^+ consisting of non-decreasing functions F defined on $[0, \infty]$ that satisfy $F(0) = 0$ and $F(\infty) = 1$. Moreover, $\mathcal{D}^+$ then consists of non-decreasing functions F defined on $[0, \infty)$ that satisfy $F(0) = 0$ and $\lim_{x\to\infty} F(x) = 1$. The class $\mathcal{D}^+$ will play the important role in the probabilistic fixed point theorems.

Now we will introduce an operation τ on Δ^+ that makes (Δ^+, τ) a semigroup.

Definition 2.8 *A* triangle function τ *is a binary operation on Δ^+ that is commutative, associative, and non-decreasing in each place, and has H_0 as identity.*

$(\Delta^+, \leq)$ is a complete lattice. We have $H_a \in \Delta^+$ for $a \in [0, \infty]$. We remark that H_∞ is the null element of τ. Namely, for every $F \in \Delta^+$ we have

$$H_\infty \leq \tau(H_\infty, F) \leq \tau(H_\infty, H_0) = H_\infty .$$

Similarly, as we have done for t-norms, we can compare triangle functions in the following way:
If τ_1 and τ_2 are triangle functions then τ_1 is weaker than τ_2 (or τ_2 is stronger than τ_1), $\tau_1 \leq \tau_2$, if for all F, G in Δ^+ and all x in $\mathbb{R}^+$

$$\tau_1(F, G)(x) \leq \tau_2(F, G)(x) .$$

Example 2.9 Let T be a left-continuous t-norm. Then the function $\mathbf{T} : \Delta^+ \times \Delta^+ \to \Delta^+$ defined by

$$\mathbf{T}(F, G)(x) = T(F(x), G(x))$$

is a triangle function.

Example 2.10 $\mathbf{T_M}$ defined by

$$\mathbf{T_M}(F, G)(x) = T_{\mathbf{M}}(F(x), G(x))$$

is the maximal triangle function. Indeed for any triangle function τ we have

$$\tau(F, G) \leq \tau(F, H_0) = F$$

and

$$\tau(F, G) \leq \tau(H_0, G) = G .$$

Hence

$$\begin{aligned} \tau(F, G)(x) &\leq T_{\mathbf{M}}(F(x), G(x)) \\ &= \mathbf{T_M}(F, G)(x) \quad (x \in \mathbb{R}^+) . \end{aligned}$$

Example 2.11 If T is a left-continuous t-norm, then τ_T, defined by

$$\tau_T(F, G)(x) = \sup\{T(F(u), G(v)) \mid u + v = x\}$$

is a triangle function. We can also write for a t-norm T and $F, G \in \Delta^+$

$$\tau_T(F, G)(x) = \sup_{\substack{\alpha,\beta \geq 0 \\ \alpha+\beta=1}} T(F(\alpha x), G(\beta x)) \ (x \in \mathbb{R}).$$

A triangle function is *continuous*, if it is continuous in the topology of the weak convergence on Δ^+.

We can not generally characterize the semigroup (Δ^+, τ) in a similar way to the semigroup $([0, 1], T)$ up to some special cases, as, for example, in the situation which leads to Menger spaces.

Example 2.12 If $F, G \in \Delta^+$ then we define their convolution $F * G$ on $[0, \infty)$ by $(F * G)(0) = 0$, $(F * G)(\infty) = 1$ and

$$(F * G)(x) = \int_{[0,x)} F(x - t)\, dG(t) \quad \text{for } x \in (0, \infty) .$$

The convolution is a commutative and associative binary operator on Δ^+ which is non-decreasing and has H_0 as a neutral element and therefore it is a triangle function.

It is important to obtain a rich source of different triangle functions which would enable the construction of new probabilistic metric spaces. One of the useful constructions goes in the following way.

Definition 2.13 $\mathcal{L}$ is the *class of all binary operators* L *on* $[0, \infty)$ which satisfy the following conditions:

(i) L maps $[0, \infty)^2$ onto $[0, \infty)$;

(ii) L is non-decreasing in both coordinate;

(iii) L is continuous on $[0, \infty)^2$ (except possibly at the points $(0, \infty)$ and $(\infty, 0)$).

We introduce now, using the family $\mathcal{L}$, the following binary operation on Δ^+.

Definition 2.14 For a t-norm T and $L \in \mathcal{L}$, the function $\tau_{T,L}$ defined on $\Delta^+ \times \Delta^+$ and with values in $[0, \infty)$ is given by

$$\tau_{T,L}(F, G)(x) = \sup \{T(F(u), G(v)) \mid L(u, v) = x\} .$$

In the special case $L(x, y) = x + y$ we obtain $\tau_{T,L} = \tau_T$. The following theorem guarantees the triangularity of $\tau_{T,L}$ under mild conditions.

Theorem 2.15 *If T is left-continuous t-norm and L from $\mathcal{L}$ is commutative, associative, has 0 as identity and satisfy the condition*

$$\text{if } u_1 < u_2 \text{ and } v_1 < v_2 \text{ then } L(u_1, v_1) < L(u_2, v_2) , \tag{2.1}$$

then $\tau_{T,L}$ is a triangle function.

Condition (2.1) is weaker than strict increasingness of L in each place. Therefore min and max satisfies it, although they are not strictly increasing in each place.

Example 2.16 $\tau_{T,\max}$ is a triangle function for any left-continuous t-conorm T.

The left-continuity of T in the preceding theorem is only sufficient but not necessary condition.

Example 2.17 $T_{\mathbf{D}}$ is not left-continuous but since we can represent $\tau_{T_{\mathbf{D}}}$ in the following way

$$\tau_{T_{\mathbf{D}}}(F,G)(x) = \max\left(F\left(\max\left(x - G^{(-1)}(1),0\right)\right), G\left(\max\left(x - F^{(-1)}(1),0\right)\right)\right)$$

we obtain that $\tau_{T_{\mathbf{D}}}$ is a triangle function on $\mathcal{D}^+$, and $G^{(-1)}$ is the pseudo-inverse of G.

It is interesting to note that the semigroup (Δ^+, τ_T) is not cancellative, but the semigroup $(\mathcal{D}^+, \tau_{T_{\mathbf{M}}})$ is cancellative, i.e., $\tau_{T_{\mathbf{M}}}(F,G) = \tau_{T_{\mathbf{M}}}(F,R)$ implies $F = 0$ or $G = R$.

Definition 2.18 *Let C be a copula and let $L : [0,\infty]^2 \to [0,\infty]$ be a surjective and continuous function on $[0,\infty]^2 \setminus \{(0,\infty),(\infty,0)\}$ and for each $x \in [0,\infty)$ the set $L_x = \{(u,v) \in [0,\infty]^2 \mid L(u,v) < x\}$ is bounded and $([0,\infty], L, \leq)$ is a partially ordered semigroup. The function $\sigma_{C,L} : \Delta^+ \times \Delta^+ \to \Delta^+$ is defined by*

$$\sigma_{C,L}(F,G)(x) = \begin{cases} 0 & \text{if } x \in [-\infty, 0], \\ \displaystyle\int_{L_x} dC(F(u),G(v)) & \text{if } x \in (0,\infty), \\ 1 & \text{if } x = \infty, \end{cases}$$

where the integral is of Lebesgue–Stieltjes type.

$\sigma_{C,L}$ is a triangle function if and only if $C = (< (a_\alpha, e_\alpha), T_{\mathbf{P}} >)_{\alpha \in A}$, see [268], Corollary 7.4.4.

2.2 Definitions of probabilistic metric and related spaces

Now we are coming to the appropriate probabilistic generalization of metric spaces. The following definition is proposed by Schweizer and Sklar [263].

Definition 2.19 *A probabilistic metric space in the sense of Schweizer and Sklar is an ordered pair $(S,\mathcal{F})$, where S is a nonempty set and $\mathcal{F} : S \times S \to \Delta^+$, if and only if the following conditions are satisfied ($\mathcal{F}(p,q) = F_{p,q}$, for every $(p,q) \in S \times S$) :*

1. *for every $(p,q) \in S \times S$, $F_{p,q}(0) = 0$;*
2. *for every $(p,q) \in S \times S$, $F_{p,q} = F_{q,p}$;*

3. $F_{p,q}(u) = 1$, *for every* $u > 0 \iff p = q$;
4. *for every* $(p, g, r) \in S \times S \times S$ *and for every* $x, y > 0$,

$$F_{p,q}(x) = 1,\ F_{q,r}(y) = 1 \Rightarrow F_{p,r}(x + y) = 1.$$

The principal inconvenience of the implication 4 in Definition 2.19 is that it is vacuous in all spaces in which the functions $F_{p,q}$, for $p \neq q$, never attain the value 1.

In the paper [273] there was introduced a probabilistic generalization of the triangle inequality for metrics. Namely, extended an idea belonging to Wald, Šerstnev suggested the use of a triangle function.

Definition 2.20 (a) *A probabilistic metric space in the sense of Šerstnev is a triple* $(S, \mathcal{F}, \tau)$ *where* S *is a nonempty set,* $\mathcal{F} : S \times S \to \Delta^+$ *is given by* $(p, q) \mapsto F_{p,q}$, τ *is a triangle function, such that the following conditions are satisfied for all* p, q, r *in* S :

(i) $F_{p,p} = H_0$;

(ii) $F_{p,q} \neq H_0$ *for* $p \neq q$;

(iii) $F_{p,q} = F_{q,p}$;

(iv) $F_{p,r} \geq \tau(F_{p,q}, F_{q,r})$.

(b) $(S, \mathcal{F}, \tau)$ *is proper if* $\tau(H_a, H_b) \geq H_{a+b}$ $(a, b \in [0, \infty))$.

Definition 2.21 *If only (i) and (iii) in Definition 2.20 hold then the pair* $(S, \mathcal{F})$ *is a probabilistic premetric space.*

Definition 2.22 *If only (i),(iii) and (iv) in Definition 2.20 hold then* $(S, \mathcal{F}, \tau)$ *is a probabilistic pseudo-metric space and* $(S, \mathcal{F})$ *is a probabilistic premetric space under* τ.

Definition 2.23 *If only (i),(ii) and (iii) in Definition 2.20 hold then the pair* $(S, \mathcal{F})$ *is a probabilistic semi-metric space.*

When we say, in what follows, that "$(S, \mathcal{F})$ is a probabilistic metric space", we mean that $(S, \mathcal{F})$ is a probabilistic metric space in the sense of Schweizer and Sklar, and that $(S, \mathcal{F}, \tau)$ is a probabilistic metric space in the sense of Šerstnev. In the following chapters related to the fixed point theorems in the probabilistic metric spaces $(S, \mathcal{F}, \tau)$ we shall always assume that

$$\text{Range}(\mathcal{F}) \subset \mathcal{D}^+.$$

Hicks and Sharma [134] introduced the following definition.

Definition 2.24 *A probabilistic semi-metric space $(S, \mathcal{F})$ is called a Hicks space if for every $\varepsilon > 0$ there exists $\delta > 0$ such that for every $p, q, r \in S$ the implication*

$$(F_{p,q}(\delta) > 1 - \delta, F_{q,r}(\delta) > 1 - \delta) \quad \Rightarrow \quad F_{p,r}(\varepsilon) > 1 - \varepsilon \tag{2.2}$$

holds.

V. Radu [250] generalized the implication (2.2) in the following way.

Definition 2.25 *Let $f : [0,1] \to [0,\infty)$ be a continuous strictly decreasing function with $f(0) = 1$. A probabilistic semi-metric space $(S, \mathcal{F})$ is called a probabilistic f-metric structure if the following implication holds:*

$$(\forall\varepsilon > 0)(\exists\delta > 0)(\forall p, q, r \in S)(f \circ F_{p,q}(\delta) < \delta, f \circ F_{q,r}(\delta) < \delta \quad \Rightarrow \quad f \circ F_{p,r}(\varepsilon) < \varepsilon)$$

In [55] R. Egbert introduced the notion of the product of probabilistic metric spaces. Let $(M_1, \mathcal{F}_1)$ and $(M_2, \mathcal{F}_2)$ be probabilistic metric spaces and τ a triangle function. The *τ-product* of $(M_1, \mathcal{F}_1)$ and $(M_2, \mathcal{F}_2)$ is the ordered pair $(M_1 \times M_2, \mathcal{F}_1\tau\mathcal{F}_2)$ such that $\mathcal{F}_1\tau\mathcal{F}_2$ is a function defined on $(S_1 \times S_2)^2$ and with values in $\mathcal{D}^+$ defined by

$$(\mathcal{F}_1\tau\mathcal{F}_2)(\mathbf{p}, \mathbf{q}) = \tau(\mathcal{F}_1(p, q), \mathcal{F}_2(p, q))$$

for every $\mathbf{p} = (p_1, p_2)$ and $\mathbf{q} = (q_1, q_2)$ from $S_1 \times S_2$.

To obtain that the τ-product of two probabilistic metric spaces is again probabilistic metric space we need some additional conditions. Extending Definition 1.60 for a triangle function we obtain the following definition.

Definition 2.26 *A triangle function τ_2 dominates a triangle function $\tau_1, \tau_2 \gg \tau_1$, if for each $F, G, H, K \in \Delta^+$,*

$$\tau_2(\tau_1(F, G), \tau_1(H, K)) \geq \tau_1(\tau_2(F, H), \tau_2(G, K)).$$

We have that if triangle function τ_1 dominates triangle function τ, $\tau_1 \gg \tau$, then τ_1-product of two probabilistic metric spaces $(M_1, \mathcal{F}_1, \tau)$ and $(M_2, \mathcal{F}_2, \tau)$ is again a probabilistic metric space. Specially this is true for $\tau = \tau_T$ and $\tau_1 = \mathbf{T}$ for a continuous t-norm T.

2.3 Some classes of probabilistic metric spaces

Now we will list some important classes of probabilistic metric spaces.

2.3.1 Menger and Wald spaces

Definition 2.27 *Let $(S, \mathcal{F}, \tau)$ be a probabilistic metric space and $\tau = \tau_T$, where*

$$\tau_T(F, G)(x) = \sup \{T(F(u), G(v)) \mid u + v = x\}$$

for a t-norm T. Then $(S, \mathcal{F}, \tau)$ is the so called Menger space, which will be denoted by $(S, \mathcal{F}, T)$.

Remark 2.28 If the t-norm T is left continuous then τ_T of Definition 2.27 is a triangle function.

Then we have

$$F_{p,r}(x + y) \geq T\left(F_{p,q}(x), F_{q,r}(y)\right) \tag{2.3}$$

for all p, q, r in S and x, y real numbers. This inequality also implies (iv). Namely, taking $x \in [0, \infty)$ we have for all $u, v \in [0, \infty)$ such that $u + v = x$

$$F_{p,r}(x) \geq T\left(F_{p,q}(u), F_{q,r}(v)\right).$$

Hence

$$F_{p,r}(x) \geq \left(\tau_T(F_{p,q}, F_{q,r})\right)(x).$$

We can interpret the inequality (2.3) in the way of the classical metric spaces that the third side in a triangle depends on the other two sides in the sense that if the knowledge of two sides increases then also the knowledge of third side increases or that knowing the upper bounds of two sides we have an upper bound for the third side.

Remark 2.29 If $(S, \mathcal{F}, T_{\mathbf{L}})$ is a Menger space then it is also a Hicks space, since (2.2) is satisfied for $\delta = \varepsilon/2$. Namely, if

$$F_{p,q}\left(\frac{\varepsilon}{2}\right) > 1 - \frac{\varepsilon}{2}, \quad F_{q,r}\left(\frac{\varepsilon}{2}\right) > 1 - \frac{\varepsilon}{2},$$

then

$$\begin{aligned} F_{p,r}(\varepsilon) &\geq T_{\mathbf{L}}\left(F_{p,q}\left(\frac{\varepsilon}{2}\right), F_{q,r}\left(\frac{\varepsilon}{2}\right)\right) \\ &\geq F_{p,q}\left(\frac{\varepsilon}{2}\right) + F_{q,r}\left(\frac{\varepsilon}{2}\right) - 1 \\ &> 1 - \varepsilon. \end{aligned}$$

If the inequality $F_{x,z}(u + v) \geq T(F_{x,y}(u), F_{y,z}(v))$ is replaced by the inequality

$$F_{p,r}(\max(x, y)) \geq T(F_{p,q}(x), F_{q,r}(y))$$

the triplet $(S, \mathcal{F}, T)$ is a *non-Archimedean Menger space.*

As a very special case of a Menger space we obtain the classical metric space.

Example 2.30 If we suppose, that there exists a function d, $d : M \times M \to [0, \infty)$, such that

$$F_{p,q}(x) = H_{d(p,q)} \qquad (p, q \in M,\ x \in \mathbb{R}) \tag{2.4}$$

then we obtain that the Menger space $(M, \mathcal{F}, \tau_T)$, for $\mathcal{F}(p, q) = H_{d(p,q)}$, for any t-norm T is a classical metric space. Namely, we have to prove only the classical triangle inequality, since all other properties follow trivially. If we have for $p, q, r \in M$ that $d(p, q) < x$ and $d(q, r) < y$ for some $x, y > 0$, then by (2.4) it follows $F_{p,q}(x) = 1$ and $F_{q,r}(y) = 1$. Then by (2.3) and the boundary property of the t-norm T we obtain $d(p, r) < x + y$, which gives the desired inequality.

If we start from a metric space (M, d), then taking $F_{p,q}$ defined by (2.4) we obtain that for any t-norm T the function $F_{p,q}$ is a probability distribution function such that conditions (i)–(iv) in Definition 2.20 are satisfied for τ_T. In this way we have proved that $(M, \mathcal{F}, \tau_T)$ for $\mathcal{F}(p, q) = H_{d(p,q)}$ is a Menger space if and only if (M, d) is a classical metric space.

A very important class of probabilistic metric spaces is given in the following example, see [54].

Example 2.31 (Drossos) Let (M, d) be a separable metric space and $(\Omega, \mathcal{A}, P)$ a probability measure space. We shall denote by S the set of all the equivalence classes of measurable mappings $X : \Omega \to M$. If $\widehat{X}, \widehat{Y} \in S$ and $x \in \mathbb{R}$ then $F_{\widehat{X},\widehat{Y}}(x)$ is defined in the following way:

$$F_{\widehat{X},\widehat{Y}}(x) = P(\{\omega \mid \omega \in \Omega,\ d(X(\omega), Y(\omega)) < x\}) \quad (X \in \widehat{X}, Y \in \widehat{Y}).$$

It is obvious that $F_{X,Y}$ has the following properties:

(i) $F_{\widehat{X},\widehat{Y}}(0) = 0$ for every $\widehat{X}, \widehat{Y} \in S$;

(ii) $F_{\widehat{X},\widehat{Y}} = F_{\widehat{Y},\widehat{X}}$ for every $X, Y \in S$;

(iii) $F_{\widehat{X},\widehat{Y}}(x) = 1$ for every $\widehat{X}, \widehat{Y} \in S$, and every $x > 0 \Rightarrow X = Y$;

(iv) For every $\widehat{X}, \widehat{Y}, \widehat{Z} \in S$ and every $x, y \in \mathbb{R}$ the following inequality holds:

$$F_{\widehat{X},\widehat{Z}}(x + y) \geq T_{\mathbf{L}}(F_{\widehat{X},\widehat{Y}}(x), F_{\widehat{Y},\widehat{Z}}(y)). \tag{2.5}$$

In order to prove (2.5) we prove

$$P(\{\omega \mid \omega \in \Omega,\ d(X(\omega), Z(\omega)) < x + y\}) \geq P(\{\omega \mid \omega \in \Omega, d(X(\omega), Y(\omega)) < x\})$$

$$+P(\{\omega \mid \omega \in \Omega, d(Y(\omega), Z(\omega)) < y\}) - 1. \tag{2.6}$$

Since

$$d(X(\omega), Z(\omega)) \leq d(X(\omega), Y(\omega)) + d(Y(\omega), Z(\omega)), \quad \omega \in \Omega,$$

it follows that

$$\{\omega \mid \omega \in \Omega, d(X(\omega), Y(\omega)) < x\} \cap \{\omega \mid \omega \in \Omega, d(Y(\omega), Z(\omega)) < y\}$$

$$\subseteq \{\omega \mid \omega \in \Omega, d(X(\omega), Z(\omega)) < x + y\}.$$

Let $A = \{\omega \mid \omega \in \Omega, d(X(\omega), Y(\omega)) < x\}$ and $B = \{\omega \mid \omega \in \Omega, d(Y(\omega), Z(\omega)) < y\}$. Since $P(A \cap B) = P(A) + P(B) - P(A \cup B)$ it follows that $P(A \cap B) \geq P(A) + P(B) - 1$ and therefore $P(A \cap B) \geq \max\{P(A) + P(B) - 1, 0\}$. Hence (2.5) holds. The ordered triple $(S, \mathcal{F}, T_{\mathbf{L}})$, where $\mathcal{F}(p, q) = F_{p,q}$, for every $(p, q) \in S \times S$, is a Menger space, which is the so called *E-space over the metric space* (M, d).

A more general example, [224], will be given in section 2.8.2.

Let M be a nonempty set, $\mathcal{D}$ a collection of metrics on M, and μ a measure on $\mathcal{D}$, and the following conditions be satisfied:

1. For every $(p, q) \in M \times M$ and every $x \in X$

$$\{d \mid d \in \mathcal{D}, d(p, q) < x\}$$

is μ-measurable.

2. $\mu(\mathcal{D}) = 1$.

For every $(p, q) \in M \times M$ and every $x \in \mathbb{R}$ let

$$F_{p,q}(x) = \mu(\{d \mid d \in \mathcal{D}, d(p, q) < x\}).$$

Then $(M, \mathcal{F}, T_{\mathbf{L}})$ is a Menger space which is *metrically generated*, see [289].

Theorem 2.32 *Every metrically generated probabilistic metric space $(X, \mathcal{F})$ is isometric to an E-space, i.e., there exists an E-space $(X', \mathcal{F}')$ and a bijection $\pi : X \to X'$ such that for every $p, q \in X$*

$$\mathcal{F}'_{\pi(p),\pi(q)}(x) = \mathcal{F}_{p,q}(x) \text{ for every } x \in \mathbb{R}.$$

There exist E-spaces which are not metrically generated.

Definition 2.33 *A probabilistic metric space $(S, \mathcal{F}, \tau)$ for which τ is a convolution is called* Wald space.

We can prove in a way analogous for the general Menger space, using now the equality $H_a * H_b = H_{a+b}$, that for a function $d : M \times M \to [0, \infty)$ and $\mathcal{F}(p, q) = H_{d(p,q)}$ the triple $(M, \mathcal{F}, *)$ is a Wald space if and only if (M, d) is the classical metric space.

Theorem 2.34 *A probabilistic metric space $(S, \mathcal{F}, \tau)$, which is a Wald space, is a Menger space $(S, \mathcal{F}, T_{\mathbf{P}})$.*

Proof. In a Wald space, for any $x, y \geq 0$ and $p, q, r \in S$ we have

$$\begin{aligned} F_{p,r}(x+y) &\geq \int_0^{x+y} F_{p,q}(x+y-z)\, dF_{q,r}(z) \\ &= \int_0^{x+y} \left(\int_0^{x+y-z} dF_{p,q}(t) \right) dF_{q,r}(z) \\ &= \iint\limits_{\substack{t,z\geq 0 \\ t+z\leq x+y}} dF_{p,q}(t)\, dF_{q,r}(z). \end{aligned}$$

Since

$$\iint\limits_{\substack{t,z\geq 0 \\ t+z\leq x+y}} dF_{p,q}(t)\, dF_{q,r}(z) \geq \iint\limits_{\substack{0\leq t\leq x \\ 0\leq z\leq y}} dF_{p,q}(t)\, dF_{q,r}(z)$$

and

$$\begin{aligned} \iint\limits_{\substack{0\leq t\leq x \\ 0\leq z\leq y}} dF_{p,q}(t)\, dF_{q,r}(z) &= \int_0^x \int_0^y dF_{p,q}(t)\, dF_{q,r}(z) \\ &= \int_0^x dF_{p,q}(t) \int_0^y dF_{q,r}(z) \\ &= F_{p,q}(x) \cdot F_{q,r}(y), \end{aligned}$$

we obtain that

$$F_{p,r}(x+y) \geq F_{p,q}(x) \cdot F_{q,r}(y).$$

□

Let (S, d) be a metric space, $G \in \mathcal{D}^+$, and $\alpha > 0$. If $p = q$ $(p, q \in S)$ let $F_{p,q} = H_0$ and if $p \neq q$ let

$$F_{p,q}(x) = G\Big(\frac{x}{d(p,q)^\alpha}\Big) \text{ for every } x \in \mathbb{R}.$$

It can be proved [264] that $(S, \mathcal{F}, T)$ is a Menger space for an arbitrary t-norm T. This space is the so called *α-simple space* and it is usually denoted by (S, G, d, α). Simple spaces are used to construct a phenomenological theory which describes hystereses in large-scale physical systems, see [268].

2.3.2 Transformation-generated spaces

Let (S, d) be metric space and $\psi : S \to S$. For any $p, q \in S$ and any $n \in \mathbb{N}$, let

$$F_{p,q}^{(n)} = \frac{1}{n} \sum_{m=0}^{n-1} H_{d(\psi^m(p), \psi^m(q))},$$

where $\psi^m(p)$ is the value of the m-th iterate of the mapping ψ at p.

If $\mathcal{F}^{(n)}$ is defined on $S \times S$ by

$$\mathcal{F}^{(n)}(p,q) = F^{(n)}_{p,q}$$

then for every $(p,q) \in S \times S$

$$\mathcal{F}^{(n)}(p,p) = H_0, \qquad \mathcal{F}^{(n)}(p,q) = \mathcal{F}^{(n)}(q,p),$$

i.e., $(S, \mathcal{F}^{(n)})$ is a probabilistic premetric space and $(S, \mathcal{F}^{(n)}, \tau_{T_L})$ is a probabilistic pseudo-metric space. The following theorem of Sherwood holds, see [277].

Theorem 2.35 *Let (S,d) be a metric space and $\psi : S \to S$. For every $(p,q) \in S \times S$ let $\varphi_{p,q}$ be defined on $\mathbb{R}^+$ by*

$$\varphi_{p,q}(x) = \liminf_{n\to\infty} F^{(n)}_{p,q}(x) = \lim_{n\to\infty} (\inf\{F^{(m)}_{p,q}(x) \mid m \geq n\})$$

and $\mathcal{F}$ be defined on $S \times S$ by

$$F_{p,q} = \ell^- \varphi_{p,q}.$$

Then $(S, \mathcal{F}, \tau_{T_L})$ is a probabilistic pseudo-metric space.

The space $(S, \mathcal{F}, \tau_{T_L})$ is called *the transformation generated space* determined by the metric space (S,d) and the transformation ψ, and it is denoted by $[S, d, \psi]$.

If $(S, \mathcal{F})$ is the transformation generated space $[S, d, \psi]$ then Sherwood proved that

$$\mathcal{F}(\psi(p), \psi(q)) = \mathcal{F}(p,q) \text{ for every } (p,q) \in S \times S.$$

The transformation generated spaces are important in the ergodic theory, see [268, 277].

The preceding approach play an important role in chaos theory, see [269], leading to a theory of distributional chaos.

2.3.3 E-processes and Markov chains.

Definition 2.36 (R. Moynihan) *Let $(\Omega, \mathcal{A}, P)$ be a probability measure space, (M,d) a metric space and $Z \in \{[0,\infty), \mathbb{Z}_+\}$.*

Let $S : Z \times \Omega \to M$ be such that the following conditions hold:

1. For every $p \in S$ and $t \in Z$ the function $p_t : \Omega \to M$, defined by

$$p_t(\omega) = p(t,\omega), \ \omega \in \Omega$$

is P-measurable with respect to a suitable σ-algebra in (M,d);

2. For all $p, q \in S$, $t \in Z$, and $x \in \mathbb{R}^+$

$$D(p,q,t,x) = \{\omega | \omega \in \Omega, d(p_t(\omega), q_t(\omega)) < x\} \in \mathcal{A};$$

3. *For each* $p, q \in S$ *and* $t \in Z$, $F_{p,q}^{(t)} \in \Delta^+$, *where*

$$F_{p,q}^{(t)}(x) = P(D(p, q, t, x));$$

4. *For each* $t \in Z$, $\mathcal{F}^{(t)} : S \times S \to \Delta^+$, *where*

$$\mathcal{F}^{(t)}(p, q) = F_{p,q}^{(t)}.$$

If $\mathcal{F}^*(t) = \mathcal{F}^{(t)}$, $t \in Z$, *then* $(S, \mathcal{F}^*)$ *is an E-space process with base* $(\Omega, \mathcal{A}, P)$ *and target* (M, d).

Definition 2.37 *Let* M *be a denumerable set and* $Z \in \{[0, \infty), \mathbb{Z}_+\}$. *Let* $(P_{ij})_{i,j\in M}$ *be a family of functions from* Z *into* $[0, 1]$ *such that for any* $s, t \in Z$

$$\sum_{j\in M} P_{ij}(t) = P_{ii}(0) = 1,$$

$$P_{ik}(s + t) = \sum_{j\in M} P_{ij}(s)P_{jk}(t).$$

Then $(P_{ij})_{i,j\in M}$ *is a family of Markov transition probabilities on* M.

Definition 2.38 *An ME-chain is an E-process* $(S, \mathcal{F}^*)$ *with base* $(\Omega, \mathcal{A}, P)$ *and target* (M, d) *such that the following conditions are satisfied:*

(i) M *is denumerable.*
(ii) *For each* $p \in S$ *there is an* $i \in M$ *such that*

$$p_0 = i \quad a.s.$$

(iii) *There is a family* $(P_{ij})_{i,j\in M}$ *of Markov transition probabilities on* M *such that for every* $p \in S$, $j \in M$ *and* $t \in Z$

$$P(\{\omega | \omega \in \Omega,\ p_t(\omega) = j\}) = P_{ij}(t),$$

where i *is given in (ii).*

In any ME-chain, when the family $(P_{ij})_{i,j\in M}$ is specified, it becomes unnecessary to refer the probability measure space $(\Omega, \mathcal{A}, P)$.

The following two theorems hold, see [268].

Theorem 2.39 *Let* $(S, \mathcal{F}^*)$ *be an ME-chain with state space* (M, d) *and Markov matrix* $(P_{ij})_{i,j\in M}$. *For any* $i, j \in M$, $t \in Z$ *and* $x \in \mathbb{R}_+$ *let*

$$G_{ij}^{(t)}(x) = \sum_{k\in M} P_{ik}(t) \sum_{\substack{\ell\in M \\ d(k,\ell)<x}} P_{j\ell}.$$

Then the function $G_{ij}^{(t)}$ *is in* $\mathcal{D}^+$ *and for distinct points* $p, q \in S$

$$F_{p,q}^{(t)} = G_{ij}^{(t)} \quad \text{for all } t \in Z,$$

where $i = p_0$ *and* $j = q_0$.

Theorem 2.40 *Let* $(S, \mathcal{F}^*)$ *be an ME-chain with state space* (M, d) *and Markov matrices* $(P_{ij})_{i,j\in M}$. *Suppose that for every* $i, j \in M$ *there is a function* $G_{ij} \in \Delta^+$ *such that one of the following two conditions hold:*

(i) For all $i, j \in M$

$$G_{ij}^{(t)} \xrightarrow{w} G_{ij};$$

(ii) $Z = \mathbb{Z}_+$ *and for all* $i, j \in M$

$$\frac{1}{n}\sum_{m=1}^{n} G_{ij}^{(m)} \xrightarrow{w} G_{ij}.$$

Then $(S, \mathcal{F}, \tau_{T_L})$ *is a probabilistic premetric space, where* $\mathcal{F}$ *is defined by*

$$\mathcal{F}(p,q) = \begin{cases} H_0 & \text{if } p = q, \\ G_{ij} & \text{if } p \neq q, i = p_0, j = q_0. \end{cases}$$

2.4 Topology, uniformity, metrics and semi-metrics on probabilistic metric spaces

Different kinds of topologies can be introduced in a probabilistic metric space, see [268]. The *strong topology* is introduced by a *strong neighbourhood system* $\mathcal{N} = \bigcup_{p\in S} \mathcal{N}_p$, where $\mathcal{N}_p = \{N_p(t) \mid t > 0\}$ and

$$N_p(t) = \{q \mid F_{p,q}(t) > 1 - t\} \text{ for } t > 0 \text{ and } p \in S.$$

In [268] the following theorem is proved.

Theorem 2.41 *Let* $(S, \mathcal{F}, \tau)$ *be a probabilistic metric space. If* τ *is continuous, then the strong neighbourhood system* $\mathcal{N}$ *determines a Hausdorff topology on* S.

It is of special interest the (ε, λ)-topology on $(S, \mathcal{F}, \tau)$ which is introduced by a family of neigbourhoods $(N_p(\varepsilon, \lambda))_{p\in S, \varepsilon>0, \lambda\in(0,1)}$, where

$$N_p(\varepsilon, \lambda) = \{q \mid q \in S, F_{p,q}(\varepsilon) > 1 - \lambda\}.$$

Since $N_p(t,t) = N_p(t)$ for $t > 0$, and

$$N_p(\min(\varepsilon, \lambda)) \subseteq N_p(\varepsilon, \lambda) \text{ for every } \varepsilon > 0, \lambda \in (0,1)$$

the strong neighbourhood system is equivalent to the (ε, λ)-neighbourhood system.

If $(S, \mathcal{F}, T)$ is a Menger space and $\sup_{a<1} T(a,a) = 1$, then the family $(\mathcal{N}_p)_{p\in S}$ defines on S a metrizable topology. A sequence $(p_n)_{n\in\mathbb{N}}$ in S converges in the (ε, λ)-topology to $p \in S$ if for every $\varepsilon > 0$ and $\lambda \in (0,1)$ there exists $n_0(\varepsilon, \lambda) \in \mathbb{N}$ such that

$$F_{p_n,p}(\varepsilon) > 1 - \lambda \text{ for every } n \geq n_0(\varepsilon, \lambda)$$

(see [268]). It is known that if $(S, \mathcal{F}, T)$ is a Menger space, where $\sup_{a<1} T(a,a) = 1$, then the family $\{N(\varepsilon, \lambda) \mid \varepsilon > 0,\ \lambda \in (0,1)\}$ is a base for the Hausdorff uniformity $\mathcal{U}_{\mathcal{F}}$ in S, see [264], where

$$N(\varepsilon, \lambda) = \{(p,q) \mid (p,q) \in S \times S, F_{p,q}(\varepsilon) > 1 - \lambda\}.$$

A sequence $(p_n)_{n\in\mathbb{N}}$ in S is a Cauchy sequence if for every ε and $\lambda \in (0,1)$ there exists $n_0(\varepsilon, \lambda) \in \mathbb{N}$ such that $F_{p_n,p_m}(\varepsilon) > 1 - \lambda$, for every $n, m \geq n_0(\varepsilon, \lambda)$, and S is complete if every Cauchy sequence in S converges to a point in S.

It was J.F.C. Kingman [159] who firstly constructed a deterministic metric for a Wald space, which is a Menger space $(S, \mathcal{F}, \tau_{T_{\mathbf{P}}})$. If $(S, \mathcal{F}, *)$ is any Wald space

$$d(p,q) = -\log \int_0^\infty e^{-x}\, dF_{p,q}(x)$$

is a metric on S. The uniformity derived from the metric d, is equivalent to the uniformity $\mathcal{U}_{\mathcal{F}}$.

G. L. Cain and R. Kasriel introduced the (ε, λ)-topology in $(S, \mathcal{F}, \tau_{T_{\mathbf{M}}})$ by a family of pseudo-metrics $(d_\alpha)_{\alpha\in(0,1)}$, where

$$d_\alpha(p,q) = \sup\{x \mid x \in \mathbb{R}, F_{p,q}(x) \leq 1 - \alpha\} \quad (p, q \in S).$$

In the paper [200] a formula for a metric on Menger space under Archimedean t-norms is given.

The problem of metrization of probabilistic metric spaces is investigated in [268, 312] and a family of deterministic metrics on a Menger space is given by V. Radu [246] in the following way.

Let $\mathcal{M}$ be the family of all the mappings $m : \overline{\mathbb{R}} \to \overline{\mathbb{R}}$ ($\overline{\mathbb{R}} = [0, \infty]$) such that the following conditions are satisfied:

a) For every $t, s \geq 0 : m(t+s) \geq m(t) + m(s)$;

b) $m(t) = 0 \iff t = 0$;

c) m is continuous.

Let $(S, \mathcal{F}, \tau_T)$ be a Menger space with an Archimedean t-norm T which has an additive generator f. If $m_1, m_2 \in \mathcal{M}$, then the metric d_{m_1,m_2}, given by

$$d_{m_1,m_2}(p,q) = \sup\{t \mid t \geq 0,\ m_1(t) \leq f \circ F_{p,q}(m_2(t))\} \quad (p, q \in S)$$

defines the (ε, λ)-topology on S.

In a special case when $T = T_{\mathbf{L}}$, then $d_{m_1,m_2} : S \times S \to [0, \infty)$ $(m_1, m_2 \in \mathcal{M})$ is defined by

$$d_{m_1,m_2}(p,q) = \sup\{s \mid s \geq 0, m_1(s) \leq 1 - F_{p,q}(m_2(s))\} \quad (p, q \in S).$$

If $m_1(s) = m_2(s) = s$, for every $s \in \mathbb{R}$ we obtain the following metric:

$$d(p,q) = \sup\{s \mid s \in \mathbb{R}, s \leq 1 - F_{p,q}(s)\}(p, q \in S).$$

The following result, obtained in [263], is useful in the fixed point theory.

Proposition 2.42 *Let T be a continuous t-norm which is not of H-type. Then if φ_{a_T} is from Theorem 1.45 we have:*

(i) If $(S, \mathcal{F}, T_{\mathbf{P}})$ is a Menger space, then $(S, \varphi_{a_T}^{-1} \circ \mathcal{F}, T)$ is a Menger space with the same (ε, λ)-uniformity.

(ii) If $(S, \mathcal{F}, T)$ is a Menger space, then $(S, \varphi_{a_T}^{-1} \circ \mathcal{F}, T_{\mathbf{P}})$ is a Menger space with the same (ε, λ)-uniformity.

(iii) If $(S, \mathcal{F}, T_{\mathbf{L}})$ is a Menger space, then $(S, e^{\mathcal{F}-1}, T_{\mathbf{P}})$ is a Menger space with the same (ε, λ)-uniformity.

Proof. Let $(S, \mathcal{F}, T)$ be a Menger space and T a continuous t-norm which is not of H-type. Then by Theorem 1.45 there exists $a_T \in [0, 1)$ such that $T(a_T, a_T) = a_T$ and $T(x, x) < x$ for every $x \in (a_T, 1]$. Moreover $T|_{[a_T,1]}$ is isomorphic either with $T_{\mathbf{L}}$ or with $T_{\mathbf{P}}$. Suppose that $T_0 \in \{T_{\mathbf{L}}, T_{\mathbf{P}}\}$. Then

$$T(x, y) = \varphi_{a_T}^{-1}(T_0(\varphi_{a_T}(a), \varphi_{a_T}(b))), \text{ for every } a, b \geq a_T.$$

Let $\tilde{\mathcal{F}} : S \times S \to \mathcal{D}^+$ be defined by

$$\tilde{F}_{x,y}(t) = \begin{cases} 0 & \text{if } F_{x,y}(t) \leq a_T, \\ \varphi_{a_T}(F_{x,y}(t)) & \text{if } F_{x,y}(t) \geq a_T. \end{cases}$$

Then $(S, \tilde{\mathcal{F}}, T_0)$ is a Menger space, which is complete if $(S, \mathcal{F}, T)$ is complete.

If $(S, \tilde{\mathcal{F}}, T_0)$ is a complete Menger space, then $(S, \mathcal{F}, T)$ is a complete Menger space, where

$$F_{x,y}(t) = \begin{cases} 0 & \text{if } \tilde{F}_{x,y}(t) \leq a_T, \\ \varphi_{a_T}^{-1}(\tilde{F}_{x,y}(t)) & \text{if } \tilde{F}_{x,y}(t) \geq a_T. \end{cases}$$

□

Theorem 2.43 *Let $(b_n)_{n\in\mathbb{N}}$ be an increasing sequence from the interval $[0,1)$ and $\lim_{n\to\infty} b_n = 1$. If $(S, \mathcal{F}, T)$ is a Menger space such that $T(b_n, b_n) = b_n$ for every $n \in \mathbb{N}$, then $r_n : S \times S \to [0, \infty)$ $(n \in \mathbb{N})$ defined by*

$$r_n(x, y) = \inf\{t \mid t > 0, F_{x,y}(t) \geq b_n\}$$

is a pseudo-metric on S and if the uniformity $\mathcal{U}_1$ is equal to the uniformity $\mathcal{U}_{\mathcal{R}}$, given by the family $\mathcal{R} = (r_n)_{n\in\mathbb{N}}$, then $\mathcal{U}_1 = \mathcal{U}_{\mathcal{F}}$.

Proof. Since $F_{x,x}(t) = 1$, for every $t > 0$, we have that

$$r_n(x, x) = \inf\{t \mid t > 0, F_{x,x}(t) \geq b_n\} = \inf\{t \mid t > 0\} = 0.$$

Suppose that $r_n(x, y) = 0$, for every $n \in \mathbb{N}$. Since the condition $F_{x,y}(t) \geq b_n$, for all $t > 0$, is equivalent with $r_n(x, y) = 0$, we obtain by $\lim_{n\to\infty} b_n = 1$ that $F_{x,y}(+0) = 1$ and therefore $x = y$. It is obvious that r_n is symmetric. It remains to prove that r_n satisfies the triangle inequality for every $n \in \mathbb{N}$. Let $x, y, z \in S$. In order to prove that for every $n \in \mathbb{N}$

$$r_n(x, y) \leq r_n(x, z) + r_n(z, y) \tag{2.7}$$

we suppose that $r_n(x, z) < a$ and $r_n(z, y) < b$. Then $F_{x,z}(a) \geq b_n$, $F_{z,y}(b) \geq b_n$, and therefore

$$F_{x,z}(a + b) \geq T(F_{x,z}(a), F_{z,y}(b)) \geq T(b_n, b_n) = b_n.$$

Hence $r_n(x, z) \leq a + b$ and (2.7) is proved. We shall prove that $\mathcal{U}_1 = \mathcal{U}_{\mathcal{F}}$, i.e., that the family of pseudo-metrics $(r_n)_{n\in\mathbb{N}}$ induces the uniformity $\mathcal{U}_{\mathcal{F}}$. It is obvious that $r_n \leq r_{n+1}$, for every $n \in \mathbb{N}$, and that for every $n \in \mathbb{N}$ we have that $r_n(x, y) < a$ implies $F_{x,y}(a) \geq b_n$. On the other hand

$$F_{x,y}(\varepsilon) > 1 - \lambda,\ 1 - \lambda \geq b_n \text{ implies } r_n(x, y) < \varepsilon.$$

□

2.5 Random normed and para-normed spaces

A very important class of Menger spaces are random normed spaces, introduced by Šerstnev [274]. We denote by $\mathcal{K}$ the real field $\mathbb{R}$ or the complex field $\mathbb{C}$.

Definition 2.44 *Let S be a vector space over $\mathcal{K}, \mathcal{F} : S \to \mathcal{D}^+$ and t-norm T is such that $T \geq T_{\mathbf{L}}$. The ordered triple $(S, \mathcal{F}, T)$ is a random normed space if and only if the following conditions are satisfied, where $\mathcal{F}(p) = F_p$, for every $p \in S$:*

1. *$F_p(0) = 0$ for every $p \in S$ and $F_p = H_0 \Longleftrightarrow p = 0$;*
2. *$F_{\lambda p}(x) = F_p\Big(\frac{x}{|\lambda|}\Big)$ for every $p \in S$, $x > 0$, $\lambda \in \mathcal{K} \setminus \{0\}$;*
3. *$F_{p+q}(x + y) \geq T(F_p(x), F_q(y))$, for every $p, q \in S$ and every $x, y > 0$.*

Every normed space $(S, \|\cdot\|)$ is a random normed space $(S, \mathcal{F}, T_{\mathbf{L}})$, where

$$\mathcal{F}_p(x) = \begin{cases} 1 & \text{if } \|p\| < x, \\ 0 & \text{if } \|p\| \geq x. \end{cases}$$

Let us remark that these conditions are generalizations of the conditions for a norm.

If $(S, \mathcal{F}, T)$ is a random normed space and for every $(p, q) \in S \times S$, $\overline{\mathcal{F}} : S \times S \to \mathcal{D}^+$ is defined by

$$\overline{\mathcal{F}}(p, q) = F_{p-q},$$

then $(S, \overline{\mathcal{F}}, T)$ is a Menger space.

Example 2.45 Let $(M, \|\cdot\|)$ be a separable normed space, $(\Omega, \mathcal{A}, P)$ a probability measure space and $S(\Omega, \mathcal{A}, P)$ the set of all the equivalence classes of measurable mappings $X : \Omega \to M$. If $\mathcal{F} : S \to \mathcal{D}^+$ is defined by

$$F_{\widehat{X}}(\varepsilon) = P(\{\omega \mid \omega \in \Omega, \|X(\omega)\| < \varepsilon\}) \quad (\varepsilon > 0, \widehat{X} \in S, X \in \widehat{X}),$$

then $(S(\Omega, \mathcal{A}, P), \mathcal{F}, T_{\mathbf{L}})$ is a random normed space.

Every random normed space $(S, \mathcal{F}, T)$, with a continuous t-norm T, is a topological vector space with a denumerable base of neighbourhoods of zero in S. Then there exists an F-norm $\|\cdot\| : S \to [0, \infty)$ ($(F1) - (F4)$) hold) such that the (ε, λ)–topology and the topology induced by $\|\cdot\|$ coincides, where:

$(F1)$ for every $x \in S : \|x\| \geq 0$ and $\|x\| = 0 \Longleftrightarrow x = 0$;
$(F2)$ for every $x \in S$ and every $a \in \mathcal{K}$ ($\mathcal{K}$ is the scalar field)

$$|a| \leq 1 \Rightarrow \|ax\| \leq \|x\|;$$

$(F3)$ for every $(x, y) \in S \times S : \|x + y\| \leq \|x\| + \|y\|$;
$(F4)$ for every $x \in S$ and every $(a_n) \in \mathcal{K}^{\mathbf{N}}$

$$\lim_{n\to\infty} a_n = 0 \Rightarrow \lim_{n\to\infty} \|a_n x\| = 0.$$

Lemma 2.46 *Let $(S, \mathcal{F}, T)$ be a random normed space and $d : S \times S \to [0, \infty)$ be defined by*

$$d(x, y) = \sup\{t \mid F_{x-y}(t) \leq 1 - t\} \quad (x, y \in S). \tag{2.8}$$

Then d is a metric on S which induces the (ε, λ)-topology and $\|x\|^* = d(0, x), x \in S$ is an F-norm.

In a random normed space the (ε, λ)-topology can be introduced, see [238], by another equivalent F-norm

$$\nu(p) = \inf_{s>0}\{s + 1 - F_p(s)\} \quad (p \in S).$$

Lemma 2.47 *If $(S, \mathcal{F}, T)$ is a random normed space and T is a continuous t-norm of the H-type, the (ε, λ)-topology in S can be introduced by the family of semi-norms $(s_n)_{n \in \mathbb{N}}$*

$$s_n(p) = \inf\{u \mid u \in \mathbb{R}, F_p(u) \geq b_n\} \ (p \in S, \ n \in \mathbb{N}),$$

which induces the structure of a locally convex topological vector space in S. Here $b_n \uparrow 1$ and $T(b_n, b_n) = b_n (n \in \mathbb{N})$.

We shall give a different proof that a random normed space $(S, \mathcal{F}, T)$, where T is a continuous t-norm T of H-type, is a locally convex space. The proof is based on the following result, [174]:

Theorem 2.48 *Let $(X, \|\cdot\|)$ be an F-normed space such that*

$$(\forall (x_i) \in X^{\mathbb{N}}) \left(\lim_{n \to \infty} x_n = 0 \quad \Rightarrow \quad \text{co}\{x_i \mid i \in \mathbb{N}\} \text{ is bounded}\right).$$

Then X is a locally convex space.

In [86] the following proposition is proved.

Proposition 2.49 *Let $(S, \mathcal{F}, T)$ be a random normed space with a continuous t-norm T of the H-type. Then in the (ε, λ)-topology S is a locally convex topological vector space.*

Proof. Let $(x_i) \in S^{\mathbb{N}}$ and $\lim_{n \to \infty} x_n = 0$. We shall prove that the set $\text{co}\{x_n \mid n \in \mathbb{N}\}$ is bounded, i.e., that for every $\varepsilon > 0$ and $\lambda \in (0, 1)$ there exists $\rho(\varepsilon, \lambda) > 0$ such that $\text{co}\{x_n \mid n \in \mathbb{N}\} \subseteq \rho(\varepsilon, \lambda) N(\varepsilon, \lambda)$, i.e.,

$$F_x(\rho(\varepsilon, \lambda)\varepsilon) > 1 - \lambda \ \text{ for every } x \in \text{co } \{x_n \mid n \in \mathbb{N}\}. \tag{2.9}$$

Choose $\varepsilon > 0$ and $\lambda \in (0, 1)$ and let $\delta = \delta(\varepsilon, \lambda) \in (0, 1)$ be such that

$$u \in (\delta, 1) \Rightarrow u_T^{(n)} > 1 - \lambda \text{ for every } n \in \mathbb{N}.$$

Such an element $\delta(\varepsilon, \lambda)$ exists since T is of the H-type. Since the sequence $(x_n)_{n\in\mathbb{N}}$ tends to zero, there exists $n(\varepsilon, \lambda) \in \mathbb{N}$ such that

$$F_{x_n}(\varepsilon) > \delta \quad \text{for every} \quad n \geq n(\varepsilon, \lambda).$$

Let $M_1 = \{x_n \mid n \geq n(\varepsilon, \lambda)\}$ and $M_2 = \{x_n \mid n < n(\varepsilon, \lambda)\}$. Since F_x is a distribution function there exists $\rho'(\varepsilon, \lambda) \geq 1$ such that

$$F_x(\rho'(\varepsilon, \lambda)\varepsilon) > \delta \quad \text{for every} \quad x \in M_2.$$

We shall prove that (2.9) holds for $\rho(\varepsilon, \lambda) = 2\rho'(\varepsilon, \lambda)$.

If $x \in \text{co}\{x_n \mid n \in \mathbb{N}\}$ then

$$x = \sum_{k=1}^{n} r_k x_{i_k} + \sum_{s=1}^{m} p_s x_{j_s},$$

for some $r_k, p_s \geq 0$ ($k \in \{1, 2, \ldots, n\}, s \in \{1, 2, \ldots, m\}$), where

$$\sum_{k=1}^{n} r_k + \sum_{s=1}^{m} p_s = 1,$$

and $x_{i_k} \in M_1 (k \in \{1, 2, \ldots, n\})$, $x_{j_s} \in M_2$ $(s \in \{1, 2, \ldots, m\})$.

Suppose that $\sum\limits_{k=1}^{n} r_k < 1$ and $\sum\limits_{s=1}^{m} p_s < 1$. If $\sum\limits_{k=1}^{n} r_k = 1$ or $\sum_{s=1}^{m} p_s = 1$ the proof is similar. We have that

$$\begin{aligned}
F_x(2\rho'(\varepsilon, \lambda)\varepsilon) &= F_{\sum\limits_{k=1}^{n} r_k x_{i_k} + \sum\limits_{s=1}^{m} p_s x_{j_s}}(2\rho'(\varepsilon, \lambda)\varepsilon) \\
&\geq T\left(F_{\sum\limits_{k=1}^{n} r_k x_{i_k}}(\rho'(\varepsilon, \lambda)\varepsilon), F_{\sum\limits_{s=1}^{m} p_s x_{j_s}}(\rho'(\varepsilon, \lambda)\varepsilon)\right).
\end{aligned}$$

Further on

$$\begin{aligned}
F_{\sum\limits_{k=1}^{n} r_k x_{i_k}}(\rho'(\varepsilon, \lambda)\varepsilon) &= F_{\sum\limits_{k=1}^{n} r_k x_{i_k}}\left(\rho'(\varepsilon, \lambda)\left(\sum_{k=1}^{n} r_k + 1 - \sum_{k=1}^{n} r_k\right)\varepsilon\right) \\
&\geq \underbrace{T\Big(T\Big(\ldots T\Big(T}_{\text{n-times}}\Big(F_{x_{i_1}}(\rho'(\varepsilon, \lambda)\varepsilon), F_{x_{i_2}}(\rho'(\varepsilon, \lambda)\varepsilon), \\
&\qquad \ldots, F_{x_{i_n}}(\rho'(\varepsilon, \lambda)\varepsilon), F_0\Big(\rho'(\varepsilon, \lambda)\Big(1 - \sum_{k=1}^{n} r_k\Big)\varepsilon\Big)\Big) \\
&\geq \delta_T^{(n)},
\end{aligned}$$

and similarly $F_{\sum\limits_{s=1}^{m} p_s x_{j_s}}(\rho'(\varepsilon,\lambda)\varepsilon) \geq \delta_T^{(m)}$. Hence

$$F_x(2\rho'(\varepsilon,\lambda)\varepsilon) \geq T(\delta_T^{(n)}, \delta_T^{(m)}) = \delta_T^{(m+n+1)} > 1-\lambda.$$

□

Definition 2.50 *Let $(S,\mathcal{F})$ be a probabilistic semi-metric space and $\varepsilon > 0, \lambda \in (0,1)$. The space $(S,\mathcal{F})$ is called (ε,λ)-chainable if for each $p,q \in S$ there exists a finite sequence of points in S, $p = p_0, p_1, \ldots, p_n = q$, such that*

$$F_{p_{i+1},p_i}(\varepsilon) > 1-\lambda \text{ for every } i \in \{0,1,\ldots,n-1\}.$$

Proposition 2.51 *Let $(S,\mathcal{F},T)$ be a random normed space. Then for every $\varepsilon > 0$ and $\lambda \in (0,1)$, $(S,\mathcal{F})$ is an (ε,λ)-chainable probabilistic semi-metric space.*

Proof. Since $T \geq T_{\mathbf{L}}$ it follows that $\sup_{x<1} T(x,x) = 1$ and therefore for every $\lambda \in (0,1)$ there exists $\lambda' \in (0,1)$ such that $T(\lambda',\lambda') > 1-\lambda$. Let $p,q \in S$, $\varepsilon > 0$ and $\lambda \in (0,1)$. Since $F_p, F_q \in \mathcal{D}^+$ there exists $\delta > 0$ such that $\frac{\varepsilon}{2\delta} < 1$ and

$$F_p(\delta) > \lambda', \ F_q(\delta) > \lambda'.$$

If $p_i = (1-\lambda_i)p + \lambda_i q, i \in \{0,1,\ldots,n\}$, where $n = \left[\dfrac{2\delta}{\varepsilon} + 1\right]$ and $\lambda_i = i\dfrac{\varepsilon}{2\delta}, i \in \{0,1,\ldots,n-1\}, \lambda_n = 1$. Then

$$\begin{aligned}
F_{p_{i+1},p_i}(\varepsilon) &\geq T\left(F_p\left(\frac{\varepsilon}{2(\lambda_{i+1}-\lambda_i)}\right), F_q\left(\frac{\varepsilon}{2(\lambda_{i+1}-\lambda_i)}\right)\right) \\
&\geq T\left(F_p\left(\frac{\varepsilon}{2}\cdot\frac{2\delta}{\varepsilon}\right), F_q\left(\frac{\varepsilon}{2}\cdot\frac{2\delta}{\varepsilon}\right)\right) \\
&\geq T(\lambda',\lambda') \\
&> 1-\lambda
\end{aligned}$$

for every $i \in \{0,1,\ldots,n-1\}$. □

In [98] we introduced the notion of a random para-normed space , as a generalization of the notion of a para-normed space.

Definition 2.52 *Let E be a vector space over $\mathcal{K}$ and $p : E \to [0,\infty)$ so that the following conditions are satisfied:*

(i) $p(x) = 0 \iff x = 0$;
(ii) $p(x) = p(-x)$ *for every* $x \in E$;
(iii) $p(x+y) \leq p(x) + p(y)$ *for every* $x,y \in E$;
(iv) if $\lim\limits_{n\to\infty} \lambda_n = \lambda$ $(\lambda_n, \lambda \in \mathcal{K})$ *and* $\lim\limits_{n\to\infty} p(x_n - x) = 0$ $(x_n, x \in E)$ *then* $\lim\limits_{n\to\infty} p(\lambda_n x_n - \lambda x) = 0$.

Then (E,p) is a para-normed space and p a para-norm.

A para-normed space (E,p) is a topological vector space if the fundamental system of neighbourhoods of zero is given by $\mathcal{V} = (V_\varepsilon)_{\varepsilon>0}$, where $V_\varepsilon = \{x \mid x \in E,\ p(x) < \varepsilon\}$.

The space $S(0,1)$ of all classes of real measurable mappings on $(0,1)$ is a para-normed space if the para-norm $p : S(0,1) \to [0,\infty)$ is given by

$$p(\widehat{x}) = \int_0^1 \frac{|x(t)|}{1+|x(t)|}\mathrm{m}_0(dt)\ (\{x(t)\} \in \widehat{x}). \tag{2.10}$$

The para-norm p defined by (2.10) is not homogeneous. More general, if $(\Omega,\mathcal{A},P)$ is a probability measure space and X a P-measurable mapping from Ω into $\mathbb{R}^n$ then a para-norm on $S(\Omega,\mathcal{A},P)$ is defined by

$$p(\widehat{X}) = \int_\Omega \frac{\|X(\omega)\|_{\mathbb{R}^n}}{1+\|X(\omega)\|_{\mathbb{R}^n}}dP.$$

Definition 2.53 *An ordered triple $(E,\mathcal{F},T)$, where E is a real vector space, $\mathcal{F} : E \to \mathcal{D}^+$ and $T \geq T_{\mathbf{L}}$, is a random para-normed space if $\mathcal{F} : E \to \mathcal{D}^+$ satisfies 1.–4., where:*

1. $F_p = H_0 \iff p = 0;$
2. $F_{-x} = F_x$ *for every* $x \in E$;
3. $F_{x+y}(u_1+u_2) \geq T(F_x(u_1), F_y(u_2))$ *for every* $x, y \in E$ *and every* $u_1, u_2 \geq 0$;
4. *if* $\lambda_n \to \lambda$ $(\lambda_n, \lambda \in \mathbb{R})$ *and* $\lim_{n\to\infty} F_{x_n - x}(\varepsilon) = 1$ *for every* $\varepsilon > 0$ $(x_n, x \in E)$ *then* $\lim_{n\to\infty} F_{\lambda_n x_n - \lambda x}(\varepsilon) = 1$ *for every* $\varepsilon > 0$.

Every para-normed space (E,p) is also a random para-normed space if

$$F_x(\varepsilon) = \begin{cases} 1 & \text{if } p(x) < \varepsilon, \\ 0 & \text{if } p(x) \geq \varepsilon. \end{cases}$$

The topology in a random para-normed space is the (ε,λ)-topology, as in a Menger space. A random para-normed space is a topological vector space in the (ε,λ)-topology, if the t-norm T is continuous.

2.6 Fuzzy metric spaces

Fuzzy metric spaces, introduced by Kramosil and Michalek [172], are in some sense generalizations of probabilistic metric spaces. Using the notion of a fuzzy number

Kaleva and Seikkala [155] defined the notion of a fuzzy metric space and investigated the relationship between probabilistic metric and fuzzy metric spaces.

A *fuzzy number* x is defined as a mapping $x : \mathbb{R} \to [0,1]$. By E we shall denote the set of all fuzzy numbers such that the following conditions hold:

a) $x(t) \geq \min\{x(s), x(r)\}$ $(s \leq t \leq r)$;
b) there exists $t_0 \in \mathbb{R}$ such that $x(t_0) = 1$;
c) x is an upper semi-continuous mapping.

From a)–c) it follows that for every $\alpha \in (0,1]$ the set

$$[x]_\alpha = \{t \mid t \in \mathbb{R}, x(t) \geq \alpha\}$$

is a closed interval $[a^\alpha, b^\alpha]$ if $a^\alpha, b^\alpha \in \mathbb{R}$, and if $a^\alpha = -\infty$ or $b^\alpha = \infty$, it is semi-open interval $(-\infty, b^\alpha]$ or $[a^\alpha, \infty)$, respectively.

If, for every $\alpha \in (0,1]$, $[x]_\alpha$ is convex, then x is a *convex* fuzzy number, i.e., a) holds.

A fuzzy number x is *non-negative* if $x(t) = 0$ for $t < 0$. The set of all non-negative fuzzy numbers from E will be denoted by G.

Let $[x]_\alpha = [a_1^\alpha, b_1^\alpha], [y_\alpha] = [a_2^\alpha, b_2^\alpha]$, for every $\alpha \in (0,1]$. A *partial ordering* $\leq$ in G is defined by

$$x \leq y \iff \text{for every } \alpha \in (0,1) : a_1^\alpha \leq a_2^\alpha, b_1^\alpha \leq b_2^\alpha.$$

A sequence $(x_n)_{n\in\mathbb{N}}$ in G is called *α-level convergent* to $x \in G$ if and only if

$$\lim_{n\to\infty} a_n^\alpha = a^\alpha, \qquad \lim_{n\to\infty} b_n^\alpha = b^\alpha,$$

where

$$[x_n]_\alpha = [a_n^\alpha, b_n^\alpha], [x_\alpha] = [a^\alpha, b^\alpha] \text{ for every } \alpha \in (0,1].$$

Let X be a nonempty set, $d : X \times X \to G$, and $L, R : [0,1] \times [0,1] \to [0,1]$. It is assumed that L and R are symmetric, associative, non-decreasing in both arguments, and

$$L(0,0) = 0, \qquad R(1,1) = 1.$$

Let

$$[d(x,y)]_\alpha = [\lambda_\alpha(x,y), \rho_\alpha(x,y)] \; (x, y \in X, \; 0 < \alpha \leq 1).$$

The ordered quadruple (X, d, L, R) is a *fuzzy metric space* and d is a *fuzzy metric* if and only if the following conditions hold :

(i) $d(x,y) = \bar{0} \iff x = y$;
(ii) $d(x,y) = d(y,x)$ for every $x, y \in X$;
(iii) For every $x, y \in X$:
a)

$$d(x,y)(s+t) \geq L(d(x,z)(s), d(z,y)(t))$$

if

$$s \leq \lambda_1(x,z), t \leq \lambda_1(z,y), s+t \leq \lambda_1(x,y);$$

b)

$$d(x,y)(s+t) \leq R(d(x,z)(s), d(z,y)(t))$$

if

$$s \geq \lambda_1(x,z), t \geq \lambda_1(z,y), s+t \geq \lambda_1(x,y).$$

Every metric space is a fuzzy metric space since $x^+ \in G$, where $x^+(t) = 1$ for $t \geq 0$, and $x^+(t) = 0$ for $t < 0$. The mappings L and R are given by

$$L(a,b) = 0, \ R(a,b) = \begin{cases} 0 & \text{if } a = b = 0, \\ 1 & \text{otherwise}. \end{cases}$$

If $R = \max$, (iii)b) is equivalent to the triangle inequality

$$\rho_\alpha(x,y) \leq \rho_\alpha(x,z) + \rho_\alpha(z,y) \text{ for every } \alpha \in (0,1] \text{ and every } x,y,z \in X.$$

Analogously, if $L = \min$, (iii)a) is equivalent to the triangle inequality

$$\lambda_\alpha(x,y) \leq \lambda_\alpha(x,z) + \lambda_\alpha(z,y) \text{ for every } \alpha \in (0,1] \text{ and every } x,y,z \in X.$$

Let $(S, \mathcal{F}, T)$ be a Menger space and let $d : S \times S \to G$ be defined in the following way:

$$d(p,q)(s) = \begin{cases} 0 & \text{if } s < s_{p,q} = \sup\{s \mid F_{p,q}(s) = 0\}, \\ 1 - F_{p,q}(s) & \text{if } s \geq s_{p,q}. \end{cases}$$

Then $d(p,q) \in G$. Let $R : [0,1] \times [0,1] \to [0,1]$ be defined by

$$R(a,b) = 1 - T(1-a, 1-b) \ (a, b \in [0,1]).$$

Then from the inequality

$$F_{p,r}(u+v) \geq T(F_{p,q}(u), \ F_{q,r}(v))$$

it follows that

$$\begin{aligned} 1 - F_{p,r}(u+v) &\leq 1 - T(F_{p,q}(u), \ F_{q,r}(v)) \\ &= R(1 - F_{p,q}(u), \ 1 - F_{q,r}(v)). \end{aligned}$$

If $L \equiv 0$, then (S, d, L, R) is a fuzzy metric space.

Let (X,d,L,R) be a fuzzy metric space such that R is associative, $R(a,0) = a$ for every $a \in [0,1]$ and $\lim_{s\to\infty} d(p,q)(s) = 0$ for every $p,q \in X$.

If for every $p,q \in S$,

$$F_{p,q}(s) = \begin{cases} 0 & \text{if } s \leq \lambda_1(p,q), \\ 1 - d(p,q)(s) & \text{if } s \geq \lambda_1(p,q), \end{cases}$$

and $T(a,b) = 1 - R(1-a, 1-b)$, for every $a,b \in [0,1]$ then $\mathcal{F} : X \times X \to \mathcal{D}^+$ and T is a t-norm. The triple $(X,\mathcal{F},T)$ is a Menger space, which is easy to verify. The convergence in (X,d,L,R) is defined by

$$\lim_{n\to\infty} x_n = x \iff \lim_{n\to\infty} d(x_n, x) = \bar{0},$$

where

$$\bar{0}(t) = \begin{cases} 1 & \text{if } t = 0, \\ 0 & \text{if } t \neq 0. \end{cases}$$

If (X,d,L,R) is a fuzzy metric space and $\lim_{a\to 0_+} R(a,a) = 0$, then the family $\mathcal{U} = \{U(\varepsilon,\alpha)\}_{\varepsilon>0,\alpha\in(0,1]}$ defined by

$$U(\varepsilon,\alpha) = \{(p,q) \mid (p,q) \in S \times S,\ \rho_\alpha(p,q) < \varepsilon\}$$

is a base of the Hausdorff uniformity on $X \times X$. The topology induced by $\mathcal{U}$ is metrizable.

If a fuzzy metric space (X,d,L,R) is given, then we call the above defined Menger space $(X,\mathcal{F},T)$ the associated Menger space.

If the family $(R_n(u))_{n\in\mathbb{N}}$, $R_1(u) = R(u,u)$, $R_n(u) = R(R_{n-1}(u),u)$ $(n \geq 2,\ u \in [0,1])$ is equicontinuous at the point $u = 0$, then from the relation $u_T^{(n)} = 1 - R_n(1-u)$ $(u \in [0,1])$ it follows that the family $(u_T^{(n)})_{n\in\mathbb{N}}$ is equicontinuous at the point $u = 1$.

J.X. Fang introduced in [62] the notion of a topological space of F-type and proved that some classes of fuzzy metric and Menger spaces belong to the class of topological spaces of F-type.

Let $D = (D, \prec)$ be a directed set. In [62], the following definition is introduced.

Definition 2.54 *A topological space X is said to be of F-type if it is Hausdorff and for each $x \in X$ there exists a neighbourhood base $\mathcal{U}_x = \{U_x(\lambda,t) \mid \lambda \in D,\ t > 0\}$ of x with the following properties:*

$(F-1)$ *If $y \in U_x(\lambda,t)$ then $x \in U_y(\lambda,t)$;*

$(F-2)$ $U_x(\lambda, t) \subset U_x(\mu, s)$, *for* $\lambda \prec \mu, t \leq s$;
$(F-3)$ *For every* $\lambda \in D$ *there exists* $\mu \in D$ *with* $\lambda \prec \mu$ *such that*

$$U_x(\mu, t_1) \cap U_y(\mu, t_2) \neq \emptyset \Rightarrow y \in U_x(\lambda, t_1 + t_2);$$

$(F-4)$ $X = \bigcup_{t>0} U_x(\lambda, t)$ *for each* $\lambda \in D$ *and* $x \in X$.

In [62] it is proved that the topology of an F-type topological space X can be generated by a family $M = \{d_\lambda \mid \lambda \in D\}$ of quasi-metrics on X such that (P-1)–(P-4) hold, where:

$(P-1)$ $d_\lambda(x, y) = 0$ for all $\lambda \in D \iff x = y$.
$(P-2)$ $d_\lambda(x, y) = d_\lambda(y, x)$ for all $\lambda \in D$, and $x, y \in X$.
$(P-3)$ $d_\lambda(x, y) \leq d_\mu(x, y)$ for all $\lambda \prec \mu$ and $x, y \in X$.
$(P-4)$ For every $\lambda \in D$ there exists $\mu \in D$ with $\lambda \prec \mu$ such that

$$d_\lambda(x, y) \leq d_\mu(x, z) + d_\mu(z, y),$$

for all $x, y, z \in X$.

This means that the topology of X is equal to the topology defined by the neigbourhood base for x, $\mathcal{B} = \{\mathcal{B}_x\}_{x \in X}$ where

$$\mathcal{B}_x = \{B_x(\lambda, t) \mid \lambda \in D, t > 0\}$$

is a neighbourhood base for $x \in X$ and for $\lambda \in D$ and $t > 0$

$$B_x(\lambda, t) = \{y \mid y \in X, \ d_\lambda(x, y) < t\}.$$

As it was proved in [62], Hausdorff topological vector spaces and some classes of Menger probabilistic metric spaces belong to the class of F-type topological spaces. The class of F-type topological spaces contains the class of fuzzy metric spaces (X, d, L, R), where $\lim_{a \to 0_+} R(a, a) = 0$ and $\lim_{t \to \infty} d(x, y)(t) = 0$ for every $x, y \in X$.

Theorem 2.55 *Let* (X, d, L, R) *be a fuzzy metric space such that* $\lim_{a \to 0_+} R(a, a) = 0$, $\lim_{t \to \infty} d(x, y)(t) = 0$. *Then* X *is an* F*-type topological space with the topology induced by the family of quasi-metrics* $M = \{\rho'_\alpha \mid \alpha \in [0, 1)\}$, $\rho'_\alpha = \rho_{1-\alpha} (\alpha \in [0, 1))$.

Proof. From the condition $\lim_{t \to \infty} d(x, y)(t) = 0$ it follows that $\rho_\alpha(x, y) < \infty$, for every $\alpha \in (0, 1]$ and $x, y \in X$. Since ρ_α is non-increasing in α, ρ'_α is non-decreasing in $\alpha \in [0, 1)$, and therefore (P-3) is satisfied. (P-1) and (P-2) are obviously satisfied. It is easy to prove that for $\alpha, \mu \in (0, 1]$:

$$R(\mu, \mu) < \alpha \quad \Rightarrow \quad \rho_\alpha(x, y) \leq \rho_\mu(x, z) + \rho_\mu(z, y),$$

for every $x, y, z \in X$. Since $\lim_{a \to 0_+} R(a, a) = 0$, for every $\alpha \in (0, 1]$, there exists $\mu < \alpha$ such that $R(\mu, \mu) < \alpha$. Hence

$$\rho_\alpha(x, y) \leq \rho_\mu(x, z) + \rho_\mu(z, y)$$

and since $\mu < \alpha$, if $\alpha = 1 - \alpha'$, $\mu = 1 - \mu'$, we have that $\alpha' < \mu'$ and

$$\rho'_{\alpha'}(x, y) \leq \rho'_{\mu'}(x, z) + \rho'_{\mu'}(z, y)$$

which means that (P-4) holds. □

It can be proved analogously that every Menger space $(S, \mathcal{F}, T)$, where t-norm T is such that $\sup_{a<1} T(a, a) = 1$, is a topological space of F-type, where d_λ, $\lambda \in (0, 1)$ is defined by

$$d_\lambda(x, y) = \sup\{t \mid t > 0, F_{x,y}(t) \leq 1 - \lambda\} \text{ for every } x, y \in S.$$

2.7 Functions of non-compactness

In [45] G. Darbo defined a new class of mappings which contains completely continuous mappings and contractions as well, and for such a class of mappings he proved some fixed point theorems. He generalized the Schauder fixed point theorem, using the function $\alpha : B(M) \to [0, \infty)$, where (M, d) is a metric space, $B(M)$ is the family of all bounded subsets of M and for every $A \in B(M)$, $\alpha(A)$ is defined by

$$\alpha(A) = \inf \Big\{ \varepsilon \mid \varepsilon > 0, \exists \text{ finite } (A_j)_{j \in J} \text{ such that } A \subset \bigcup_{j \in J} A_j, \ \mathrm{diam} A_j < \varepsilon, \forall j \in J \Big\}.$$

The function α, introduced by Kuratowski [175], is called Kuratowski's measure of non-compactness. Darbo considered the class of mappings $f : M \to M$, where $(M, \|\cdot\|)$ is a normed space, such that for every $A \in B(M)$

$$\alpha(f(A)) \leq q\alpha(A), \tag{2.11}$$

where $q \in [0, 1)$. In some sense (2.11) means that $f(A)$ is 'more' compact than A and f, which satisfies (2.11), is the so called q-set contraction mapping, see [260]. There exists a large literature devoted to this subject with many generalizations and applications in fixed point theory, functional analysis, integral and differential equations, game theory and extremum problems.

In this section we shall present probabilistic generalizations of Kuratowski's and Hausdorff's measure of non-compactness together with some of their properties , see [17], which we shall use in the fixed point theory in probabilistic metric spaces.

Definition 2.57 and some applications of Kuratowski's function are given in [17]. First, we shall give the definition of the *Kuratowski function* α_A of the set A ($A \subseteq S$, $(S, \mathcal{F})$ is a probabilistic metric space).

Let $(S, \mathcal{F})$ be a probabilistic metric space and A a nonempty subset of S. The probabilistic diameter of A is defined by

$$D_A(x) = \sup_{u<x} \inf_{p,q\in A} F_{p,q}(u) \quad (x \in \mathbb{R})$$

and the set A is probabilistic bounded if and only if $\sup_x D_A(x) = 1$, which means that D_A is in $\mathcal{D}^+$.

Lemma 2.56 *Let $(S, \mathcal{F}, T)$ be a Menger space such that $\sup_{x<1} T(x,x) = 1$ and let A be a compact subset of S. Then A is a probabilistic bounded subset of S.*

Proof. We have to prove that

$$\sup_{x>0} D_A(x) = 1,$$

i.e., that for every $\lambda \in (0,1)$ there exists x_λ such that $D_A(x_\lambda) > 1 - \lambda$. Since A is compact in the (ε, λ)-topology, for every $\varepsilon > 0$ and $\lambda' \in (0,1)$ there exists a finite subset $U = \{u_1, u_2, \ldots, u_m\} \subseteq S$ such that

$$A \subseteq \bigcup_{i=1}^{m} N_{u_i}(\varepsilon, \lambda'). \tag{2.12}$$

Relation (2.12) means that for every $p \in A$ there exists u_i such that

$$F_{p,u_i}(\varepsilon) > 1 - \lambda'.$$

Let $\lambda \in (0,1)$ be given. Since $\sup_{x<1} T(x,x) = 1$ there exists $\lambda' \in (0,1)$ such that

$$T(T(1-\lambda', 1-\lambda'), 1-\lambda') > 1 - \lambda.$$

Let ε be an arbitrary positive number and $\{u_1, u_2, \ldots, u_m\}$ be such that (2.12) holds.

Let $\overline{\varepsilon}$ be such that

$$F_{u_i,u_j}(\overline{\varepsilon}) > 1 - \lambda' \quad \text{for every } i, j \in \{1, 2, \ldots, m\}.$$

Let $s(\varepsilon, \lambda') = \max\{3\varepsilon, 3\overline{\varepsilon}\}$. Then we have for every $p, q \in S$, $u_i, u_j \in U$,

$$\begin{aligned} F_{p,q}(s(\varepsilon,\lambda')) &\geq T\Big(T\Big(F_{p,u_i}\Big(\frac{s(\varepsilon,\lambda')}{3}\Big), F_{q,u_j}\Big(\frac{s(\varepsilon,\lambda')}{3}\Big)\Big), F_{u_i,u_j}\Big(\frac{s(\varepsilon,\lambda')}{3}\Big)\Big) \\ &\geq T(T(F_{p,u_i}(\varepsilon), F_{q,u_i}(\varepsilon)), F_{u_i,u_j}(\overline{\varepsilon})). \end{aligned}$$

If u_i and u_j are such that

$$F_{p,u_i}(\varepsilon) > 1 - \lambda', \qquad F_{q,u_j}(\varepsilon) > 1 - \lambda'$$

then

$$\begin{aligned} F_{p,q}(s(\varepsilon, \lambda')) &\geq T(T(1-\lambda', 1-\lambda), 1-\lambda') \\ &> 1 - \lambda, \end{aligned}$$

for every $p, q \in A$.

Hence

$$\inf_{p,q \in A} F_{p,q}(s(\varepsilon, \lambda')) \geq 1 - \lambda,$$

which implies that for $x_\lambda > s(\varepsilon, \lambda')$

$$\sup_{\delta < x_\lambda} \inf_{p,q \in A} F_{p,q}(\delta) \geq 1 - \lambda.$$

This means that $D_A(x_\lambda) \geq 1 - \lambda$ and A is a probabilistic bounded set. □

Definition 2.57 *Let $(S, \mathcal{F})$ be a probabilistic metric space and A a probabilistic bounded subset of S. Then the Kuratowski function $\alpha_A : \mathbb{R} \to [0,1]$ of the set A is defined for $x \in \mathbb{R}$ by*

$$\alpha_A(x) = \sup\Big\{\varepsilon \mid \varepsilon > 0, \exists\, (A_j)_{j \in J}, J \text{ is finite}, \ A \subseteq \bigcup_{j \in J} A_j, D_{A_j}(x) \geq \varepsilon, \forall j \in J\Big\}.$$

If (M, d) is a metric space and $(M, \mathcal{F}, T_{\mathrm{M}})$ is the associated Menger space, i.e., $\mathcal{F}$ is defined by $\mathcal{F}_{p,q}(x) = H_{d(p,q)}$ then for every bounded subset A of M

$$\alpha_A(x) = H_0(x - \alpha(A)).$$

It is easy to see that the Kuratowski function has the following properties:

a) $\alpha_A \in \mathcal{D}^+$;

b) A is probabilistic bounded $\Rightarrow \alpha_A \geq D_A$;

c) A and B are probabilistic bounded $\Rightarrow \alpha_{A \cup B} = \min\{\alpha_A, \alpha_B\}$.

The next theorem is proved in [13].

Theorem 2.58 *Let $(S, \mathcal{F}, T)$ be a random normed space and A and B two probabilistic bounded subsets of S. Then:*

1) $\alpha_{\lambda A}(x) = \alpha_A\left(\frac{x}{|\lambda|}\right)$, $x \in \mathbb{R}$, $\lambda \neq 0$ (λ is in the scalar field);

2) If T is continuous then $\alpha_{A+B}(x+y) \geq T(\alpha_A(x), \alpha_B(y))$ for every $(x, y) \in \mathbb{R} \times \mathbb{R}$;

3) If $T = T_{\mathbf{M}}$ then $\alpha_A = \alpha_{\text{co } A}$;

4) If T is continuous then $\alpha_A = \alpha_{\bar{A}}$, where $\bar{A}$ is the closure of A in the (ε, λ)-topology.

Proof. In order to prove 1) we shall prove that

$$D_{\lambda A}(x) = D_A\left(\frac{x}{|\lambda|}\right),$$

for every $x \in \mathbb{R}$ and $\lambda \neq 0$ (λ is in the scalar field).

Let $D'_A(x) = \inf_{p,q \in A} F_{p-q}(x)$. Then $D_A(x) = D'_A(x - 0)$. Hence

$$\begin{aligned} D'_{\lambda A} &= \inf_{p,q \in \lambda A} F_{p-q}(x) \\ &= \inf_{p',q' \in A} F_{\lambda(p'-q')}(x) \\ &= \inf_{p',q' \in A} F_{p'-q'}\left(\frac{x}{|\lambda|}\right) \\ &= D'_A\left(\frac{x}{|\lambda|}\right). \end{aligned}$$

This implies that $D_{\lambda A}(x) = D_A\left(\frac{x}{|\lambda|}\right)$. Let

$$K_{A,x} = \left\{\varepsilon \mid \varepsilon > 0, \exists (A_j)_{j \in J},\ J \text{ finite, such that } D_{A_j}(x) \geq \varepsilon, \forall j \in J, A \subseteq \bigcup_{j \in J} A_j\right\}.$$

We prove that for every $\lambda \neq 0$, $K_{\lambda A,x} \subseteq K_{A,x/|\lambda|}$. If $\varepsilon \in K_{\lambda A,x}$ then there exist $A_1, A_2, ..., A_n$ from S such that

$$\lambda A \subseteq \bigcup_{j=1}^{n} A_j \quad \text{and} \quad D_{A_j}(x) \geq \varepsilon \quad \text{for every} \quad j \in \{1, 2, \ldots, n\}.$$

Let $B_j = \frac{1}{\lambda} A_j$, for every $j \in \{1, 2, \ldots, n\}$. Then from $\lambda A \subseteq \bigcup_{j=1}^n A_j$ it follows that $A \subseteq \bigcup_{j=1}^n B_j$ and from $D_{A_j}(x) = D_{B_j}\left(\frac{x}{|\lambda|}\right)$ we have that $D_{B_j}\left(\frac{x}{|\lambda|}\right) \geq \varepsilon$ for every $j \in \{1, 2, \ldots, n\}$. This implies that $\varepsilon \in K_{A,x/|\lambda|}$ and therefore

$$K_{\lambda A,x} \subseteq K_{A,x/|\lambda|}. \tag{2.13}$$

From (2.13) we obtain that

$$\alpha_{\lambda A}(x) \le \alpha_A\left(\frac{x}{|\lambda|}\right) = \alpha_{\frac{1}{\lambda}\cdot\lambda A}\left(\frac{x}{|\lambda|}\right) \le \alpha_{\lambda A}(x),$$

which implies 1).

In order to prove that 2) holds observe that $D'_{A+B}(x+y) \ge T(D'_A(x), D'_B(y))$. By the continuity of t-norm T it follows that

$$D_{A+B}(x+y) \ge T(D_A(x), D_B(y)). \tag{2.14}$$

Since A and B are probabilistic bounded it follows from (2.14) that $A+B$ is probabilistic bounded.

Let $\varepsilon \in K_{A,x}$ and $\eta \in K_{B,y}$. Then there exist $(A_j)_{j\in J_1}$ and $(B_j)_{j\in J_2}$ such that J_1 and J_2 are finite and

$$A \subseteq \bigcup_{j\in J_1} A_j, \qquad B \subseteq \bigcup_{j\in J_2} B_j, \qquad D_{A_j}(x) \ge \varepsilon,\ j \in J_1, \qquad D_{B_j}(y) \ge \eta,\ j \in J_2.$$

Then

$$A+B \subseteq \bigcup_{(i,j)\in J_1\times J_2} C_{i,j} \text{ where } C_{i,j} = A_i + B_j,\ (i,j) \in J_1 \times J_2$$

and

$$D_{C_{i,j}}(x+y) = D_{A_i+B_j}(x+y) \ge T(D_{A_i}(x), D_{B_j}(y)) \ge T(\varepsilon,\eta). \tag{2.15}$$

(2.15) means that $T(\varepsilon,\eta) \in K_{A+B,x+y}$ which implies that $T(\varepsilon,\eta) \le \alpha_{A+B}(x+y)$, for every ε in $K_{A,x}$ and every η in $K_{B,y}$. Since T is continuous it follows that 2) is satisfied.

We prove that $\alpha_A = \alpha_{\text{co } A}$ if t-norm T is equal to $T_{\mathbf{M}}$. First, we shall show that

$$D_A = D_{\text{co}A}. \tag{2.16}$$

The set $\mathcal{D}^+$ is ordered by

$$F, G \in \mathcal{D}^+, \qquad F \le G \Longleftrightarrow F(x) \le G(x) \quad \text{for every } x > 0.$$

For $F \in \mathcal{D}^+$ and $p \in S$ let

$$S_p(F) = \{q \mid q \in S,\ F_{p-q} \ge F\}.$$

It is easy to see that $S_p(F)$ is a convex set, since for $q_1, q_2 \in S_p(F)$ and $u \in [0,1]$ we have

$$\begin{aligned} F_{p-uq_1-(1-u)q_2}(x) &\ge \min\{F_{u(p-q_1)}(ux), F_{(1-u)(p-q_2)}((1-u)x)\} \\ &\ge \min\{F(x), F(x)\} \\ &= F(x). \end{aligned}$$

For probabilistic bounded sets A and B such that $A \subseteq B$, we have that $D_B \leq D_A$ and therefore $D_{\text{co}A} \leq D_A$. In order to prove (2.16) it remains to prove that $D_{\text{co}A} < D_A$ leads to a contradiction. If $D_{\text{co}\,A} < D_A$ there exist p_1 and p_2 in co$A$ such that $F_{p_1-p_2} < D_A$. First, suppose that A is not contained in $S_{p_1}(D_A)$, which means that there exists $q \in A \setminus S_{p_1}(D_A)$. From the definition of D_A it follows that

$$A \subseteq S_q(D_A)$$

since $q \in A$, and for every $p \in A$

$$F_{p-q}(x) \geq \sup_{u<x} \inf_{q_1,q_2 \in A} F_{q_1-q_2}(u).$$

From the convexity of $S_q(D_A)$ it follows that co$A \subseteq S_q(D_A)$. Further, from $q \notin S_{p_1}(D_A)$, we obtain that $F_{q-p_1} < D_A$ and so $p_1 \notin S_q(D_A)$, which is a contradiction since $p_1 \in$ co A.

Suppose that $A \subseteq S_{p_1}(D_A)$. Then co$A \subseteq S_{p_1}(D_A)$. But from $F_{p_1-p_2} < D_A$ it follows that $p_2 \in \text{co}A \setminus S_{p_1}(D_A)$ and this is a contradiction. Hence, we have proved that $D_A = D_{\text{co}A}$. Let

$K_A = \{F \mid F \in \mathcal{D}^+, \exists (A_j)_{j\in J},\ J \text{ is finite}, A \subseteq \bigcup_{j\in J} A_j,\ D_{A_j} \geq F, \forall j \in J\}$.

Let $F \in K_A$ and $B_j = \text{co}A_j$, $j \in J$, $J = \{1, 2, \ldots, n\}$. From $A \subseteq \bigcup_{j\in J} A_j$ it follows that

$$\text{co}A \subseteq \left\{ p \mid p \in S,\ p = \sum_{j=1}^{n} \lambda_j p_j,\ p_j \in B_j,\ \lambda_j \geq 0,\ j \in \{1, 2, \ldots, n\},\ \sum_{j=1}^{n} \lambda_j = 1 \right\}.$$

Let

$$L = \left\{ \lambda \mid \lambda \in \mathbb{R}^n,\ \lambda = (\lambda_1, \lambda_2, \ldots, \lambda_n),\ \lambda_i \geq 0,\ i \in \{1, 2, \ldots, n\},\ \sum_{i=1}^{n} \lambda_i = 1 \right\}.$$

By $\|\lambda\|$ we shall denote $\max\limits_{i\in\{1,2,\ldots,n\}} |\lambda_i|$. Since L is compact, for every $\varepsilon > 0$ there exists a finite set

$$L_m = \{\lambda^1, \lambda^2, \ldots, \lambda^m\} \subseteq L, \quad \lambda^j = (\lambda_1^j, \ldots, \lambda_n^j),\ j \in \{1, \ldots, m\}$$

such that for every $\lambda \in L$ there exists $\lambda^j \in L_m$ such that $\|\lambda - \lambda^j\| < \varepsilon$.

Let $F_\varepsilon(x) = G\left(\dfrac{x}{n\varepsilon} - 0\right)$, where

$$G(x) = \inf \left\{ F_p(x) \mid p \in \bigcup_{i=1}^{n} B_i \right\}.$$

It is obvious that $F_\varepsilon \in \mathcal{D}^+$.

We prove that for every $p \in \mathrm{co}A$ there exist $j \in \{1, 2, \ldots, m\}$ and $q \in S_j$ such that $F_{p-q} \geq F_\varepsilon$, where

$$S_j = \lambda_1^j B_1 + \cdots + \lambda_n^j B_n \ (j \in \{1, 2, \ldots, m\}). \tag{2.17}$$

In order to prove $F_{p-q} \geq F_\varepsilon$ we shall show that $F_{p-q}(x) \geq G\Big(\dfrac{x}{n\varepsilon}\Big)$, for some $q \in S_j$.

From $p \in \mathrm{co}A$ it follows that

$$p = \sum_{i=1}^{n} \lambda_i p_i, \text{ where } \lambda = (\lambda_1, \ldots, \lambda_n) \in L \text{ and } p_i \in B_i \text{ for every } i \in \{1, 2, \ldots, n\}.$$

Let $\lambda^j \in L_m$ such that $\|\lambda - \lambda^j\| < \varepsilon$ and let $q = \sum_{i=1}^{n} \lambda_i^j p_i$. Then from (2.17) it follows that $q \in S_j$. Further

$$\begin{aligned} F_{p-q}(x) &= F_{\sum_{i=1}^n (\lambda_i - \lambda_i^j) p_i}\Big(\frac{x}{n} + \frac{n-1}{n} x\Big) \\ &\geq \min\left(F_{p_1}\Big(\frac{x}{n|\lambda_1 - \lambda_1^j|}\Big), F_{p_2}\Big(\frac{x}{n|\lambda_2 - \lambda_2^j|}\Big), \ldots, F_{p_n}\Big(\frac{x}{n|\lambda_n - \lambda_n^j|}\Big)\right). \end{aligned}$$

From the definition of G we obtain that

$$F_{p-q}(x) \geq \min\Big\{G\Big(\frac{x}{n\varepsilon}\Big), \ldots, G\Big(\frac{x}{n\varepsilon}\Big)\Big\} = G\Big(\frac{x}{n\varepsilon}\Big),$$

which implies that $F_{p-q} \geq F_\varepsilon$.

Let

$$V = \{p \mid p \in S, F_p \geq F_\varepsilon\}, \qquad C_j = V + S_j \ \text{ for every } j \in \{1, 2, \ldots, m\}.$$

We shall prove that $\mathrm{co}A \subseteq \bigcup_{j=1}^{m} C_j$. If $p \in \mathrm{co}A$ and $q \in S_j$ is defined as above, it follows that $p = (p - q) + q \in V + S_j = C_j$ since $F_{p-q} \geq F_\varepsilon$ means that $p - q \in V$. Then $D_V(x) \geq F_\varepsilon(x/2)$.

This follows from

$$\begin{aligned} D'_V(x) &= \inf_{q_1, q_2 \in V} F_{q_1 - q_2}(x) \\ &\geq \inf_{q_1, q_2 \in V} \min\{F_{q_1}(x/2), F_{q_2}(x/2)\} \\ &\geq F_\varepsilon(x/2). \end{aligned}$$

since

$$F_{q_1}(x/2) \geq F_\varepsilon(x/2), \quad F_{q_2}(x/2) \geq F_\varepsilon(x/2).$$

Further, $D_{S_j}(x) \geq F(x)$, for every $j \in \{1,2,\ldots,m\}$, since from the definition of S_j it follows that

$$\begin{aligned} D_{S_j}(x) &= D_{\lambda_1^j B_1+\cdots+\lambda_n^j B_n}\left(\sum_{i=1}^{n} \lambda_i^j x\right) \\ &\geq \min(D_{B_1}(x), D_{B_2}(x), \ldots, D_{B_n}(x)) \\ &= \min(D_{A_1}(x), D_{A_2}(x), \ldots, D_{A_n}(x)) \\ &\geq \min(F(x), F(x), \ldots, F(x)) \\ &= F(x). \end{aligned}$$

For a positive real number y denote by $F_{\varepsilon,y}$ the function

$$F_{\varepsilon,y}(x) = \min\left(F_\varepsilon\left(\frac{x-y}{2}\right), F(y)\right),$$

and prove that $F_{\varepsilon,y} \in K_{\mathrm{co}A}$. Since $\{C_1, C_2, \ldots, C_m\}$ is a cover of co$A$, it is enough to prove that $D_{C_j}(x) \geq F_{\varepsilon,y}(x)$ $(j \in \{1,2,\ldots,m\})$.

If $x - y \leq 0$, then

$$F_{\varepsilon,y}(x) = \min(0, F(y)) = 0 \leq D_{C_j}(x) \quad (j \in \{1,2,\ldots,m\}).$$

If $x > y$, the inequality $D_{C_j}(x) \geq F_{\varepsilon,y}(x)$ follows from

$$\begin{aligned} D_{C_j}(x) &\geq \min(D_V(x-y), D_{S_j}(y)) \\ &\geq \min\left(F_\varepsilon\left(\frac{x-y}{2}\right), F(y)\right) \\ &= F_{\varepsilon,y}(x). \end{aligned}$$

Therefore we have that $F_{\varepsilon,y} \in K_{\mathrm{co}A}$. It is easy to show that this implies that $F_{\varepsilon,y} \leq \alpha_{\mathrm{co}A}$, for every $\varepsilon > 0$ and $y > 0$.

From the definition of the function $F_{\varepsilon,y}$, for fixed $x > 0$ and $y \in (0,x)$, we obtain that

$$\lim_{\varepsilon\to 0_-} F_{\varepsilon,y}(x) = \min(1, F(y)) = F(y), \tag{2.18}$$

since $\lim\limits_{\varepsilon\to 0_-} F_\varepsilon\left(\frac{x-y}{2}\right) = 1$. Relation (2.18) implies that $F(y) \leq \alpha_{\mathrm{co}A}(x)$, for every $y \in (0,x)$ and therefore $F(x) \leq \alpha_{\mathrm{co}A}(x)$. Since F is an arbitrary element from K_A, it follows that $\alpha_A \leq \alpha_{\mathrm{co}A}$.

We prove that $\alpha_A = \alpha_{\bar{A}}$, where the closure of A is in the $(\varepsilon,\lambda)-$ topology. From $A \subseteq \bar{A}$ it follows that $\alpha_A \geq \alpha_{\bar{A}}$. It remains to prove that $\alpha_A \leq \alpha_{\bar{A}}$, which implies that $\alpha_A = \alpha_{\bar{A}}$. We shall show that $\varepsilon \in K_{A,x}$ implies that $\varepsilon \in K_{\bar{A},x}$. If $\varepsilon \in K_{A,x}$, then there exists $(A_i)_{i\in I}$ (I is finite) such that $A \subseteq \bigcup_{i\in I} A_i$ and $D_{A_i}(x) \geq \varepsilon$, for every $i \in I$. Since $\bar{A} \subseteq \bigcup_{i\in I} \bar{A}_i$ and $D_{\bar{A}_i}(x) = D_{A_i}(x) \geq \varepsilon$, see [55], it follows that $\varepsilon \in K_{\bar{A},x}$. This implies that $\alpha_A(x) \leq \alpha_{\bar{A}}(x)$. □

Definition 2.59 *Let $(S, \mathcal{F})$ be a probabilistic metric space and $A \subseteq S$. We say that A is a probabilistic precompact set if for every $\varepsilon > 0, \lambda \in (0,1)$ there exists a finite cover of A, $(A_i)_{i \in I}$, such that*

$$D_{A_i}(\varepsilon) > 1 - \lambda \quad \textit{for every } i \in I.$$

The above definition and the following four theorems 2.60, 2.61, 2.62 and 2.63 are given in [17].

Theorem 2.60 *If $A \subseteq S$ and A is a probabilistic precompact set, then for every $\varepsilon > 0, \lambda \in (0,1)$ there exists a finite subset $A_{\varepsilon,\lambda}$ of A such that for each $p \in A$ there exists $q = q(p) \in A_{\varepsilon,\lambda}$ such that $F_{p,q}(\varepsilon) > 1 - \lambda$.*

Proof. Take that $\varepsilon = 1/n$ and $\lambda = 1/m$. Then, from Definition 2.59 it follows that there exists a finite cover $(A_i^{mn})_{i \in I}$ of A such that $D_{A_i^{mn}}(1/n) > 1 - 1/m$ $(i \in I)$. Since for every $x \in \mathbb{R}$, $D_A(x) \leq F_{p,q}(x)$ $(p, q \in A)$, it follows that $F_{p,q}(1/n) > 1 - 1/m$, for every $p, q \in A_i^{mn}$.

Let p_i^{mn} be an element from A_i^{mn} and $A_{mn} = \{p_i^{mn} \mid i \in I\}$.

Let $r \in A$. Then there exists $i_0 \in I$ such that $r \in A_{i_0}^{mn}$. Then, from $p_{i_0}^{mn} \in A_{i_0}^{mn}$ it follows that

$$F_{r,p_{i_0}^{mn}}\left(\frac{1}{n}\right) > 1 - \frac{1}{m}.$$

For $\varepsilon > 0, \lambda \in (0,1)$ it is enough to take that $A_{\varepsilon,\lambda} = A_{mn}$ where $\frac{1}{n} < \varepsilon$ and $\frac{1}{m} < \lambda$. □

Theorem 2.61 *Let $(S, \mathcal{F}, T)$ be a Menger space with the t-norm T such that $\sup_{x<1} T(x, x) = 1$ and $A \subseteq S$. If for every $\varepsilon > 0$, $\lambda \in (0,1)$, there exists a finite subset $A_{\varepsilon,\lambda}$ of A, as in Theorem 2.60, then A is a probabilistic precompact set.*

Proof. Let $\varepsilon > 0, \lambda \in (0,1)$ and $\varepsilon' = \varepsilon/2$, $\lambda' \in (0,1)$ such that $T(1-\lambda', 1-\lambda') > 1 - \frac{\lambda}{2}$. Since $A_{\varepsilon',\lambda'}$ is finite, we have $A_{\varepsilon',\lambda'} = (p_i)_{i \in I}$, I is finite. We prove that there exists a finite cover $(A_i)_{i \in J}$ of A such that $D_{A_i}(\varepsilon) > 1 - \lambda$, for every $i \in J$. Let $p \in A$. Then there exists $q \in A_{\varepsilon',\lambda'}$ such that $F_{p,q}(\varepsilon') > 1 - \lambda'$. Thus, if

$$A_p(\varepsilon, \lambda) = \{q \mid q \in A, F_{p,q}(\varepsilon) > 1 - \lambda\},$$

then $A = \bigcup_{i \in I} A_i$, where $A_i = A_{p_i}(\varepsilon', \lambda')$ $(i \in I)$. It remains to prove that $D_{A_i}(\varepsilon) > 1 - \lambda$. Let $r, q \in A_i$. Then

$$\begin{aligned} F_{r,q}(\varepsilon) &\geq T\left(F_{r,p_i}(\varepsilon/2),\, F_{p_i,q}(\varepsilon/2)\right) \\ &\geq T(1-\lambda', 1-\lambda') \\ &> 1 - \frac{\lambda}{2}, \end{aligned}$$

which implies that $D_{A_i}(\varepsilon) > 1 - \lambda$, for every $i \in I$. □

Theorem 2.62 *Let $A \subseteq S$, $A \neq \varnothing$, where $(S, \mathcal{F}, T)$ is a Menger space such that $\sup_{x<1} T(x,x) = 1$. Then A is probabilistic precompact if and only if it is precompact with respect to the uniformity $\mathcal{U}_{\mathcal{F}}$.*

Proof. It is known that A is precompact relative to $\mathcal{U}_{\mathcal{F}}$ if and only if for every $\varepsilon > 0$ and $\lambda \in (0,1)$ there exists $\{p_1, p_2, \ldots, p_n\} = A_{\varepsilon,\lambda}$ such that $A = \bigcup_{i=1}^{n} A_{p_i}(\varepsilon, \lambda)$. Let $\varepsilon > 0$, $\lambda \in (0,1)$ and $\lambda' < \lambda$. Since $\sup_{x<1} T(x,x) = 1$, there exists $\eta > 0$ such that $T(1-\eta, 1-\eta) > 1-\lambda'$ and let $\{p_1, p_2, \ldots, p_n\}$ be such that $A = \bigcup_{i=1}^{n} A_{p_i}(\varepsilon/4, \eta)$. Let $A_i = A_{p_i}(\varepsilon/4, \eta)$ $(i \in \{1,2,\ldots,n\})$. We shall prove that $D_{A_i}(\varepsilon) > 1-\lambda$, for every $i \in \{1,2,\ldots,n\}$. For $u \in \left(\varepsilon/2, \varepsilon\right)$ we have that $(p, q \in A_i)$

$$\begin{aligned} F_{p,q}(u) &\geq T\left(F_{p,p_i}(u/2), F_{p_i,q}(u/2)\right) \\ &\geq T\left(F_{p,p_i}(\varepsilon/4), F_{p_i,q}(\varepsilon/4)\right) \\ &\geq T(1-\eta, 1-\eta) \\ &> 1-\lambda' \end{aligned}$$

and since

$$D_{A_i}(\varepsilon) = \sup_{u<\varepsilon} \inf_{p,q\in A_i} F_{p,q}(u) = \sup_{\frac{\varepsilon}{2}<u<\varepsilon} \inf_{p,q\in A_i} F_{p,q}(u),$$

it follows that $D_{A_i}(\varepsilon) \geq 1-\lambda' > 1-\lambda$, for every $i \in \{1,2,\ldots,n\}$. From Theorem 2.60 it follows that if A is probabilistic precompact it is precompact relative to $\mathcal{U}_{\mathcal{F}}$. □

Theorem 2.63 *Let $A \subseteq S$, where $(S, \mathcal{F})$ is a probabilistic metric space. Then A is a probabilistic precompact subset if and only if $\alpha_A = H_0$.*

Proof. First, we shall suppose that A is a probabilistic precompact and prove that $\alpha_A(x) = 1$, for every $x > 0$. Let $\eta \in (0,1)$ and $\varepsilon = x > 0$, $\lambda = 1-\eta$. Then there exists a finite cover $(A_i)_{i\in I}$ of A such that

$$D_{A_i}(\varepsilon) > 1-\lambda = \eta,$$

which means that $\eta \in K_{A,x}$. Therefore we have that $\eta \leq \alpha_A(x)$ and this implies that $1 \leq \alpha_A(x)$, for every $x > 0$. Using the relation $\alpha_A(x) \in \mathcal{D}^+$ we obtain that $\alpha_A = H_0$. Let us suppose that $\alpha_A = H_0$ and let $\varepsilon > 0, \lambda \in (0,1)$. Then $\alpha_A(\varepsilon) = 1$ and let $\varepsilon_\lambda \in K_{A,\varepsilon}$ be such that $\alpha_A(\varepsilon) - \lambda < \varepsilon_\lambda$ $\quad (1-\lambda < \varepsilon_\lambda)$. Since $\varepsilon_\lambda \in K_{A,\varepsilon}$ means that there exists a finite cover $(A_i)_{i\in I}$ of the set A such that $D_{A_i}(\varepsilon) > \varepsilon_\lambda > 1-\lambda$, $i \in I$, it follows that A is precompact relative to the uniformity $\mathcal{U}_{\mathcal{F}}$. □

In [300] the function of non-compactness $\beta_A(u)$ is defined in the following way:

$$\beta_A(u) = \sup\{\varepsilon \mid \varepsilon \geq 0, \text{ there exists a finite subset } A_f \text{ of } S \text{ such that } \mathbf{F}_{A,A_f}(u) \geq \varepsilon\},$$

where

$$\mathbf{F}_{A,B}(u) = \sup_{s<u} \inf_{x\in A} \sup_{y\in B} F_{x,y}(s)$$

(A and B are probabilistic bounded subsets of S) . It is proved in [300] that β_A has properties 1)–6) as the function α_A.

Further information on probabilistic generalizations of Kuratowski's and Hausdorff's measures of non-compactness can be found in [14, 15, 16, 18, 19].

2.8 Probabilistic metric spaces related to decomposable measure

Special non-additive measures the so called pseudo-additive (decomposable) measures and the corresponding integrals give a base for the so called pseudo-analysis with important applications in optimization, nonlinear partial differential equations, nonlinear difference equations, optimal control, fuzzy systems. For the range of a set function instead of the field of real numbers it is taken a semiring on a real interval $[a, b] \subset [-\infty, +\infty]$ (in this book we restrict on the interval $[0, 1]$), denoting the corresponding operations as $\oplus$ (pseudo-addition) and $\odot$ (pseudo-multiplication) (in this book t-conorm $\mathbf{S}$ and t-norm T, respectively). This structure is applied for solving nonlinear equations (ODE, PDE, difference equations, etc.) using now the pseudo-linear principle, see [171, 183, 217, 223]. Based on semiring structure it is developed in [164, 171, 183, 201, 217, 218, 219, 223] the pseudo-analysis in an analogous way as classical analysis, introducing $\oplus$-measure, pseudo-integral, pseudo-convolution, pseudo-Laplace transform, etc.

For further results and applications see [164, 171, 183, 201, 217, 218, 219, 223].

2.8.1 Decomposable measures

Let $\mathcal{A}$ be a σ-algebra of subsets of a given set Ω. A classical measure is a set function m defined on $\mathcal{A}$ and with values in $[0, \infty]$ such that $m(\varnothing) = 0$ and

$$m\left(\bigcup_{i=1}^{\infty} A_i\right) = \sum_{i=1}^{\infty} m(A_i)$$

for every sequence $(A_i)_{i\in\mathbb{N}}$ of pairwise disjoint sets from $\mathcal{A}$. In a special case, when the range of m is the interval $[0, 1]$ and $m(\Omega) = 1$ holds, we will call m a probability measure and we denote it by P.

Each measure is also finitely additive, in the sense that the sequence $(A_i)_{i\in\mathbb{N}}$ consists of finite number of nonempty elements. The opposite statement is true if we additionally suppose the continuity of m.

As a generalization of the classical notion of finitely additiveness for a set function $m : \mathcal{A} \to [0, 1]$ we introduce for $A, B \in \mathcal{A}$ and $A \cap B = \varnothing$ a functional dependence of $m(A \cup B)$ on $m(A)$ and $m(B)$, i.e., that there exists a function $F : [0, 1]^2 \to [0, 1]$ such that we have

$$m(A \cup B) = F(m(A), m(B)) \tag{2.19}$$

for every $A, B \in \mathcal{A}$ such that $A \cap B = \emptyset$. For finitely additive set function this functional dependence has the form $F(x,y) = x + y$, supposing the normalization condition $m(\Omega) = 1$.

Going back to the general case we have that the empty set plays the role of the neutral element with respect to the operation of the union of sets and it is natural to ask that $m(\emptyset) = 0$. But this property immediately implies that the function F in (2.19) have to satisfies $F(x, 0) = x$. The commutativity and associativity of the union implies by (2.19) that F as a binary operation have to be commutative and associative. And finally, if we want to save monotonicity property of m which was a characteristic property of a measure we have to suppose that F is monotone. Summarizing all these properties of the function F which appears in (2.19) we meet t-conorms. In this way we introduce (see [164, 217]) the following definition.

Definition 2.64 *Let* $\mathbf{S}$ *be a t-conorm. A* $\mathbf{S}$*-decomposable measure* m *is a set function* $m : \mathcal{A} \to [0,1]$ *such that* $m(\emptyset) = 0$ *and*

$$m(A \cup B) = \mathbf{S}(m(A), m(B))$$

whenever $A, B \in \mathcal{A}$ *and* $A \cap B = \emptyset$.

Example 2.65 Taking $\mathbf{S_L}$ t-conorm, $\Omega = \mathbb{N}$, $\mathcal{A} = 2^{\mathbb{N}}$ and $m(E) = \min(|E|/N, 1)$ for a fixed natural number N, where $|E|$ is the cardinal number of E, we obtain that m is $\mathbf{S_L}$-decomposable measure.

Definition 2.66 *Let* $\mathbf{S}$ *be a left-continuous t-conorm. A set function* $m : \mathcal{A} \to [0,1]$ *is* σ-$\mathbf{S}$*-decomposable measure if* $m(\emptyset) = 0$ *and*

$$m\left(\bigcup_{i=1}^{\infty} A_i\right) = \mathop{\mathbf{S}}_{i=1}^{\infty} m(A_i)$$

for every sequence $(A_i)_{i \in \mathbb{N}}$ *from* $\mathcal{A}$ *whose elements are pairwise disjoint set.*

Set function considered in Example 2.65 is σ-$\mathbf{S_L}$-decomposable.

Example 2.67 Let P, $P : \mathcal{A} \to [0,1]$, be a probability measure. Then m_λ defined by

$$m_\lambda(E) = \frac{1}{\lambda}\left((1+\lambda)^{P(E)} - 1\right)$$

is a σ-$\mathbf{S}_\lambda^{\mathbf{SW}}$-decomposable measure.

We have two kinds of monotone continuity of a set function.

We will say that m, $m : \mathcal{A} \to [0,1]$, is continuous from below if for every sequence $(E_n)_{n\in\mathbf{N}}$ from $\mathcal{A}$ such that $E_1 \subset E_2 \subset \ldots$ we have

$$\lim_{n\to\infty} m(E_n) = m\left(\bigcup_{n=1}^{\infty} E_n\right) .$$

We will say that m, $m : \mathcal{A} \to [0,1]$, is continuous from above if for every sequence $(E_n)_{n\in\mathbf{N}}$ from $\mathcal{A}$ such that $E_1 \supset E_2 \supset \ldots$ we have

$$\lim_{n\to\infty} m(E_n) = m\left(\bigcap_{n=1}^{\infty} E_n\right) .$$

As is well known in the Classical Measure Theory, a set function is a measure (σ-additive positive set function) if and only if it is additive and continuous from below or continuous from above. For decomposable measures we have the same result only for one part.

Theorem 2.68 *A set function m, $m : \mathcal{A} \to [0,1]$, is σ-$\mathbf{S}$-decomposable if and only if m is $\mathbf{S}$-decomposable and continuous from below.*

It can happen that a decomposable measure is σ-$\mathbf{S}$-decomposable, even with respect to a continuous t-conorm $\mathbf{S}$, but then it is not continuous from above.

Example 2.69 Let $\mathcal{B}$ be the σ-algebra of Borel subsets of the unit interval and let f, $f : [0,1] \to [0,1]$, be a continuous function such that $f(0) \neq 0$. The set function m, $m : \mathcal{B} \to [0,1]$, defined by

$$m(E) = \sup_{x\in E} f(x)$$

is σ-sup-decomposable measure but it is not continuous from above at $\varnothing$. Namely if we take a sequence $((0,1/n))_{n\in\mathbf{N}}$ of open intervals, then obviously $\left(0,\frac{1}{n}\right) \downarrow \varnothing$, but

$$\lim_{n\to\infty} m\left((0,1/n) \right) = \lim_{n\to\infty} \sup_{x\in(0,1/n)} f(x) = f(0) \neq 0 .$$

But for special class of t-norms we have the following theorem.

Theorem 2.70 *Let $\mathbf{S}$ be a continuous Archimedean t-conorm. Then a $\mathbf{S}$-decomposable measure is σ-$\mathbf{S}$-decomposable if and only if it is continuous from above for all sequences $(E_n)_{n\in\mathbf{N}}$ such that $E_n \downarrow \varnothing$ and $m(E_1) < 1$.*

The representation of a continuous Archimedean t-conorm $\mathbf{S}$ by an additive generator $\mathbf{s}$ gives us in this case the following classification of a σ-$\mathbf{S}$-decomposable measures (see [164, 217, 321]):

(A) $\mathbf{s} \circ m$ is a σ-additive measure on $\mathcal{A}$;

(P) $\mathbf{s} \circ m$ is a pseudo-σ-additive measure on $\mathcal{A}$ in the sense that there exists a sequence $(A_n)_{n\in\mathbb{N}}$ of disjoint sets in $\mathcal{A}$ such that

$$(\mathbf{s} \circ m)\left(\bigcup_{n\in\mathbb{N}} A_n\right) = \mathbf{s}(1) < \sum_{n\in\mathbb{N}} (\mathbf{s} \circ m)(A_n) .$$

The case **(A)** occurs for strict t-conorm **S** or if **S** is non-strict but with finite σ-additive $\mathbf{s} \circ m$.

Example 2.71 $\mathbf{S_L}$-decomposable measure from Example 2.65 is of type **(P)**. Namely, if we take $A, B \in 2^{\mathbb{N}}$ such that $A \cap B = \emptyset$ and $|A| + |B| > N$, then we obtain

$$(\mathbf{s} \circ m)(A) + (\mathbf{s} \circ m)(B) = \min\left(\frac{|A|}{N}, 1\right) + \min\left(\frac{|B|}{N}, 1\right) > 1 = \mathbf{s}(1) .$$

Proposition 2.72 *Let P be probability measure on $\mathcal{A}$ and $\mathbf{s}$ some additive generator of an Archimedean t-conorm* **S**. *Then $m = \mathbf{s}^{(-1)} \circ P$ is an* **S***-decomposable measure on $\mathcal{A}$. m is of type* **(P)** *if and only if $\mathbf{s}(1) < P(\Omega) = 1$ and there exists $E \in \mathcal{A}$ such that $0 < P(E) < P(\Omega) = 1$.*

Continuity at zero of a t-conorm **S** ensures a topological connection with a submeasure η. A set function η, $\eta : \mathcal{A} \to [0, \infty]$, is a submeasure if it satisfies the conditions:

(i) $\eta(\emptyset) = 0$;

(ii) $\eta(A) \leq \eta(B)$ for $A \subseteq B$, $A, B \in \mathcal{A}$;

(iii) $\eta(A \cup B) \leq \eta(A) + \eta(B)$ for $A, B \in \mathcal{A}$ and $A \cap B = \emptyset$.

Namely the following theorem holds, see [217].

Theorem 2.73 *Let m, $m : \mathcal{A} \to [0, 1]$, be a* **S***-decomposable measure with respect to a continuous at zero t-conorm* **S**. *Then there exists a submeasure η on $\mathcal{A}$ such that $\lim_{n\to\infty} m(E_n) = 0$ if and only if $\lim_{n\to\infty} \eta(E_n) = 0$.*

This key theorem enables us to transfer many known theorems on submeasures to decomposable measures, see [217]. We remark that decomposable measure can be subadditive but also superadditive, i.e., $m(A \cup B) \geq m(A) + m(B)$ for disjoint sets A and B from $\mathcal{A}$.

Example 2.74 σ-S_λ^{SW}-decomposable measure from Example 2.67 is subadditive for $\lambda > 0$ and superadditive for $\lambda < 0$.

Special decomposable measure is very important in many applications, [164, 183, 201, 217, 218, 219].

Definition 2.75 *A* max-*decomposable measure m, $m : \mathcal{A} \to [0, \infty]$, i.e., m satisfies the condition*

$$m(A \cup B) = \sup(m(A), m(B))$$

whenever $A, B \in \mathcal{A}$ and $A \cup B \in \mathcal{A}$, is called a maxitive measure. m is completely maxitive if for any family $(E_i)_{i\in I}$ from $\mathcal{A}$ with $\cup_{i\in I} E_i \in \mathcal{A}$ we have

$$m\left(\bigcup_{i\in I} E_i\right) = \sup_{i\in I} m(E_i) .$$

If I is countable, then m is σ-maxitive.

Theorem 2.76 *Every completely maxitive measure m such that there exists $E \in \mathcal{A}$ with $m(E) < \infty$ is continuous from below.*

As we have seen in Example 2.69 completely maxitive measure may not be continuous from above.

Example 2.77 A set function $m : 2^{\mathbb{R}} \to [0, 1]$ defined by

$$m(A) = \begin{cases} 1 & \text{if } A \neq \varnothing, \\ 0 & \text{if } A = \varnothing \end{cases}$$

for $A \in 2^{\mathbb{R}}$ is completely maxitive measure but it is not continuous from above. Namely, taking $E_n = (0, 1/n)$ $(n \in \mathbb{N})$, we obtain a decreasing sequence $(E_n)_{n\in\mathbb{N}}$ with $E_n \searrow \varnothing$, but $1 = \lim_{n\to\infty} m(E_n) \neq m(\varnothing) = 0$.

Example 2.78 Every function $f : \Omega \to [0, \infty)$ determines a completely maxitive measure m in the following way, $m(A) = \sup_{x\in A} f(x)$ $(A \in \mathcal{A})$. Conversely, to any maxitive measure defined on 2^Ω we can make correspond a function f defined on Ω (density function) by $f(x) = m(\{x\})$ $(x \in \Omega)$.

Definition 2.79 *A completely maxitive measure $\pi : 2^\Omega \to [0, 1]$ with $m(\Omega) = 1$ is called a possibility measure . A set function $m : 2^\Omega \to [0, 1]$ is called a necessity measure if for every family $(E_i)_{i\in I}$ from 2^Ω we have $m\left(\bigcap_{i\in I} E_i\right) = \inf_{i\in I} m(E_i)$ and $m(\varnothing) = 0$.*

It is easy to see that for every possibility measure π we can correspond a necessity measure by $m(A) = 1 - \pi(A^c)$ $(A \in 2^{\Omega})$.

Possibility theory is based on possibility measure and necessity measure and can be considered in terms of fuzzy sets or as a restriction of Dempster–Shafer theory to bodies of evidence that are nested. Maxitive measures also appear in the theory of fractals (see [164, 217]).

There are many different definitions of fractal dimension, depending on the examined problem. One of the oldest and most important is the Hausdorff dimension defined for arbitrary subset of $\mathbb{R}^n$. Namely, for a subset F of $\mathbb{R}^n$, a non-negative number s, and any $\delta > 0$ we introduce first the set function

$$\mathcal{H}^s_\delta(F) = \inf\left\{\sum_{i=1}^{\infty}(\text{diam } U_i)^s \mid (U_i) \text{ is a } \delta\text{-cover of } F\right\},$$

where a δ-cover of F is a collection (countable or finite) of sets of diameter at most δ that cover F. Then

$$\mathcal{H}^s(F) = \lim_{\delta\to 0}\mathcal{H}^s_\delta(F)$$

is called the s-dimensional Hausdorff measure of F. Now the Hausdorff dimension of the set F is given by

$$\dim_{\mathbf{H}} F = \inf\{s \mid \mathcal{H}^s(F) = 0\} = \sup\{s \mid \mathcal{H}^s(F) = \infty\}.$$

The Hausdorff dimension is a σ-maxitive measure on $2^{\mathbb{R}^n}$ (in the theory of fractals this property is usually called countable stability).

The classical integration theory (Riemann, Lebesgue, Choquet) is based on the common arithmetical operation of addition and multiplication. If we restrict ourselves to the case of probability measures and functions with range in the unit interval, we can replace the addition of reals by the Łukasiewicz t-conorm $\mathbf{S_L}$, and the multiplication is then just the product t-norm $T_{\mathbf{P}}$. As we have seen, several generalizations of classical measures with respect to the additivity replaced by some t-conorm (pseudo-addition) have been suggested and investigated. The same idea in integration theory leads to the generalization based on pseudo-additions (t-conorms) and pseudo-multiplications (specially t-norms). On the other hand, such functions correspond to membership functions of (measurable) fuzzy sets. This allows us to define a measure on fuzzy subsets as an integral of corresponding membership functions. So, for example, let $(\Omega, \mathcal{A}, P)$ be a probability space and let $\mathcal{X}$ be the system of all fuzzy subsets of Ω with $\mathcal{A}$-measurable membership function μ_A (so called fuzzy events). Using the common Lebesgue integral, we can define a measure m of fuzzy events [326], $m : \mathcal{X} \to [0,1]$, putting

$$m(A) = \int_{\Omega} \mu_A \, dP .$$

Amongst several applications of above mentioned integrals, recall e.g. non-linear differential equations [217] and multicriteria decision making.

2.8.2 Related probabilistic metric spaces

An **S**-decomposable measure m is monotone, which means that $A, B \in \mathcal{A}$, $A \subseteq B$ implies $m(A) \leq m(B)$. A measure m is of (NSA)-*type* (see [321]) if and only if $\mathbf{s} \circ m$ is a finite additive measure, where $\mathbf{s}$ is an additive generator of the t-conorm **S** (see [321]), which is continuous, non-strict, and Archimedean, and with respect to which m is decomposable ($\mathbf{s}(1) = 1$).

If m is **S**-decomposable then m is a **S**-*valuation*, which means that

$$\mathbf{S}(m(A \cup B), m(A \cap B)) = \mathbf{S}(m(A), m(B)).$$

Lemma 2.80 *Let m be of (NSA)-type, where $\mathbf{s}$ strictly increases. Then for every $A, B \in \mathcal{A}$:*

$$m(A \cap B) \geq T(m(A),\ m(B)),$$

where T is a t-norm with an additive generator $\mathbf{t} = 1 - \mathbf{s}$.

If, in addition,

$$\mathbf{s}(x) + \mathbf{s}(1 - x) = 1 \text{ for every } x \in [0, 1], \tag{2.20}$$

then

$$T(x, y) = 1 - \mathbf{S}(1 - x, 1 - y), \text{ for every } x, y \in [0, 1],$$

i.e., T is a dual t-norm to the t-conorm **S**.

Proof. Since m is of (NSA)-type, $\mathbf{s} \circ m$ is a finite additive measure and so

$$\begin{aligned} m(A \cap B) &= \mathbf{s}^{-1}(\mathbf{s}(m(A \cap B))) \\ &\geq \mathbf{s}^{-1}(\max(0, \mathbf{s} \circ m(A) + \mathbf{s} \circ m(B) - 1)) \\ &= \mathbf{t}^{-1}(\min(1, \mathbf{t} \circ m(A) + \mathbf{t} \circ m(B))) \\ &= T(m(A), m(B)), \end{aligned}$$

where $\mathbf{t} = 1 - \mathbf{s}$ is an additive generator of t-norm T.

If $\mathbf{s}$ satisfies (2.20) we have that $\mathbf{s}^{-1}(1 - \mathbf{s}(u)) = 1 - u$ ($u \in [0, 1]$). For

$$u = \mathbf{s}^{-1}(\mathbf{s}(1 - x) + \mathbf{s}(1 - y)) \quad (x, y \in [0, 1])$$

we obtain that

$$\mathbf{s}^{-1}(1 - \mathbf{s}(\mathbf{s}^{-1}(\mathbf{s}(1 - x) + \mathbf{s}(1 - y)))) = 1 - u$$

and therefore

$$\mathbf{s}^{-1}(1 - \mathbf{s}(1 - x) - \mathbf{s}(1 - y)) = 1 - \mathbf{s}^{-1}(\mathbf{s}(1 - x) + \mathbf{s}(1 - y)).$$

Since $\mathbf{s}(1 - x) = 1 - \mathbf{s}(x)$, $\mathbf{s}(1 - y) = 1 - \mathbf{s}(y)$, we have that

$$\mathbf{s}^{-1}(\mathbf{s}(x) + \mathbf{s}(y) - 1) = 1 - \mathbf{s}^{-1}(\mathbf{s}(1 - x) + \mathbf{s}(1 - y)).$$

Hence $T(x, y) = 1 - \mathbf{S}(1 - x, 1 - y)$, which means that T and **S** are in the duality.□

Example 2.81 (i) For the family of Yager's t-conorms $(\mathbf{S}^{\mathbf{Y}}_\lambda)_{\lambda\in(0,\infty)}$, with the additive generators $\mathbf{s}^{\mathbf{Y}}_\lambda(x) = x^\lambda$:

$$\mathbf{S}^{\mathbf{Y}}_\lambda(x,y) = \min(1,(x^\lambda + y^\lambda)^{1/\lambda}),$$

the family of t-norms $(T_\lambda)_{\lambda\in(0,\infty)}$ from Lemma 2.80 is given by

$$T_\lambda(x,y) = (\max(x^\lambda + y^\lambda - 1, 0))^{1/\lambda}.$$

If $\lambda \neq 1$, $T^{\mathbf{Y}}_\lambda$ and T_λ are not in the duality.

(ii) For the family of Schweizer–Sklar t-conorms $(\mathbf{S}^{\mathbf{SS}}_\lambda)_{\lambda\in(0,\infty)}$,

$$\mathbf{S}^{\mathbf{SS}}_\lambda(x,y) = 1 - \max(0,((1-x)^\lambda + (1-y)^\lambda - 1)^{1/\lambda}),$$

with the additive generators $\mathbf{s}^{\mathbf{SS}}_\lambda(x) = 1-(1-x)^\lambda$, the family of the t-norms $(T_\lambda)_{\lambda\in(0,\infty)}$ from Lemma 2.80, is the family of Yager's t-norms

$$T^{\mathbf{Y}}_\lambda(x,y) = 1 - \min(1,((1-x)^\lambda + (1-y^\lambda))^{1/\lambda}).$$

(iii) For the family of Sugeno–Weber's t-conorms $(\mathbf{S}^{\mathbf{SW}}_\lambda)_{\lambda\in(-1,\infty)}$, with the additive generators

$$\mathbf{s}^{\mathbf{SW}}_\lambda(x) = \begin{cases} x & \text{if } \lambda = 0, \\ \dfrac{\log(1+\lambda x)}{\log(1+\lambda)} & \text{if } \lambda \in (-1,\infty)\setminus\{0\}, \end{cases}$$

the corresponding family of t-norms from Lemma 2.80 is $(T^{\mathbf{SW}}_\lambda)_{\lambda\in(-1,\infty)}$ with corresponding additive generators $\mathbf{t}^{\mathbf{SW}}_\lambda = 1 - \mathbf{s}^{\mathbf{SW}}_\lambda$. $T^{\mathbf{SW}}_\lambda$ and $\mathbf{S}^{\mathbf{SW}}_\mu$ are in duality for $\mu = -\lambda/(\lambda+1)$.

If $(\Omega,\mathcal{A},m)$ is a measure space and (M,d) is a separable metric space, by S we shall denote the set of all the equivalence classes of measurable mappings $X:\Omega\to M$. An element from S will be denoted by $\widehat{X}$ if $\{X(\omega)\}\in\widehat{X}$.

Proposition 2.82 *Let $(\Omega,\mathcal{A},m)$ be a measure space, where m is a continuous* **S**-*decomposable measure of (NSA)-type with monotone increasing generator* $\mathbf{s}$. *Then $(S,\mathcal{F},T)$ is a Menger space, where $\mathcal{F}$ and t-norm T are given in the following way $(\mathcal{F}(\widehat{X},\widehat{Y}) = F_{\widehat{X},\widehat{Y}})$:*

$$F_{\widehat{X},\widehat{Y}}(u) = m\{\omega \mid \omega\in\Omega,\ d(X(\omega),Y(\omega)) < u\} = m\{d(X,Y) < u\}$$

(for every $\widehat{X},\widehat{Y}\in S, u\in\mathbb{R}$),

$$T(x,y) = \mathbf{s}^{-1}(\max(0,\mathbf{s}(x)+\mathbf{s}(y)-1)),\ \textit{for every } x,y\in[0,1].$$

Proof. We have to prove that the mapping $\mathcal{F}$ maps $S \times S$ into $\mathcal{D}^+$, such that the following conditions are satisfied:

(i) $F_{\widehat{X},\widehat{Y}}(u) = 1$ for every $u > 0 \Rightarrow \widehat{X} = \widehat{Y}$.

(ii) $F_{\widehat{X},\widehat{Y}} = F_{\widehat{Y},\widehat{X}}$ for every $(\widehat{X}, \widehat{Y}) \in S \times S$.

(iii) $F_{\widehat{X},\widehat{Z}}(u+v) \geq T(F_{\widehat{X},\widehat{Y}}(u), F_{\widehat{Y},\widehat{Z}}(v))$ for every $(\widehat{X}, \widehat{Y}, \widehat{Z}) \in S \times S \times S$ and every $u, v > 0$.

It is obvious that $F_{\widehat{X},\widehat{Y}} : \mathbb{R} \to [0,1]$ and since m is monotone, $F_{\widehat{X},\widehat{Y}}$ is monotone nondecreasing. Since $F_{\widehat{X},\widehat{Y}}(0) = m(\emptyset) = 0$, in order to prove that $F_{\widehat{X},\widehat{Y}} \in \mathcal{D}^+$ it remains to prove that

$$\lim_{u\to\infty} F_{\widehat{X},\widehat{Y}}(u) = 1.$$

Let $A_n = \{d(X,Y) < n\}$ $(n \in \mathbb{N})$. Then $A_1 \subseteq A_2 \subseteq \cdots$ and from the continuity of m and $m(\Omega) = 1$ it follows that

$$m\left(\bigcup_{n\in\mathbb{N}} A_n\right) = m(\Omega) = 1 = \lim_{n\to\infty} F_{\widehat{X},\widehat{Y}}(n) = \lim_{u\to\infty} F_{\widehat{X},\widehat{Y}}(u).$$

We shall prove that (i) holds. Since $F_{\widehat{X},\widehat{Y}}(u) = 1$, for every $u > 0$ implies that $F_{\widehat{X},\widehat{Y}}(1/n) = 1$, for every $n \in \mathbb{N}$ we have that $m(A_n) = 1$ $(n \in \mathbb{N})$, where $A_n = \{d(X,Y) < 1/n\}$. From $\{d(X,Y) = 0\} = \bigcap_{n\in\mathbb{N}} A_n$ and the continuity of m it follows that

$$m\{d(X,Y) = 0\} = \lim_{n\to\infty} m(A_n) = 1$$

and therefore $\widehat{X} = \widehat{Y}$.

Let $(\widehat{X}, \widehat{Y}, \widehat{Z}) \in S \times S \times S$ and $u, v > 0$. We shall prove that (iii) holds.

Let $A = \{d(X,Y) < u\}$, $B = \{d(Y,Z) < v\}$, $C = \{d(X,Z) < u+v\}$. Then $A \cap B \subseteq C$ and since m is monotone we have that $m(A \cap B) \leq m(C)$. Lemma 2.80 implies that $m(C) \geq m(A \cap B) \geq T(m(A), m(B))$, where T is a t-norm given by

$$T(x,y) = \mathsf{s}^{-1}(\max(0, \mathsf{s}(x) + \mathsf{s}(y) - 1)),\ x, y \in [0,1].$$

□

Remark 2.83 If the additive generator for $\mathbf{S}$ is $\mathsf{s}(x) = x$ $(x \in [0,1])$ then $T(x,y) = T_{\mathbf{L}}(x,y) = \max(x+y-1, 0)$. The probability measure $m = P$ is $\mathbf{S}_{\mathbf{L}}$-decomposable, where $\mathbf{S}_{\mathbf{L}}(x,y) = \min(x+y, 1)$. Hence Proposition 2.82 is a generalization of Drossos result from [54].

Proposition 2.84 *Let $(\Omega, \mathcal{A}, m)$ be as in Proposition 2.82 and (M,d) be a complete separable metric space. Then $(S, \mathcal{F}, T)$ from Proposition 2.82 is a complete probabilistic metric space.*

Proof. Suppose that $(\widehat{X}_n)_{n\in\mathbb{N}}$ is a Cauchy sequence in S in the (ε, λ)-topology. This means that for every $\varepsilon > 0$ and $\lambda \in (0,1)$ there exists $n_0(\varepsilon, \lambda) \in \mathbb{N}$ such that

$$m\{d(X_n, X_{n+p}) < \varepsilon\} = F_{\widehat{X}_n, \widehat{X}_{n+p}}(\varepsilon) > 1 - \lambda$$

for every $n \geq n_0(\varepsilon, \lambda)$ and every $p \in \mathbb{N}$.

Let $(\varepsilon_k)_{k\in\mathbb{N}}$ be a sequence from $(0,1)$ such that $\sum\limits_{k\in\mathbb{N}} \varepsilon_k$ is convergent. Since $\mathbf{s}$ is continuous and $\mathbf{s}(1) = 1$, it follows that there exists $\lambda_k \in (0,1)$ $(k \in \mathbb{N})$ such that $\mathbf{s}(1 - \lambda_k) > 1 - \varepsilon_k$ $(k \in \mathbb{N})$. Let $N_s \in \mathbb{N}$ $(s \in \mathbb{N})$ be such that $N_1 < N_2 < \ldots$

$$m\{d(X_n, X_{n+p}) < \varepsilon_s\} > 1 - \lambda_s, \; n \geq N_s, \; p \in \mathbb{N}.$$

Then $\mathbf{s} \circ m\{d(X_{N_s}, X_{N_{s+1}}) < \varepsilon_s\} > \mathbf{s}(1 - \lambda_s) > 1 - \varepsilon_s$, for every $s \in \mathbb{N}$. Since the measure $\mathbf{s} \circ m$ is finite additive we have that

$$\mathbf{s} \circ m\{d(X_{N_s}, X_{N_{s+1}}) > \varepsilon_s\} < 1 - \mathbf{s}(1 - \lambda_s) < \varepsilon_s \text{ for every } s \in \mathbb{N}.$$

If

$$B_s = \{d(X_{N_s}, X_{N_{s+1}}) > \varepsilon_s\} \quad (s \in \mathbb{N})$$

we have that $\mathbf{s} \circ m(B_s) < \varepsilon_s$ $(s \in \mathbb{N})$. Hence $\sum\limits_{s\in\mathbb{N}} \mathbf{s} \circ m(B_s)$ is convergent. Borel–Cantelli's lemma implies that

$$\mathbf{s} \circ m(\{\omega \mid \omega \in \Omega, \sum_{s\in\mathbb{N}} d(X_{N_s}, X_{N_{s+1}}) < \infty\}) = 1.$$

Since (M, d) is a complete metric space, there exists $\widehat{X} \in S$, $\Omega_0 \in \mathcal{A}$, $\mathbf{s} \circ m(\Omega_0) = 1$ such that

$$\lim_{s\to\infty} X_{N_s}(\omega) = X(\omega), \; \omega \in \Omega_0.$$

Since $\mathbf{s} \circ m$ is a probability measure it follows that $\widehat{X}_n$ converges to $\widehat{X}$ in the probability $\mathbf{s} \circ m$. This follows from $(\widehat{X}_n)_{n\in\mathbb{N}}$ being a Cauchy sequence with respect to $\mathbf{s} \circ m$ as well.

But the convergence in $\mathbf{s} \circ m$ implies the convergence in m and therefore the proof is complete. □

Chapter 3

Probabilistic B-contraction principles for single-valued mappings

The first fixed point theorem in probabilistic metric spaces was proved by Sehgal and Barucha-Reid [272] for mappings $f : S \to S$, where $(S, \mathcal{F}, T_{\mathbf{M}})$ is a Menger space. Further development of the fixed point theory in a more general Menger space $(S, \mathcal{F}, T)$ was connected with investigations of the structure of the t-norm T. Very soon the problem was in some sense completely solved. Namely, if we restrict ourselves to complete Menger spaces $(S, \mathcal{F}, T)$, where T is a continuous t-norm, then any probabilistic q-contraction $f : S \to S$ has a fixed point if and only if the t-norm is of H-type.

A very interesting new approach to the fixed point theory in probabilistic metric spaces is given in Tardiff's paper [310], where some additional growth conditions for the mapping $\mathcal{F} : S \times S \to \mathcal{D}^+$ are assumed, and $T \geq T_{\mathbf{L}}$. V. Radu introduced a stronger growth condition for $\mathcal{F}$ than in Tardiff's paper. We prove in section 3.1 a fixed point theorem for a probabilistic q-contraction $f : S \to S$, where $(S, \mathcal{F}, T)$ is a complete Menger space, $\mathcal{F}$ satisfies Radu's condition, and T is a q-convergent t-norm. In section 3.2 two special classes of probabilistic q-contractions are investigated, and without any growth conditions on $\mathcal{F}$ some fixed point results are obtained. Section 3.3 contains some generalizations of probabilistic B-contraction principles, and in section 3.4 we prove some fixed point theorems of Caristi's type in F-topological spaces with applications to fuzzy metric and Menger spaces. Common fixed point theorems for some classes of mappings are obtained in section 3.5.

When we say, in what follows, that '$(S, \mathcal{F})$ is a probabilistic metric space', we mean that $(S, \mathcal{F})$ is a probabilistic metric space in the sense of Schweizer and Sklar, and that $(S, \mathcal{F}, \tau)$ is a probabilistic metric space in the sense of Šerstnev. We assume in the whole chapter for the probabilistic metric spaces $(S, \mathcal{F}, \tau)$ that Range$(\mathcal{F}) \subset \mathcal{D}^+$.

3.1 Probabilistic B-contraction principles

Let (M, d) be a metric space and $f : M \to M$. If there exists a $q \in [0, 1)$ such that

$$d(fx, fy) \le qd(x, y) \text{ for every } x, y \in M,$$

then f is the so called q-contraction.

Every q-contraction $f : M \to M$ on a complete metric space (M, d) has one and only one fixed point. This is the well known Banach contraction principle (shortly, B-contraction principle).

Sehgal and Bharucha-Reid introduced in 1972 the notion of a probabilistic q-contraction ($q \in (0, 1)$) in a probabilistic metric space [272].

Definition 3.1 *Let $(S, \mathcal{F})$ be a probabilistic metric space. A mapping $f : S \to S$ is a probabilistic q-contraction ($q \in (0, 1)$) if*

$$F_{fp_1, fp_2}(x) \ge F_{p_1, p_2}\left(\frac{x}{q}\right) \tag{3.1}$$

for every $p_1, p_2 \in S$ and every $x \in \mathbb{R}$.

Remark 3.2 It is obvious that $f : S \to S$ is a probabilistic q-contraction if and only if for every $p_1, p_2 \in S$ and every $x \in \mathbb{R}$ the following implication holds

$$(\forall \alpha \in (0, 1))(F_{p_1, p_2}(x) > 1 - \alpha \quad \Rightarrow \quad F_{fp_1, fp_2}(qx) > 1 - \alpha).$$

Inequality (3.1) is a generalization of the inequality

$$d(fp_1, fp_2) \le qd(p_1, p_2), \tag{3.2}$$

where $f : M \to M$ and (M, d) is a metric space. In order to prove that (3.2) implies (3.1) recall that every metric space (M, d) is also a Menger space $(M, \mathcal{F}, T_{\mathbf{M}})$, if $\mathcal{F}$ is defined in the following way:

$$F_{p_1, p_2}(x) = \begin{cases} 1 & \text{if } d(p_1, p_2) < x, \\ 0 & \text{if } d(p_1, p_2) \ge x \end{cases} \qquad (x \in \mathbb{R}).$$

Suppose that $f : M \to M$ is such that (3.2) holds and prove that (3.1) is satisfied, i.e., that for every $x > 0$

$$F_{p_1, p_2}\left(\frac{x}{q}\right) = 1 \quad \Rightarrow \quad F_{fp_1, fp_2}(x) = 1.$$

If $F_{p_1, p_2}\left(\frac{x}{q}\right) = 1$, then $d(p_1, p_2) < \frac{x}{q}$ and (3.2) implies

$$d(fp_1, fp_2) < q \cdot \frac{x}{q} = x,$$

which means that $F_{fp_1, fp_2}(x) = 1$.

Example 3.3 Let $(\Omega, \mathcal{A}, P)$ be a probability measure space, (M, d) a separable metric space, and $\mathcal{B}_M$ the family of Borel subsets of M. A mapping $f : \Omega \times M \to M$ is a *random operator* if for every $C \in \mathcal{B}_M$ and every $x \in M$

$$\{\omega \mid \omega \in \Omega, f(\omega, x) \in C\} \in \mathcal{A},$$

i.e., if the mapping $\omega \mapsto f(\omega, x)$ is measurable on Ω. A random operator $f : \Omega \times M \to M$ is *continuous* if for every $\omega \in \Omega$ the mapping $x \mapsto f(\omega, x)$ is continuous on M.

If $f : \Omega \times M \to M$ is a continuous random operator then (by [11]) for every measurable mapping $X : \Omega \to M$ the mapping $\omega \mapsto f(\omega, X(\omega))$ is measurable on Ω.

Let S be the set of all equivalence classes of measurable mappings $X : \Omega \to M$ and let f be a continuous random operator. The mapping $\hat{f} : S \to S$, defined by

$$(\hat{f}\widehat{X})(\omega) = f(\omega, X(\omega)) \text{ for every } \widehat{X} \in S \quad (\omega \in \Omega, X \in \widehat{X}),$$

is the so called *Nemytskij operator* of f. If $f : \Omega \times M \to M$ is a random operator then a measurable mapping $X : \Omega \to M$ is a *random fixed point* of the mapping f if

$$X(\omega) = f(\omega, X(\omega)) \text{ a.e. .} \tag{3.3}$$

Many authors investigated the problem of the existence of a random fixed point of a random operator, see [40]. If f is a continuous random operator then (3.3) holds if and only if $\widehat{X} = \widehat{f}\widehat{X}, \quad X \in \widehat{X}$. Hence in this case the problem of the existence of a random fixed point of a continuous random operator f reduces to the problem of the existence of a fixed point of the Nemytskij operator $\widehat{f}$ of f.

Let for every $\widehat{X}, \widehat{Y} \in S$ and every $x > 0$

$$\begin{aligned} &P(\{\omega \mid \omega \in \Omega, d(f(\omega, X(\omega)), f(\omega, Y(\omega)) < qx\}) \\ &\geq P(\{\omega \mid \omega \in \Omega, d(X(\omega), Y(\omega)) < x\}), \end{aligned} \tag{3.4}$$

where $q \in (0, 1)$. $(S, \mathcal{F}, T_{\mathbf{L}})$ is a Menger space, where

$$F_{\widehat{X},\widehat{Y}}(x) = P(\{\omega \mid \omega \in \Omega, d(X(\omega), Y(\omega)) < x\})$$

for every $\widehat{X}, \widehat{Y} \in S$ and $x \in \mathbb{R}$. Then (3.4) implies

$$F_{\widehat{f}\widehat{X},\widehat{f}\widehat{Y}}(qx) \geq F_{\widehat{X},\widehat{Y}}(x).$$

Hence $\widehat{f}$ is a probabilistic q-contraction if and only if (3.4) holds.

Let $(S, \mathcal{F})$ be a probabilistic metric space. We recall that a sequence $(x_n)_{n \in \mathbb{N}}$ in S is a Cauchy sequence if and only if for every $\varepsilon > 0$ and $\lambda \in (0, 1)$ there exists $n_0(\varepsilon, \lambda) \in \mathbb{N}$ such that $F_{x_{n+p}, x_n}(\varepsilon) > 1 - \lambda$ for every $n \geq n_0(\varepsilon, \lambda)$ and every $p \in \mathbb{N}$.

If a probabilistic metric space $(S, \mathcal{F})$ is such that every Cauchy sequence $(x_n)_{n \in \mathbb{N}}$ in S converges in S, then $(S, \mathcal{F})$ is a complete space.

The first fixed point theorem in probabilistic metric spaces was proved by Sehgal and Bharucha-Reid in [272].

Theorem 3.4 *Let $(S, \mathcal{F}, T_{\mathbf{M}})$ be a complete Menger space and $f : S \to S$ a probabilistic q-contraction. Then there exists a unique fixed point x of the mapping f and $x = \lim_{n\to\infty} f^n p$ for every $p \in S$.*

Proof. G.L. Cain and R. Kasriel [25] proved Theorem 3.4 by introducing the family of semi-metrics $(d_b)_{b\in(0,1)}$ defined by

$$d_b(p,q) = \sup\{x \mid x \in \mathbb{R},\ F_{p,q}(x) \le b\}, \quad b \in (0,1),\ p,q \in S.$$

The family of semi-metrics $(d_b)_{b\in(0,1)}$ generates the (ε,λ)-topology in S. We obtain by (3.1) that for every $p_1, p_2 \in S$ and every $b \in (0,1)$

$$d_b(fp_1, fp_2) \le q d_b(p_1, p_2).$$

Namely, if $d_b(p_1,p_2) < r$ then $F_{p_1,p_2}(r) > b$, and (3.1) implies $F_{fp_1,fp_2}(qr) > b$, which means that $d_b(fp_1, fp_2) < qr$.

Using an analogous procedure as in the case of metric spaces they proved a generalization of Banach contraction principle for uniform space $(S, (d_b)_{b\in(0,1)})$. □

For every $x_0 \in S$ let

$$O(x_0, f) = \{f^n x_0 \mid n \in \mathbb{N} \cup \{0\}\}.$$

The set $O(x_0,f)$ is the *orbit of the mapping $f : S \to S$ at x_0*. Let $D_{O(x_0,f)} : \mathbb{R} \to [0,1]$ be the diameter of $O(x_0,f)$, i.e.,

$$D_{O(x_0,f)}(x) = \sup_{s<x} \inf_{u,v\in O(x_0,f)} F_{u,v}(s).$$

If $\sup_{x\in\mathbb{R}} D_{O(x_0,f)}(x) = 1$ then the orbit $O(x_0,f)$ is a probabilistic bounded subset of S. Hence $O(x_0,f)$ is a probabilistic bounded set if and only if $D_{O(x_0,f)} \in \mathcal{D}^+$.

The fixed point theorem of Sehgal and Bharucha-Reid can be obtained from the next theorem [276] .

Theorem 3.5 *Let $(S, \mathcal{F}, T)$ be a complete Menger space, T a t-norm such that $\sup_{a<1} T(a,a) = 1$ and $f : S \to S$ a probabilistic q-contraction. If $O(x_0,f)$ is probabilistic bounded for some $x_0 \in S$, then there exists a unique fixed point x of the mapping f and $x = \lim_{n\to\infty} f^n p$ for every $p \in S$.*

Proof. Let $x_n = f^n x_0$, $n \in \mathbb{N}$. We shall prove that $(x_n)_{n\in\mathbb{N}}$ is a Cauchy sequence.

Let $n, m \in \mathbb{N}$, $\varepsilon > 0$ and $\lambda \in (0,1)$. We have that

$$\begin{aligned} F_{x_{n+m},x_n}(\varepsilon) &= F_{f^{n+m}x_0,f^n x_0}(\varepsilon) \\ &\geq F_{f^{n+m-1}x_0,f^{n-1}x_0}\left(\frac{\varepsilon}{q}\right) \\ &\;\;\vdots \\ &\geq F_{f^m x_0,x_0}\left(\frac{\varepsilon}{q^n}\right) \\ &\geq D_{O(x_0,f)}\left(\frac{\varepsilon}{q^n}\right). \end{aligned}$$

Since $D_{O(x_0,f)}\left(\frac{\varepsilon}{q^n}\right) \to 1$, when $n \to \infty$, it follows that there exists $n_0(\varepsilon,\lambda) \in \mathbb{N}$ such that for every $n \geq n_0(\varepsilon,\lambda)$ and every $m \in \mathbb{N}$

$$F_{x_{n+m},x_n}(\varepsilon) > 1 - \lambda.$$

Hence $(x_n)_{n\in\mathbb{N}}$ is a Cauchy sequence and since S is complete, it follows the existence of $x \in S$ such that $x = \lim_{n\to\infty} x_n$. By the continuity of f and $x_{n+1} = fx_n$ for every $n \in \mathbb{N}$, we obtain that $x = fx$.

It is easy to prove that if $O(x_0, f)$ is probabilistic bounded for some $x_0 \in S$, then $O(p, f)$ is probabilistic bounded for any $p \in S$. □

Corollary 3.6 *Let $(S, \mathcal{F}, T)$ be a complete Menger space, T a t-norm of H-type and $f : S \to S$ a probabilistic q-contraction. Then there exists a unique fixed point $x \in S$ of the mapping f and $x = \lim_{n\to\infty} f^n p$ for every $p \in S$.*

Proof. It is easy to prove that for a t-norm T, which is of H-type, the orbit $O(x_0, f)$ is probabilistic bounded for every $x_0 \in S$. Indeed, let $x_0 \in S$, $m \in \mathbb{N}$ and $u \in \mathbb{R}$. Then

$$\begin{aligned} F_{f^m x_0,x_0}(u) &\geq T(F_{f^m x_0,fx_0}(qu),\, F_{fx_0,x_0}(u-qu)) \\ &\geq T(F_{f^{m-1}x_0,x_0}(u),\, F_{fx_0,x_0}(u-qu)) \\ &\;\;\vdots \\ &\geq (F_{fx_0,x_0}(u-qu))_T^{(m)} \end{aligned}$$

and since $\lim_{u\to\infty} F_{fx_0,x_0}(u-qu) = 1$, from the equicontinuity of the family $(x_T^{(m)})_{m\in\mathbb{N}}$ at the point $x = 1$, it follows that $D_{O(x_0,f)} \in \mathcal{D}^+$. □

Since the t-norm $T_{\mathbf{M}}$ is of H-type, the fixed point theorem of Sehgal and Bharucha-Reid follows from Corollary 3.6.

Let τ_T be a triangle function and $(F_i)_{i\in\mathbb{N}}$ a sequence in $\mathcal{D}^+$. Let

$$\tau_T^1(F_i) = \tau_T(F_1, F_2),\ \tau_T^n(F_i) = \tau_T(\tau_T^{n-1}(F_i), F_{n+1})\ (n \geq 2).$$

Since τ_T is nondecreasing in each place, the weak limit of $\tau_T^n(F_i)$, when $n \to \infty$, exists and will be denoted by $\tau_T^\infty(F_i)$.

The following theorem is proved in [276].

Theorem 3.7 *Let $(S, \mathcal{F}, T)$ be a complete Menger space, T a t-norm such that $\sup_{a<1} T(a,a) = 1$ and $f : S \to S$ a probabilistic q-contraction such that $\tau_T^\infty(F_i) \in \mathcal{D}^+$, where $F_i(x) = F_{p,fp}\Big(\dfrac{x}{q^i}\Big)$ $(i \in \mathbb{N},\ x \in \mathbb{R})$ and $p \in S$. Then $D_{O(p,f)} \in \mathcal{D}^+$.*

Proof. We know that

$$F_{p,r} \geq \tau_T(F_{p,w}, F_{w,r})\ (p, w, r \in S),$$

since $(S, \mathcal{F}, T)$ is a Menger space. For $p_m = f^m p, m \in \mathbb{N}$, we have that

$$\begin{aligned} F_{p,p_m} &\geq \underbrace{\tau_T(\tau_T(\ldots\tau_T}_{(m-1)-\text{times}}(F_{p,p_1}, F_{p_1,p_2}), \ldots, F_{p_{m-1},p_m}) \\ &\geq \underbrace{\tau_T(\tau_T(\ldots\tau_T}_{(m-1)-\text{times}}(F_{p,p_1}, F_{p,p_1}(j_q)), \ldots, F_{p,p_1}(j_{q^{m-1}})) \\ &= \tau_T^{m-1}(F_{p,p_1}(j_{q^i})), \end{aligned}$$

where $j_\alpha(x) = \dfrac{x}{\alpha}$ $(x \in \mathbb{R}, \alpha \in (0,1))$. Hence

$$F_{p,p_m} \geq \tau_T^\infty(F_i) \quad \text{for every} \quad m \in \mathbb{N}. \tag{3.5}$$

Inequality (3.5) implies that $D_{O(p,f)} \in \mathcal{D}^+$. □

Corollary 3.8 *Let $(S, \mathcal{F}, T)$ be a complete Menger space, T a t-norm such that $\sup_{a<1} T(a,a) = 1$, $f : S \to S$ a probabilistic q-contraction and $\tau_T^\infty(F_i) \in \mathcal{D}^+$ where $F_i(x) = F_{p,fp}\Big(\dfrac{x}{q^i}\Big)$ $(i \in \mathbb{N},\ x \in \mathbb{R})$ and $p \in S$. Then there exists a unique fixed point z of the mapping f and $z = \lim_{n\to\infty} f^n x_0$ for every $x_0 \in S$.*

Proof. The proof follows by Theorem 3.5 and Theorem 3.7. □

Definition 3.9 *Let τ_T be a triangle function, $q \in (0,1)$, and $G \in \mathcal{D}^+$. Then $G^{\tau_T,q} \in \mathcal{D}^+$ is defined by*

$$G^{\tau_T,q}(x) = \lim_{t<x} \lim_{n\to\infty} G_n^{\tau_T,q}(t),\ x > 0,$$

where

$$G_n^{\tau_T,q} = \tau_T^n(F_i),\ F_i = G(j_{q^i}),\ i \in \mathbb{N},\ n \in \mathbb{N}.$$

Sherwood [276] proved the following theorem.

Theorem 3.10 *Let T be a lower continuous t-norm and let τ_T be the lower continuous triangle function induced by T. Then there exists a complete probabilistic metric space $(S, \mathcal{F}, \tau_T)$ and a probabilistic q-contraction mapping $f : S \to S$ which has no fixed point if and only if there exists $G \in \mathcal{D}^+$ such that $G^{\tau_T, q} \notin \mathcal{D}^+$.*

Several authors have studied the problem of probabilistic boundedness of the set $O(p, f)$ and essentially two different approaches have been pursued.

One is to identify those t-norm T which are strong enough to guarantee that the sequence of iterates $(f^n p)_{n\in\mathbb{N}}$ is a Cauchy sequence and the other is to impose a growth condition on the distance distribution functions [228, 310].

Definition 3.11 *A t-norm T has the fixed point property if and only if every probabilistic q-contraction $f : S \to S$, where $(S, \mathcal{F}, T)$ is an arbitrary complete Menger space, has a fixed point.*

Hence if a t-norm T is of H-type then T has the fixed point property.

V. Radu [250] proved the following theorem.

Theorem 3.12 *Any continuous t-norm T with the fixed point property is of H-type.*

Proof. Suppose that T is not of H-type. Then there exists $\lambda \in (0, 1)$ such that for every $\delta \in (0, 1)$, there exists $b > \delta$ such that

$$b_T^{(m)} < \lambda \text{ for some } m = m_b \in \mathbb{N}.$$

Let $(b_n)_{n\in\mathbb{N}}$ be a strictly increasing sequence from $(0, 1)$ such that $\lim_{n\to\infty} b_n = 1, s_n = m_{b_n}, n \in \mathbb{N}$, such that $s_{n+1} > s_n$ for every $n \in \mathbb{N}$ and

$$(b_n)_T^{(s_n)} < \lambda \quad \text{for every } n \in \mathbb{N}.$$

Let

$$X = \mathbb{N}, \; F_{n,n} = H_0, \; F_{n,n+m}(t) = \mathop{\mathrm{T}}_{i=1}^{m+1} G(2^{n+i-1}t),$$

where

$$G(t) = \begin{cases} 0 & \text{if } t \geq 1, \\ b_1 & \text{if } t \in (1, 2^{2+s_1}], \\ b_{n+1} & \text{if } t \in (2^{2n+s_n}, 2^{2n+2+s_{n+1}}], \; n \geq 1. \end{cases}$$

Then $(X, \mathcal{F}, T)$ is a Menger space and let $f : X \to X$ be defined by $f(n) = n+1, n \in \mathbb{N}$. Then f is a probabilistic $1/2$-contraction, but $(f^n(1))_{n\in\mathbb{N}}$ is not a Cauchy sequence since

$$F_{n,n+s_n}(1) \leq (b_n)_T^{(s_n)} < \lambda.$$

Thus, no Archimedean t-norm has the fixed point property. □

From the above consideration it is clear that in order to obtain some kind of fixed point theorems for Menger spaces $(S, \mathcal{F}, T)$, where T is an Archimedean t-norm, one has to impose some additional conditions on the mapping $\mathcal{F}$.

Interesting results in this direction are obtained in Tardiff's paper [310]. We recall that a continuous t-norm T is Archimedean if for each $a \in (0,1)$ we have $T(a,a) < a$. Hence, $T_{\mathbf{P}}(a,b) = a \cdot b$ and $T_{\mathbf{L}}(a,b) = \max(a+b-1, 0)$ are Archimedean t-norms but $T_{\mathbf{M}}$ is not.

In [198] the following two theorems are proved.

Theorem 3.13 *Let $(F_i)_{i\in\mathbb{N}}$ be a sequence in $\mathcal{D}^+$ and T an Archimedean t-norm. Then either $\tau_T^\infty(F_i)$ is identically 0 or is in $\mathcal{D}^+$.*

Theorem 3.14 *Let $(F_i)_{i\in\mathbb{N}}$ be a sequence in $\mathcal{D}^+$. Then $\tau_{T_{\mathbf{P}}}^\infty(F_i)$ is identically 0 if and only if $\tau_{T_{\mathbf{L}}}^\infty(F_i)$ is identically 0.*

Lemma 3.15 *Let $F \in \mathcal{D}^+$ and $\alpha \in (0,1)$. Then*

$$G = \tau_{T_{\mathbf{P}}}^\infty(F \circ j_{\alpha^{i-1}}) \in \mathcal{D}^+ \quad \Longleftrightarrow \quad \int_1^\infty \ln(u)\, dF(u) < \infty.$$

Proof. Since $T_{\mathbf{P}}$ is an Archimedean t-norm, we can apply Theorem 3.13. In order to prove that $\tau_{T_{\mathbf{P}}}^\infty(F \circ j_{\alpha^{i-1}})$ is in $\mathcal{D}^+$, it is enough to prove that for some $x \in \mathbb{R}$, $G(x) > 0$. The function G is a non-decreasing and left continuous function and therefore G has at most countable number of discontinuities. Let $x \in \mathbb{R}$ be a point of the continuity of G. Then

$$G(x) = \lim_{n\to\infty} \sup_{\sum_{i=1}^n \beta_i = 1} \prod_{i=1}^n F\left(\frac{\beta_i x}{\alpha^{i-1}}\right).$$

Let $\delta \in (\alpha, 1)$. Since F is non-decreasing and $\sum_{i=1}^\infty (1-\delta)\delta^{i-1} = 1$, it follows that

$$\prod_{i=1}^\infty F\left(\frac{x}{\alpha^{i-1}}\right) \geq G(x) \geq \prod_{i=1}^\infty F\left(\frac{(1-\delta)x}{\left(\frac{\alpha}{\delta}\right)^{i-1}}\right).$$

Since $F \in \mathcal{D}^+$, there exists $y \in \mathbb{R}$ such that $F(y) > 0$. We shall prove that for any $\sigma \in (0,1)$

$$\prod_{i=1}^\infty F\left(\frac{y}{\sigma^{i-1}}\right) > 0 \quad \Longleftrightarrow \quad \int_1^\infty \ln(u)\, dF(u) < \infty.$$

Let $F(y) > 0$, $\sigma \in (0,1)$ and g be defined by

$$g(t) = \begin{cases} 0 & \text{if } t \leq 1, \\ y\sigma^{1-t} & \text{if } t > 1. \end{cases}$$

Since $0 < \sigma < 1$ and $F \in \mathcal{D}^+$, it follows that $F \circ g \in \mathcal{D}^+$. Further by [169]

$$\prod_{i=1}^{\infty} F \circ g(i) > 0 \iff \sum_{i=1}^{\infty} (1 - F \circ g(i)) < \infty$$
$$\iff \int_0^{\infty} (1 - F \circ g(t))\, dt < \infty$$

and by [64]

$$\int_0^{\infty} (1 - F \circ g(t))\, dt < \infty \quad \iff \quad \int_0^{\infty} t\, dF(g(t)) < \infty.$$

Hence we conclude that

$$\prod_{i=1}^{\infty} F \circ g(i) > 0 \quad \iff \quad \int_0^{\infty} t\, dF(g(t)) < \infty.$$

If $u = g(t)$ then the relation

$$\int_0^{\infty} t\, dF(g(t)) = F(y) + \int_y^{\infty} \left(1 - \frac{\ln(u) - \ln(y)}{\ln(\sigma)}\right) dF(u) \tag{3.6}$$

implies that the right side of (3.6) is finite if and only if $\int_1^{\infty} \ln(u)\, dF(u) < \infty$. □

Corollary 3.16 *Let $F \in \mathcal{D}^+$ and $\alpha \in (0,1)$. Then $\tau_{T_{\mathbf{L}}}^{\infty}(F \circ j_{\alpha^{i-1}}) \in \mathcal{D}^+$ if and only if*

$$\int_1^{\infty} \ln(u)\, dF(u) < \infty. \tag{3.7}$$

Theorem 3.17 *Let $(S, \mathcal{F}, T)$ be a complete Menger space and T a t-norm such that $T \geq T_{\mathbf{L}}$. If for every $u, v \in S$ (3.7) holds, with F replaced by $F_{u,v}$, then any probabilistic q-contraction $f : S \to S$ has a unique fixed point x and $x = \lim_{n\to\infty} f^n p$ for every $p \in S$.*

Proof. From Lemma 3.15 and Theorem 3.14 it follows that

$$\tau_{T_{\mathbf{L}}}^{\infty}(F_{p,f(p)} \circ j_{q^i-1}) \in \mathcal{D}^+$$

and since $T \geq T_{\mathbf{L}}$

$$\tau_T^{\infty}(F_{p,f(p)} \circ j_{q^i-1}) \in \mathcal{D}^+.$$

This completes the proof. □

Corollary 3.18 *Let $(S, \mathcal{F}, T)$ be a complete Menger space, T a t-norm such that $T \geq T_{\mathbf{L}}$ and $f : S \to S$ a probabilistic q-contraction. If there exists $p \in S$ such that*

$$\int_1^{\infty} \ln(u)\, dF_{p,fp}(u) < \infty \tag{3.8}$$

then there exists a unique fixed point x of the mapping f and $x = \lim_{n\to\infty} f^n p$.

Proof. The proof is identical to the proof of Theorem 3.17. □

Corollary 3.19 *Let $(S, \mathcal{F}, T)$ be a complete Menger space and T a t-norm such that $T \geq T_{\mathbf{L}}$. If for all $u, v \in S$ the first moment $F_{u,v}$ is finite, then any probabilistic q-contraction $f : S \to S$ has a unique fixed point x and $x = \lim_{n\to\infty} f^n p$ for every $p \in S$.*

Proof. Let

$$g(x) = \begin{cases} x - 1 & \text{if } x \leq 1, \\ \ln x & \text{if } x > 1. \end{cases}$$

Using the Jensen inequality, since g is concave, we have that for every $u, v \in S$

$$\int_0^{\infty} g(x)\, dF_{u,v}(x) \leq g\left(\int_0^{\infty} x\, dF_{u,v}(x)\right) < \infty,$$

and therefore (3.7) holds. □

Remark 3.20 Theorem 3.17 can be applied on a complete E space ($T = T_{\mathbf{L}}$) and the Wald space ($T_{\mathbf{P}} > T_{\mathbf{L}}$).

Some interesting results are obtained by Părău and Radu in [228] by introducing for $k > 0$ a functional ϑ_k on S.

Proposition 3.21 *Let $(S, \mathcal{F})$ be a probabilistic semi-metric space and define for every $k > 0$ and every $p, q \in S$*

$$e_k(p,q) = \delta_k(F_{p,q}) = \sup_{x>0} x^k(1 - F_{p,q}(x))e^{-x}.$$

Then

(i) e_k is a semi-metric for the strong $\mathcal{F}$-topology.

(ii) e_k generates the $\mathcal{F}$-uniformity if the later exists.

(iii) If $(S, \mathcal{F}, T_{\mathbf{L}})$ is a Menger space, then the functional ϑ_k defined by

$$(p,q) \mapsto \vartheta_k(p,q) := e_k(p,q)^{\frac{1}{k+1}}$$

gives a metric on S.

Moreover, $(S, \mathcal{F})$ is complete if and only if (S, ϑ_k) is complete.

Proof. We prove only (iii) since (i) and (ii) can be proved easily. If $(S, \mathcal{F}, T_{\mathbf{L}})$ is a Menger space we have that

$$\begin{aligned} F_{p,q}(tx + (1-t)x) &= F_{p,q}(x) \\ &\geq T_L(F_{p,r}(tx), F_{r,q}((1-t)x)) \\ &\geq F_{p,r}(tx) + F_{r,q}((1-t)x) - 1 \end{aligned}$$

and therefore

$$1 - F_{p,q}(x) \leq 1 - F_{p,r}(tx) + 1 - F_{r,q}((1-t)x) \tag{3.9}$$

for every $p, q, r \in S$, every $x \in \mathbb{R}$ and every $t \in [0,1]$.

Hence (3.9) implies that

$$x^k(1 - F_{p,q}(x))e^{-x} \leq x^k(1 - F_{p,r}(tx))e^{-tx} + x^k(1 - F_{r,q}((1-t)x))e^{-(1-t)x}$$

for every $t \in (0,1)$, and so

$$x^k(1 - F_{p,q}(x))e^{-x} \leq \frac{1}{t^k} e_k(p,r) + \frac{1}{(1-t)^k} e_k(r,q) \tag{3.10}$$

for every $t \in (0,1)$.

From (3.10) we have that

$$e_k(p,q) \leq \frac{1}{t^k} e_k(p,r) + \frac{1}{(1-t)^k} e_k(r,q),$$

for every $t \in (0,1)$, which implies that

$$(e_k(p,q))^{\frac{1}{k+1}} \leq (e_k(p,r))^{\frac{1}{k+1}} + (e_k(r,q))^{\frac{1}{k+1}}.$$

This means that ϑ_k satisfies the triangle inequality. □

Remark 3.22 If $(S, \mathcal{F}, T_{\mathbf{L}})$ is a non-Archimedean Menger space then e_k itself is a metric which generates the $\mathcal{F}$-uniformity.

Theorem 3.23 *Let $(S, \mathcal{F}, T)$ be a complete Menger space such that $T \geq T_{\mathbf{L}}$ and $f : S \to S$ a probabilistic q-contraction. Then (i) and (ii) hold, where*

(i) If for some $p \in S$, $p = fp$, then for every $k > 0$

$$E_k(p) = \sup_{x>0} x^k(1 - F_{p,fp}(x)) < \infty.$$

(ii) If there exist $p \in S$ and $k > 0$ such that

$$E_k(p) = \sup_{x>0} x^k(1 - F_{p,fp}(x)) < \infty,$$

then f has a fixed point $x \in S$ and the following error estimation holds

$$\vartheta_k(x, f^n p) \leq \left(\sum_{i=n}^{\infty} \left(q^{\frac{k}{k+1}}\right)^i\right) (E_k(p))^{\frac{1}{k+1}} \quad \textit{for every } n \in \mathbb{N}. \tag{3.11}$$

Proof. Suppose that (i) holds. Then there exists $p \in S$ such that $p = fp$, which implies that $F_{p,fp}(x) = 1$ for every $x > 0$. Then

$$E_k(p) = 0 \quad \text{for every } k > 0.$$

Let, for some $p \in S$ and $k > 0$, $E_k(p) < \infty$.

Since $\delta_k(F_{p,fp}) \leq E_k(p)$ and f is a probabilistic q-contraction we have that

$$\begin{aligned} x^k(1 - F_{fp,f^2p}(x))e^{-x} &\leq x^k\left(1 - F_{p,fp}\left(\frac{x}{q}\right)\right)e^{-x} \\ &= q^k\left(\left(\frac{x}{q}\right)^k\left(1 - F_{p,fp}\left(\frac{x}{q}\right)\right)\right)e^{-x} \\ &\leq q^k E_k(p). \end{aligned}$$

Hence

$$\vartheta_k(fp, f^2p) \leq q^{\frac{k}{k+1}}(E_k(p))^{\frac{1}{k+1}}. \tag{3.12}$$

Since, for every $p \in S$, every $n \in \mathbb{N}$ and every $x \in \mathbb{R}$

$$F_{f^np,f^{n+1}p}(x) \geq F_{p,fp}\left(\frac{x}{q^n}\right) \tag{3.13}$$

similarly as we have proved (3.12), we obtain that

$$\vartheta_k(f^np, f^{n+1}p) \leq \left(q^{\frac{k}{k+1}}\right)^n (E_k(p))^{\frac{1}{k+1}}$$

for every $n \in \mathbb{N}$. Therefore

$$\sum_{n=1}^{\infty} \vartheta_k(f^n p, f^{n+1} p) \leq \left(\sum_{n=1}^{\infty} (q^{\frac{k}{k+1}})^n \right) (E_k(p))^{\frac{1}{k+1}} < \infty,$$

from which we conclude that $(f^n p)_{n \in \mathbb{N}}$ is a Cauchy sequence. Since $(S, \mathcal{F}, T)$ is a complete Menger space it follows that there exists uniquely determined fixed point x of f. Further on we have for every $n, m \in \mathbb{N}$

$$\begin{aligned}
\vartheta_k(f^n p, f^{n+m} p) &\leq \sum_{i=n}^{n+m-1} \vartheta_k(f^i p, f^{i+1} p) \\
&\leq \left(\sum_{i=n}^{n+m-1} (q^{\frac{k}{k+1}})^i \right) (E_k(p))^{\frac{1}{k+1}} \\
&\leq \left(\sum_{i=n}^{\infty} (q^{\frac{k}{k+1}})^i \right) (E_k(p))^{\frac{1}{k+1}}.
\end{aligned}$$

Since $x = \lim_{m \to \infty} f^{n+m} p$ we obtain the error estimation (3.11). □

Example 3.24 Let $S(0,1)$ be the space of all equivalence classes of real random variables on the Lebesgue measure space $((0,1), \mathcal{L}, \mathbf{m_0})$. It is known that

$$(S(0,1), \mathcal{F}, T_{\mathbf{L}})$$

is a Menger space, where

$$F_{p,q}(x) = \mathbf{m_0}(\{t \mid t \in (0,1), \ |p(t) - q(t)| < x\})$$

for every $p, q \in S(0,1)$ and $x \in \mathbb{R}$.

Let M be any closed (for the convergence in probability) linear subspace of $S(0,1)$, which contains elements 1 and w defined by

$$w(t) = e^{\frac{1}{t}}, \quad t \in (0,1).$$

The distribution function F_w is given by

$$F_w(x) = \begin{cases} 0 & \text{if } x \leq e, \\ 1 - \dfrac{1}{\ln x} & \text{if } x > e, \end{cases}$$

since $F_w(x) = \mathbf{m_0}(\{t \mid t \in (0,1), \ e^{1/t} < x\})$.

Suppose that $a \in M$ is such that $\sup_{x>0} x^k(1 - F_{|a|}(x)) < \infty$. Let $f : M \to M$ be such that $fp = Lp + w$, $p \in M$, $L \in (0,1)$. If $\tilde{p} = \mu a + \frac{w}{1-L}$ then $f\tilde{p} = L\tilde{p} + w = L\mu a + \frac{Lw}{1-L} + w$ and therefore

$$\begin{aligned} \tilde{p} - f\tilde{p} &= \mu a + \frac{w}{1-L} - \left(L\mu a + \frac{Lw}{1-L} + w\right) \\ &= \mu(1-L)a. \end{aligned}$$

Hence $\sup_{x>0} x^k(1 - F_{|\tilde{p}-f\tilde{p}|}(x)) < \infty$. The mapping f has a fixed point $\overline{p} = \frac{w}{1-L}$ and $E_k(\overline{p}) = \infty$. The condition $\int_1^\infty \ln x \, dF_{p,q}(x) < \infty$ does not hold for $p = \mu_1\overline{p}$, $q = \mu_2\overline{p}$, $\mu_1 \neq \mu_2$.

Remark 3.25 Condition $E_k(p) < \infty$ is verified if there exists p such that $F_{p,fp}(t_p) = 1$, for some $t_p > 0$.

Corollary 3.26 *If $(S, \mathcal{F}, T)$ is a complete Menger space such that $T \geq T_{\mathbf{L}}$, then a probabilistic q-contraction $f : S \to S$ has a fixed point if and only if there exist $k > 0$ and $p \in S$ such that*

$$\int_0^\infty x^k \, dF_{p,fp}(x) < \infty. \tag{3.14}$$

Proof. It is well known that (3.14) implies $\lim_{x\to\infty} x^k(1 - F_{p,fp}(x)) = 0$ and therefore the condition $E_k(p) < \infty$ is satisfied. □

We know that a t-norm T is Archimedean if and only if there exists an increasing homeomorphism $h : [0,1] \to [0,1]$ such that

$$T_h(a,b) = h^{-1}(T_*(h(a),\ h(b))),$$

where $T_* \in \{T_L, T_P\}$. Since $a \cdot b \geq a + b - 1$, for every $a, b \in [0,1]$, we have the following result.

Theorem 3.27 *Let $(S, \mathcal{F}, T)$ be a complete Menger space such that $T \geq T_h$, for some increasing homeomorphism $h : [0,1] \to [0,1]$. Then a probabilistic q-contraction $f : S \to S$ has a fixed point if and only if for some $p \in S$ and $k > 0$*

$$\sup_{x>0} x^k(1 - h \circ F_{p,fp}(x)) < \infty.$$

Proof. The proof follows from $(S, h \circ \mathcal{F}, T_{\mathbf{L}})$ being a complete Menger space and f is a probabilistic q-contraction in $(S, h \circ \mathcal{F}, T_{\mathbf{L}})$. □

Theorem 3.28 *Let $(S, \mathcal{F}, T)$ be a complete Menger space such that $\sup_{a<1} T(a, a) = 1$ and $f : S \to S$ a probabilistic q-contraction such that for some $p \in S$ and $k > 0$*

$$\sup_{x>0} x^k(1 - F_{p,fp}(x)) < \infty. \tag{3.15}$$

If t-norm T is μ^k-convergent for some $\mu \in (q, 1)$, then there exists a unique fixed point z of the mapping f and $z = \lim_{n\to\infty} f^n p$.

Proof. Let $\delta = q/\mu < 1$. We shall prove that $(f^n p)_{n\in\mathbb{N}}$ is a Cauchy sequence. Choose $\varepsilon > 0$ and $\lambda \in (0, 1)$ and prove that there exists $n_0(\varepsilon, \lambda) \in \mathbb{N}$ such that

$$F_{f^n p, f^{n+m} p}(\varepsilon) > 1 - \lambda \text{ for every } n \geq n_0(\varepsilon, \lambda) \text{ and every } m \in \mathbb{N}.$$

Since the series $\sum_{i=1}^{\infty} \delta^i$ is convergent, there exists $n_1 = n_1(\varepsilon) \in \mathbb{N}$ such that $\sum_{i=n_1}^{\infty} \delta^i \leq \varepsilon$. Let $n > n_1$. Then we have

$$\begin{aligned}
F_{f^n p, f^{n+m} p}(\varepsilon) &\geq F_{f^n p, f^{n+m} p}\left(\sum_{i=n}^{\infty} \delta^i\right) \\
&\geq F_{f^n p, f^{n+m} p}\left(\sum_{i=n}^{n+m-1} \delta^i\right) \\
&\geq \underbrace{T\Big(T\Big(\cdots\Big(T}_{(m-1)-\text{times}}\Big(F_{f^n p, f^{n+1} p}(\delta^n), F_{f^{n+1} p, f^{n+2} p}(\delta^{n+1})\Big), \\
&\qquad \ldots, F_{f^{n+m-1} p, f^{n+m} p}(\delta^{n+m-1})\Big) \\
&\geq \underbrace{T\Big(T\Big(\cdots\Big(T}_{(m-1)-\text{times}}\Big(F_{p,fp}\Big(\frac{1}{\mu^n}\Big), F_{p,fp}\Big(\frac{1}{\mu^{n+1}}\Big)\Big), \ldots, F_{p,fp}\Big(\frac{1}{\mu^{n+m-1}}\Big)\Big).
\end{aligned}$$

Let $M > 0$ be such that

$$x^k(1 - F_{p,fp}(x)) \leq M \text{ for every } x > 0. \tag{3.16}$$

Suppose that n_2 is such that

$$1 - M(\mu^k)^n \in [0, 1) \text{ for every } n \geq n_2. \tag{3.17}$$

From (3.16) it follows that

$$F_{p,fp}\left(\frac{1}{\mu^n}\right) > 1 - M(\mu^k)^n \text{ for every } n \in \mathbb{N}$$

and by (3.17) for $n \geq \max(n_1, n_2)$

$$F_{f^np,f^{n+m}p}(\varepsilon) \geq \underbrace{T\Big(T\Big(\cdots\Big(T}_{(m-1)-\text{times}}\Big(1 - M(\mu^k)^n, 1 - M(\mu^k)^{n+1}\Big), \ldots, 1 - M(\mu^k)^{n+m-1}\Big).$$

Let s_0 be such that $M(\mu^k)^{s_0} < \mu^k$. Then for every $n \in \mathbb{N}$

$$1 - M(\mu^k)^{n+s_0} \geq 1 - (\mu^k)^{n+1}$$

and therefore for $n \geq \max(n_1, n_2)$ and $m \in \mathbb{N}$

$$\begin{aligned} F_{f^{n+s_0}p,f^{n+s_0+m}p}(\varepsilon) &\geq \underbrace{T\Big(T\Big(\cdots\Big(T}_{(m-1)-\text{times}}\Big(1 - M(\mu^k)^{n+s_0}, 1 - M(\mu^k)^{n+s_0+1}\Big), \\ &\qquad \ldots, 1 - M(\mu^k)^{n+s_0+m-1}\Big) \\ &\geq \mathop{\mathrm{T}}_{i=n+1}^{\infty} (1 - (\mu^k)^i). \end{aligned}$$

Since T is μ^k-convergent we conclude that $(f^np)_{n\in\mathbb{N}}$ is a Cauchy sequence. Let $z = \lim_{n\to\infty} f^np$. By the continuity of the mapping f it follows that $fz = z$. □

Remark 3.29 The proof of Theorem 3.28 also holds if we assume instead of the condition (3.15) that for some $p \in S$ and $\mu \in (q, 1)$

$$\lim_{n\to\infty} \mathop{\mathrm{T}}_{i=n}^{\infty} F_{p,fp}\left(\frac{1}{\mu^i}\right) = 1$$

and that T is such that $\sup_{a<1} T(a, a) = 1$.

Corollary 3.30 *Let $(S, \mathcal{F}, T)$ be a complete Menger space such that T is a strict t-norm with a multiplicative generator θ, and $f : S \to S$ a probabilistic q-contraction such that for some $k > 0$ and $p \in S$ (3.15) holds. If there exists $\mu \in (q, 1)$ such that*

$$\lim_{n\to\infty} \prod_{i=n}^{\infty} \theta(1 - (\mu^k)^i) = 1,$$

then there exists a unique fixed point x of the mapping f and $x = \lim_{n\to\infty} f^np$.

Let us recall

$$\mathcal{T}_0 = \bigcup_{\lambda\in(0,\infty)} \{T_\lambda^{\mathbf{D}}\} \bigcup \bigcup_{\lambda\in(-1,\infty]} \{T_\lambda^{\mathbf{SW}}\} \bigcup \bigcup_{\lambda\in(0,\infty)} \{T_\lambda^{\mathbf{AA}}\} \bigcup \mathcal{T}^H.$$

Corollary 3.31 *Let $(S, \mathcal{F}, T)$ be a complete Menger space such that $T \geq T_1$ for some $T_1 \in \mathcal{T}_0$ and $f : S \to S$ a probabilistic q-contraction such that for some $k > 0$ and $p \in S$ (3.15) holds. Then there exists a unique fixed point x of the mapping f and $x = \lim_{n\to\infty} f^n p$.*

Let $(\Omega, \mathcal{A}, m)$ be a measure space, (M, d) a separable metric space and $f : \Omega \times M \to M$. Let the mapping f be a continuous random operator. Then for every measurable mapping $X : \Omega \to M$, the mapping $\omega \mapsto f(\omega, X(\omega))(\omega \in \Omega)$ is measurable. If $X : \Omega \to M$ is a measurable mapping let $(\hat{f}\widehat{X})(\omega) = f(\omega, X(\omega)), \omega \in \Omega, X \in \widehat{X}$. Hence $\hat{f} : S \to S$, where S is the set of all equivalence classes of measurable mappings from Ω into M.

Corollary 3.32 *Let $(\Omega, \mathcal{A}, m)$ be a measure space, where m is a continuous* **S**-*decomposable measure of (NSA)-type , s is a monotone increasing additive generator of* **S**, *(M, d) a complete separable metric space and $f : \Omega \times M \to M$ a continuous random operator such that for some $q \in (0, 1)$*

$$m(\{\omega \mid \omega \in \Omega, d((\hat{f}\widehat{X})(\omega), (\hat{f}\widehat{Y})(\omega)) < u\}) \geq m(\{\omega \mid \omega \in \Omega, d(X(\omega), Y(\omega)) < \frac{u}{q}\})$$

for every measurable mappings $X, Y : \Omega \to M$ and every $u > 0$. If there exists a measurable mapping $U : \Omega \to M$ such that for some $k > 0$

$$\sup_{x>0} x^k(1 - m(\{d(\widehat{U}, \hat{f}\widehat{U}) < x\})) < \infty$$

and t-norm T defined by

$$T(x, y) = \mathrm{s}^{-1}(\max(0, \mathrm{s}(x) + \mathrm{s}(y) - 1), x, y \in [0, 1],$$

is μ^k-convergent for some $\mu \in (q, 1)$, then there exists a random fixed point of the operator f.

Proof. The proof follows by Propositions 2.82, 2.84 and Theorem 3.28. □

Remark 3.33 Let $(\Omega, \mathcal{A}, m)$, (M, d) and f be as in Corollary 3.32 for some t-conorm $\mathbf{S}_\lambda^{\mathbf{SW}}$, $\lambda \in (-1, \infty]$. Since the t-norms $T_\lambda^{\mathbf{SW}}$, $\lambda \in (-1, \infty]$, are q-convergent there follows the existence of a random fixed point of the random operator $f : \Omega \times M \to M$.

3.2 Two special classes of probabilistic q-contractions

In this section two special classes of probabilistic q-contractions will be considered. The first one is the so called strict q-contraction $f : S \to S$, where $(S, \mathcal{F}, T_{\mathbf{L}})$ is an

E-space and Sherwood's fixed point theorem for such a class of mappings is given. A generalization of the notion of a q-contraction of (ε, λ)-type is given and a fixed point theorem for such a class of mappings $f : S \to S$ is proved, where $(S, \mathcal{F}, T)$ is a Menger space and T belongs to the class of t-norms which has $\mathcal{T}^H \cup \{T_{\mathbf{L}}\}$ as a proper subclass.

Definition 3.34 *Let $(S, \mathcal{F}, T_{\mathbf{L}})$ be an E-space over the metric space (M, d) and let $(\Omega, \mathcal{A}, P)$ be the associated probability measure space. A mapping $f : S \to S$ is a strict q-contraction, where $q \in (0,1)$, if for every $p_1, p_2 \in S$ and every $x \in \mathbb{R}$*

$$\{\omega \mid \omega \in \Omega, d(p_1(\omega), p_2(\omega)) < x\} \subseteq \{\omega \mid \omega \in \Omega, d((fp_1)(\omega), (fp_2)(\omega) < qx\}.$$

H. Sherwood proved [276] the following fixed point theorem.

Theorem 3.35 *Every strict q-contraction mapping $f : S \to S$, where $(S, \mathcal{F}, T_{\mathbf{L}})$ is a complete E-space, has a unique fixed point.*

Proof. For the sequence $(p_m)_{m\in\mathbb{N}}$, defined by $p_m = f^m p_0, m \in \mathbb{N}$ and $p_0 \in S$, we have that for every $m \in \mathbb{N}$ and $x \in \mathbb{R}$

$$\{\omega \mid \omega \in \Omega, d(p_0(\omega), p_1(\omega)) < (1-q)x\}$$

$$\begin{aligned}
&\subseteq \{\omega \mid \omega \in \Omega, (1-q^m)d(p_0(\omega), p_1(\omega)) < (1-q)x\} \\
&\subseteq \{\omega \mid \omega \in \Omega, (1+q+\cdots+q^{m-1})d(p_0(\omega), p_1(\omega)) < x\} \\
&\subseteq \{\omega \mid \omega \in \Omega, d(p_0(\omega), p_1(\omega)) + \cdots + d(p_{m-1}(\omega), p_m(\omega)) < x\} \\
&\subseteq \{\omega \mid \omega \in \Omega, d(p_0(\omega), p_m(\omega)) < x\}.
\end{aligned}$$

Therefore $F_{p_0,p_m}(x) \geq F_{p_0,p_1}((1-q)x)$, which implies that $(f^m p_0)_{m\in\mathbb{N}}$ is probabilistic bounded. □

In [193] D. Miheţ introduced the following definition.

Definition 3.36 *Let $(S, \mathcal{F})$ be a probabilistic metric space. A mapping $f : S \to S$ is said to be a q-contraction of (ε, λ)-type, where $q \in (0,1)$, if the following implication holds for every $p_1, p_2 \in S$:*

$$(\forall \varepsilon > 0)(\forall \lambda \in (0,1))\,(F_{p_1,p_2}(\varepsilon) > 1 - \lambda \quad \Rightarrow \quad F_{fp_1,fp_2}(q\varepsilon) > 1 - q\lambda)\,. \qquad (3.18)$$

It is obvious that (3.18) implies that f is a probabilistic q-contraction, i.e., that for every $p_1, p_2 \in S$ and every $\varepsilon > 0$

$$F_{fp_1,fp_2}(q\varepsilon) \geq F_{p_1,p_2}(\varepsilon).$$

Indeed, in the opposite case for some $\varepsilon > 0$, $p_1, p_2 \in S$ and $\lambda \in (0,1)$

$$F_{fp_1,fp_2}(q\varepsilon) < 1 - \lambda < F_{p_1,p_2}(\varepsilon).$$

Since $F_{fp_1,fp_2}(q\varepsilon) > 1 - q\lambda > 1 - \lambda$ we obtain a contradiction.
In [193] the following theorem was proved.

Theorem 3.37 *Let $(S, \mathcal{F}, T_{\mathbf{L}})$ be a complete Menger space and $f : S \to S$ a q-contraction of (ε, λ)-type. Then there exists a unique fixed point $x \in S$ of the mapping f and $x = \lim\limits_{n\to\infty} f^n p$ for every $p \in S$.*

The proof is based on a fixed point theorem for uniform spaces, see [311]. Here, using a different approach, we shall prove that Miheţ's theorem 3.37 holds for a more general class of Menger spaces. First, we introduce the following definition.

Definition 3.38 *Let $(S, \mathcal{F})$ be a probabilistic metric space. A mapping $f : S \to S$ is said to be a (q, q_1)-contraction of (ε, λ)-type, where $q, q_1 \in (0,1)$, if the following implication holds for every $p_1, p_2 \in S$:*

$$(\forall\varepsilon > 0)(\forall\lambda \in (0,1))\,(F_{p_1,p_2}(\varepsilon) > 1 - \lambda \quad \Rightarrow \quad F_{fp_1,fp_2}(q\varepsilon) > 1 - q_1\lambda). \qquad (3.19)$$

Example 3.39 Let $(S, \mathcal{F})$ be a probabilistic metric space and $f : S \to S$ such that for some $q, q_1 \in (0,1)$

$$1 - F_{fp_1,fp_2}(qu) \le q_1(1 - F_{p_1,p_2}(u)) \qquad (3.20)$$

for every $p_1, p_2 \in S$ and every $u > 0$. Then f is a (q, q_1)-contraction of (ε, λ)-type. Indeed, if $F_{p_1,p_2}(u) > 1 - \lambda$ then

$$\begin{aligned} 1 - F_{fp_1,fp_2}(qu) &\le q_1(1 - F_{p_1,p_2}(u)) \\ &< q_1\lambda \end{aligned}$$

and therefore $F_{fp_1,fp_2}(qu) > 1 - q_1\lambda$.

Definition 3.40 *Let $(S, \mathcal{F})$ be a probabilistic metric space and $f : S \to S$. The mapping $f : S \to S$ is a strong probabilistic q-contraction ($q \in (0,1)$) if there exists $q_1 \in (0,1)$ such that (3.20) holds.*

Remark 3.41 It is easy to see that (3.20) implies

$$\begin{aligned} F_{f^{n+1}p, f^{n+1+m}p}(u) &\ge 1 - q_1^{n+1}\left(1 - F_{fp,f^m p}\left(\frac{u}{q^{n+1}}\right)\right) \\ &\ge 1 - q_1^{n+1} \end{aligned}$$

for every $p \in S, n, m \in \mathbb{N}$ and $u > 0$. Hence $(f^n p)_{n\in\mathbb{N}}$ is a Cauchy sequence.

Theorem 3.42 *Let $(S,\mathcal{F},T)$ be a complete Menger space such that* $\sup_{a<1} T(a,a) = 1$ *and $f : S \to S$ a (q,q_1)-contraction of (ε,λ)-type. If T is q_1-convergent, i.e.,*

$$\lim_{n\to\infty} \mathop{\mathrm{T}}_{i=n}^{\infty} (1-q_1^i) = 1, \tag{3.21}$$

then there exists a unique fixed point x of the mapping f and $x = \lim_{n\to\infty} f^n p$ *for every $p \in S$.*

Proof. Let $p \in S$ and $\delta > 0$ be such that $F_{p,fp}(\delta) > 0$. Since $F_{p,fp} \in \mathcal{D}^+$ such a δ exists. Let $\lambda_1 \in (0,1)$ be such that $F_{p,fp}(\delta) > 1-\lambda_1$. From (3.19) we have that

$$F_{fp,f^2p}(q\delta) > 1 - q_1\lambda_1$$

and, generally, for every $n \in \mathbb{N}$

$$F_{f^np,f^{n+1}p}(q^n\delta) > 1 - q_1^n\lambda_1. \tag{3.22}$$

We prove that $(f^np)_{n\in\mathbb{N}}$ is a Cauchy sequence, i.e., that for every $\varepsilon > 0$ and $\lambda \in (0,1)$ there exists $n_0(\varepsilon,\lambda) \in \mathbb{N}$ such that

$$F_{f^np,f^{n+m}p}(\varepsilon) > 1-\lambda \text{ for every } n \geq n_0(\varepsilon,\lambda) \text{ and every } m \in \mathbb{N}.$$

Let $\varepsilon > 0$ and $\lambda \in (0,1)$ be given. Since the series $\sum_{n=1}^{\infty} q^n\delta$ converges, there exists $n_0 = n_0(\varepsilon)$ such that $\sum_{n=n_0}^{\infty} q^n\delta < \varepsilon$. Then for every $n \geq n_0$

$$\begin{aligned}
F_{f^np,f^{n+m}p}(\varepsilon) &\geq F_{f^np,f^{n+m}p}\left(\sum_{n=n_0}^{\infty} q^n\delta\right) \\
&\geq F_{f^np,f^{n+m}p}\left(\sum_{i=n}^{n+m-1} q^i\delta\right) \\
&\geq \underbrace{T(\ldots(T}_{(m-1)-\text{times}}\,(F_{f^np,f^{n+1}p}(q^n\delta), F_{f^{n+1}p,f^{n+2}p}(q^{n+1}\delta)), \\
&\qquad \ldots, F_{f^{n+m-1}p,f^{n+m}p}(q^{n+m-1}\delta)).
\end{aligned}$$

Let $n_1 = n_1(\lambda) \in \mathbb{N}$ be such that $\mathop{\mathrm{T}}_{i=n_1}^{\infty} (1-q_1^i) > 1-\lambda$. Since (3.21) holds, such a number n_1 exists. Using (3.22) we obtain for every $n \geq \max(n_0,n_1)$ and every

$m \in \mathbb{N}$

$$\begin{aligned} F_{f^n p, f^{n+m} p}(\varepsilon) &\geq \overset{n+m-1}{\underset{i=n}{\mathrm{T}}} (1 - q_1^i \lambda_1) \\ &\geq \overset{n+m-1}{\underset{i=n}{\mathrm{T}}} (1 - q_1^i) \\ &\geq \overset{\infty}{\underset{i=n}{\mathrm{T}}} (1 - q_1^i) \\ &> 1 - \lambda. \end{aligned}$$

It is obvious that the mapping f is continuous. Indeed, let $\mu > 0$ and $\eta \in (0,1)$ be given. Then for $\varepsilon > 0$ and $\lambda \in (0,1)$ such that $\varepsilon = \mu/q, \lambda = \eta/q_1$ we have the implication

$$F_{p_1,p_2}(\varepsilon) > 1 - \lambda \quad \Rightarrow \quad F_{fp_1,fp_2}(q\varepsilon) = F_{fp_1,fp_2}(\mu) > 1 - q_1\lambda = 1 - \eta.$$

The relation $x = \lim_{n\to\infty} f^n p$ implies that

$$fx = f(\lim_{n\to\infty} f^n p) = \lim_{n\to\infty} f^n p = x.$$

It remains to prove the uniqueness of the fixed point x. Suppose that $y = fy, y \neq x$. If $\varepsilon > 0$ is such that $F_{x,y}(\varepsilon) > 0$ and $F_{x,y}(\varepsilon) > 1 - \lambda$ we have that $F_{fx,fy}(q\varepsilon) > 1 - q_1\lambda$ and similarly

$$F_{x,y}(q^n\varepsilon) = F_{f^n x, f^n y}(q^n \varepsilon) > 1 - q_1^n \lambda \text{ for every } n \in \mathbb{N}.$$

Therefore, $F_{x,y}(u) = 1$, for every $u > 0$, which contradicts to $x \neq y$. □

Corollary 3.43 *Let $(S, \mathcal{F}, T)$ be a complete Menger space such that T is a strict t-norm with a multiplicative generator θ and $f : S \to S$ a (q, q_1)-contraction of (ε, λ)-type. If*

$$\lim_{n\to\infty} \prod_{i=n}^{\infty} \theta(1 - q_1^i) = 1, \tag{3.23}$$

then there exists a unique fixed point x of the mapping f and $x = \lim_{n\to\infty} f^n p$ for every $p \in S$.

Proof. Since θ^{-1} is continuous we obtain by (3.23)

$$\lim_{n\to\infty} \overset{\infty}{\underset{i=n}{\mathrm{T}}} (1 - q_1^i) = \theta^{-1}\left(\lim_{n\to\infty} \prod_{i=n}^{\infty} \theta(1 - q_1^i)\right) = 1.$$

□

Corollary 3.44 *Let $(S, \mathcal{F}, T)$ be a complete Menger space such that $T \geq T_0$ for some $T_0 \in \mathcal{T}_0$ and $f : S \to S$ a (q, q_1)-contraction of (ε, λ)-type. Then there exists a unique fixed point x of the mapping f and $x = \lim_{n\to\infty} f^n p$ for every $p \in S$.*

Theorem 3.37 is a special case of Corollary 3.44 for $q = q_1$ and $T_0 = T_{\mathbf{L}}$.

Corollary 3.45 *Let $(\Omega, \mathcal{A}, m)$ be a measure space, where m is a continuous $\mathbf{S}$-decomposable measure of (NSA)-type, $\mathbf{s}$ is a monotone increasing additive generator of $\mathbf{S}$, (M, d) a complete separable metric space and $f : \Omega \times M \to M$ a random operator such that for some $q, q_1 \in (0,1)$ and every measurable mappings $X, Y : \Omega \to M$*

$$(\forall u > 0)(\forall \lambda \in (0,1))\Big(m(\{\omega \mid \omega \in \Omega, d(X(\omega), Y(\omega)) < u\}) > 1 - \lambda$$

$$\Rightarrow m(\{\omega \mid \omega \in \Omega, d((\hat{f}X), (\hat{f}Y)) < qu\}) > 1 - q_1\lambda\Big).$$

If the t-norm T defined by

$$T(x,y) = \mathbf{s}^{-1}(\max(0, \mathbf{s}(x) + \mathbf{s}(y) - 1), x, y \in [0,1],$$

is q_1-convergent, then there exists a random fixed point of the operator f.

Remark 3.46 Let $(\Omega, \mathcal{A}, m)$, (M, d) and f be as in Corollary 3.45 for some t-conorm $\mathbf{S}_\lambda^{\mathbf{SW}}, \lambda \in (-1, \infty)$. Since the t-norms $T_\lambda^{\mathbf{SW}}, \lambda \in (-1, \infty)$, are q_1-convergent, there follows the existence of the random fixed point of the random operator $f : \Omega \times M \to M$.

3.3 Some generalizations of probabilistic B-contractions principles for single-valued mappings

In this section we shall present some generalizations of Sehgal and Barucha-Reid fixed point theorem. There are a lot of number of fixed point theorems which are in some sense generalizations of Theorem 3.4. Most of the results are probabilistic versions of analogous results from the metric fixed point theory. Very often, mostly in the case when t-norm T is in the class $\mathcal{T}^H$, the proofs are similar to the proofs as in the metric fixed point theory, after introducing a metric or a family of semi-metrics which induces the (ε, λ)-topology. Hence it would be interesting to generalize some of these theorems using a more general class of t-norms with possible some additional conditions on $\mathcal{F}$.

In [193] the following definition of a strict (b_n)-probabilistic contraction is introduced.

Definition 3.47 *Let $(S,\mathcal{F})$ be a probabilistic metric space and $(b_n)_{n\in\mathbb{N}}$ an increasing sequence from $(0,1)$ such that $\lim\limits_{n\to\infty} b_n = 1$. A mapping $f : S \to S$ is a strict (b_n)-probabilistic contraction if*

$$(\forall n \in \mathbb{N})(\exists q = q_n \in (0,1))(\forall p_1, p_2 \in S)(\forall t > 0)$$

$$(F_{p_1,p_2}(t) > b_n \quad \Rightarrow \quad F_{fp_1,fp_2}(q_n t) > b_n). \tag{3.24}$$

The class of strict (b_n)-probabilistic contractions is strictly larger then the class of probabilistic q-contractions.

Example 3.48 Let $S = \{p_1, p_2, p_3\}$ and

$$F_{p_1,p_3}(x) = F_{p_3,p_1}(x) = F_{p_2,p_3}(x) = F_{p_3,p_2}(x) = \begin{cases} 0 & \text{if } x \leq 0, \\ 3/4 & \text{if } x \in (0,2], \\ 1 & \text{if } x > 2, \end{cases}$$

$$F_{p_1,p_2}(x) = F_{p_2,p_1}(x) = \begin{cases} 0 & \text{if } x \leq 0, \\ 1/2 & \text{if } x \in (0,3/2], \\ 1 & \text{if } x > 2, \end{cases}$$

and $f : S \to S$ is defined by $f(p_1) = f(p_2) = p_1$, $f(p_3) = p_2$.

Then f is not a probabilistic q-contraction, for any $q \in (0,1)$, but it is a strict probabilistic $(1 - 1/4^n)$-contraction.

Let $q \in (0,1)$ be given and $x \in (0, 3/(2q)] \cap [0,2]$. Then

$$F_{p_1,p_3}(x) = 3/4 \quad \text{and} \quad F_{fp_1,fp_3}(qx) = 1/2,$$

hence

$$F_{fp_1,fp_3}(qx) < F_{p_1,p_3}(x).$$

On the other side

$$\begin{aligned} F_{p_1,p_3}(x) > 1 - \frac{1}{4^n} \quad &\Rightarrow \quad F_{p_1,p_3}(x) > 1 - 1/4 = 3/4 \\ &\Rightarrow \quad F_{p_1,p_3}(x) = 1 \\ &\Rightarrow \quad x > 2 \\ &\Rightarrow \quad 3x/4 > 3/2 \\ &\Rightarrow \quad F_{p_1,p_2}(3x/4) = 1 > 1 - 1/4^n. \end{aligned}$$

and therefore $q_n = 3/4$ and $b_n = 1 - 1/4^n$ $(n \in \mathbb{N})$.

Every strict (b_n)-probabilistic contraction is uniformly continuous since for every $\delta > 0$ and $\lambda \in (0,1)$ the following implication holds:

$$(x,y) \in N(\frac{\delta}{q_n}, 1-b_n) \quad \Rightarrow \quad (fx,fy) \in N(\delta,\lambda)$$

where $b_n > 1-\lambda$.

Theorem 3.49 *Let $(S,\mathcal{F},T)$ be a complete Menger space, T a t-norm of H-type and $f: S \to S$ a strict (b_n)-probabilistic contraction. Then there exists a unique fixed point $x \in S$ of the mapping f and $x = \lim_{n\to\infty} f^n p$ for every $p \in S$.*

Proof. Let $p \in S, \varepsilon > 0$ and $\lambda \in (0,1)$ be given. Let $\eta \in (0,1)$ be such that $(\eta)_T^{(r)} > 1-\lambda$ for every $r \in \mathbb{N}$ and $m \in \mathbb{N}$ such that $b_m > \max(\eta, 1-\lambda)$. If $s \in \mathbb{N}$ is such that $F_{p,fp}\left(\frac{\varepsilon}{q_m^s}\right) > b_m$ then

$$F_{f^{s+k}p, f^{s+k+1}p}(q_m^k \varepsilon) > b_m > 1-\lambda \text{ for every } k \in \mathbb{N}.$$

Let $k_0 \in \mathbb{N}$ be such that $\sum_{k \geq k_0} q_m^k < 1$. Then for every $u \in \mathbb{N}$ and $r \geq 2$

$$\begin{aligned} F_{f^{s+k_0+u}p, f^{s+k_0+u+r}p}(\varepsilon) &\geq F_{f^{s+k_0+u}p, f^{s+k_0+u+r}p}\left(\sum_{i=k_0+u}^{k_0+u+r-1} q_m^i \varepsilon\right) \\ &\geq \mathop{\mathrm{T}}_{i=0}^{r-1} F_{f^{s+k_0+u+i}p, f^{s+k_0+u+i+1}p}(q_m^{k_0+u+i}) \\ &\geq (b_m)_T^{(r)} \geq (\eta)_T^{(r)} > 1-\lambda, \end{aligned}$$

which means that $(f^n p)_{n\in\mathbb{N}}$ is a Cauchy sequence. Then $x = \lim_{n\to\infty} f^n p$ is the unique fixed point of the mapping f. □

We recall that if T is a continuous t-norm then there is a finite or infinite sequence of pairwise, open subintervals $((\alpha_k, \beta_k))_{k\in K}$ of $[0,1]$ and a sequence of continuous Archimedean t-norms $(T_k)_{k\in K}$, such that

$$T(a,b) = \alpha_k + (\beta_k - \alpha_k) T_k \left(\frac{a-\alpha_k}{\beta_k - \alpha_k}, \frac{b-\alpha_k}{\beta_k-\alpha_k}\right)$$

for $a, b \in (\alpha_k, \beta_k)$ and $T(a,b) = \min(a,b)$ otherwise. We use the notation

$$T = (\langle(\alpha_k, \beta_k), T_k\rangle)_{k\in K}.$$

If T is a continuous t-norm of H-type and each T_k, $k \in K$, is strict, then there exists a sequence $(b_k)_{k\in\mathbb{N}}$ such that

$$\lim_{k\to\infty} b_k = 1 \tag{3.25}$$

and the following implication holds for every $n \in \mathbb{N}$

$$x > b_n,\ y > b_n \quad \Rightarrow \quad T(x,y) > b_n. \tag{3.26}$$

In [111] the following result is proved.

Theorem 3.50 *Let $(S,\mathcal{F},T)$ be a Menger space and T a continuous t-norm of H-type, $T = (\langle(\alpha_k,\beta_k),T_k\rangle)_{k\in K}$ and each T_k is strict. Then the family of pseudo-metrics $(\rho_s)_{s\in\mathbb{N}}$ given by*

$$\rho_s(x,y) = \sup\{u \mid u \in \mathbb{R}, F_{x,y}(u) \leq b_s\} \text{ for every } s \in \mathbb{N} \text{ and } x,y \in S. \tag{3.27}$$

is such that the (ε,λ)-topology and the topology induced by the family $(\rho_s)_{s\in\mathbb{N}}$ coincide.

Using Theorem 3.50 Hadžić and Budinčević [111] obtained the following result.

Proposition 3.51 *Let $(S,\mathcal{F},T)$ be a complete Menger space, and T satisfy conditions from Theorem 3.50 where $(b_k)_{k\in\mathbb{N}}$ is a sequence which satisfies (3.25) and (3.26). Let $f : S \to S$ be a strict (b_n)-probabilistic contraction. Then $\rho_k(fx,fy) \leq q_k\rho_k(x,y)$ for every $k \in \mathbb{N}$ and every $x,y \in S$.*

Proof. Let ρ_k be a pseudo-metric given in (3.27). Let for every $k \in \mathbb{N}$, $u_{k,x,y} = \rho_k(fx,fy)/q_k$. We shall prove that for every $x,y \in S$ and every $k \in \mathbb{N}$

$$\rho_k(fx,fy) \leq q_k\rho_k(x,y),$$

i.e., that $\rho_k(x,y) \geq u_{k,x,y}$, which means that

$$F_{x,y}(u_{k,x,y}) \leq b_k. \tag{3.28}$$

Suppose, on the contrary, that $F_{x,y}(u_{k,x,y}) > b_k$. Since f is a strict (b_n)-probabilistic contraction we have that $F_{fx,fy}(q_k u_{k,x,y}) > b_k$, and therefore

$$F_{fx,fy}(\rho_k(fx,fy)) > b_k.$$

This contradicts (3.27) since, according to (3.27),

$$F_{fx,fy}(\rho_k(fx,fy)) \leq b_k.$$

We have proved that $\rho_k(x,y) \geq \rho_k(fx,fy)/q_k$, which implies that $\rho_k(fx,fy) \leq q_k\rho_k(x,y)$ for every $k \in \mathbb{N}$ and every $x,y \in S$. □

Remark 3.52 In a similar way one can prove a fixed point result if (3.24) is replaced by (3.29)

$$m(x,y,u) > b_k \quad \Rightarrow \quad F_{fx,fy}(q_k u) > b_k, \tag{3.29}$$

where $m(x,y,u) = \min\{F_{x,y}(u),\, F_{x,fy}(u),\, F_{y,fx}(u),\, F_{x,fx}(u),\, F_{y,fy}(u)\}$, and T is as in Proposition 3.51.

We shall prove that (3.29) implies

$$\rho_k(fx, fy) \le q_k r_k(x,y) \tag{3.30}$$

for every $k \in \mathbb{N}$, where

$$r_k(x,y) = \max\{\rho_k(x,y), \rho_k(x,fy), \rho_k(y,fx), \rho_k(x,fx), \rho_k(y,fy)\}.$$

If (3.30) does not hold for some $k \in \mathbb{N}$ and $x, y \in S$, there exists $u > 0$ such that

$$\rho_k(fx, fy) > q_k u > q_k r_k(x,y). \tag{3.31}$$

Hence

$$F_{x,y}(u) > b_k, \quad F_{x,fy}(u) > b_k, \quad F_{y,fx}(u) > b_k, \quad F_{x,fx}(u) > b_k, \quad F_{y,fy}(u) > b_k,$$

and therefore

$$\min\{F_{x,y}(u), F_{x,fy}(u), F_{y,fx}(u), F_{x,fx}(u), F_{y,fy}(u)\} > b_k.$$

From (3.29) it follows that $F_{fx,fy}(q_k u) > b_k$, but (3.31) implies $\rho_k(fx,fy) > q_k u$ and therefore $F_{fx,fy}(q_k u) \le b_k$, which is a contradiction. Hence (3.30) holds, and the rest of the proof is similar to the proof as in the case of metric spaces , see [41].

The next fixed point theorem is a probabilistic generalization of a fixed point theorem for (q,n)-locally contractions proved in [77].

Definition 3.53 *Let $(S,\mathcal{F})$ be a probabilistic metric space. A mapping $f : S \to S$ is a probabilistic (q,n)-locally contraction ($q \in (0,1)$), if for every $x \in S$ there exists $n(x) \in \mathbb{N}$ such that for every $y \in S$ and $u \in \mathbb{R}$*

$$F_{f^{n(x)}x, f^{n(x)}y}(qu) \ge F_{x,y}(u).$$

Theorem 3.54 *Let $(S,\mathcal{F},T)$ be a complete Menger space, T a t-norm which is continuous at $(1,1)$, $f : S \to S$ a probabilistic (q,n)-locally contraction and $D_{O(p,f)} \in \mathcal{D}^+$ for some $p \in S$. Then there exists a unique fixed point x of the mapping f and $\lim\limits_{n\to\infty} f^n x_0 = x$ for every $x_0 \in S$.*

Proof. Let $x_n = f^{n(x_{n-1})}x_{n-1}$ for every $n \in \mathbb{N}$ and $x_0 = p \in S$. First, we shall prove that $(x_n)_{n\in\mathbb{N}}$ is a Cauchy sequence. For every $n, m \in \mathbb{N}$ and every $u > 0$ we have

$$\begin{aligned}
F_{x_{n+m},x_n}(u) &= F_{f^{n(x_{n+m-1})} f^{n(x_{n+m-2})} \cdots f^{n(x_{n-1})} x_{n-1},\, f^{n(x_{n-1})} x_{n-1}}(u) \\
&\ge F_{f^{n(x_{n+m-1})} \cdots f^{n(x_n)} x_{n-1},\, x_{n-1}}\left(\frac{u}{q}\right) \\
&\ \ \vdots \\
&\ge F_{f^{n(x_{n+m-1})} \cdots f^{n(x_n)} x_0,\, x_0}\left(\frac{u}{q^n}\right) \\
&\ge D_{O(p,f)}\left(\frac{u}{q^n}\right).
\end{aligned}$$

Since $D_{O(p,f)} \in \mathcal{D}^+$ it follows that for every $u > 0$ and every $\lambda \in (0,1)$ there exists $n_0(u,\lambda) \in \mathbb{N}$ such that

$$D_{O(p,f)}\Big(\frac{u}{q^n}\Big) > 1 - \lambda \quad \text{for every} \quad n \geq n_0(u,\lambda)$$

and therefore $(x_n)_{n\in\mathbb{N}}$ is a Cauchy sequence.

Hence from the completeness of S it follows that there exists $x = \lim_{n\to\infty} x_n$ and we shall prove that $f^{n(x)}x = x$. From the inequality

$$F_{f^{n(x)}x_n, f^{n(x)}x}(u) \geq F_{x_n,x}\Big(\frac{u}{q}\Big) \quad (u \in \mathbb{R},\ n \in \mathbb{N}),$$

it follows that $\lim_{n\to\infty} f^{n(x)}x_n = f^{n(x)}x$. For every $u \in \mathbb{R}$ and $\varepsilon \in (0, qu)$

$$\begin{aligned} F_{f^{n(x)}x_n,x_n}(qu-\varepsilon) &= F_{f^{n(x)}f^{n(x_{n-1})}x_{n-1}, f^{n(x_{n-1})}x_{n-1}}(qu-\varepsilon) \\ &\geq F_{f^{n(x)}x_{n-1},x_{n-1}}(u - q^{-1}\varepsilon) \\ &\ \vdots \\ &\geq F_{f^{n(x)}x_0,x_0}(q^{-n+1}u - q^{-n}\varepsilon). \end{aligned}$$

Hence from

$$F_{f^{n(x)}x,x}(u) \geq T(F_{f^{n(x)}x,f^{n(x)}x_n}(u-qu), T(F_{f^{n(x)}x_n,x_n}(qu-\varepsilon),\ F_{x_n,x}(\varepsilon)))$$

we obtain

$$F_{f^{n(x)}x,x}(u) \geq T(F_{x,x_n}(q^{-1}u-u), T(F_{f^{n(x)}x_0,x_0}(q^{-n+1}u - q^{-n}\varepsilon),\ F_{x_n,x}(\varepsilon)))$$

and when $n \to \infty$, by the continuity of T at the point $(1,1)$

$$1 \geq F_{f^{n(x)}x,x}(u) \geq T(1,1) = 1.$$

This implies that $f^{n(x)}x = x$. We prove that the relation $f^{n(x)}z = z$, for some $z \in S$, implies that $x = z$. Suppose that $x \neq z$. Then there exists an $s > 0$ such that $F_{x,z}(s) = a \in [0,1)$. Then for every $n \in \mathbb{N}$

$$a = F_{x,z}(s) = F_{f^{n(x)}x,f^{n(x)}z}(s) \geq F_{x,z}(q^{-1}s) \geq \cdots \geq F_{x,z}(q^{-n}s),$$

and when $n \to \infty$ we obtain $a = 1$, which contradicts $a < 1$, and therefore $x = z$. Since

$$f(x) = ff^{n(x)}x = f^{n(x)+1}x = f^{n(x)}fx,$$

it follows that $fx = x$. We shall prove that $x = \lim_{n\to\infty} f^n(x_0)$. Let $n = m \cdot n(x) + r$, where $0 \le r < n(x)$. Then for every $u > 0$

$$\begin{aligned} F_{f^n x_0, x}(u) &= F_{f^{m\cdot n(x)+r} x_0, f^{n(x)} x}(u) \\ &\ge F_{f^{(m-1)n(x)+r} x_0, x}(q^{-1}u) \\ &\ge F_{f^{(m-2)n(x)+r} x_0, x}(q^{-2}u) \\ &\vdots \\ &\ge F_{f^r x_0, x}(q^{-m}u) \quad (u \in \mathbb{R}). \end{aligned}$$

If $n \to \infty$ then $m \to \infty$ and $\lim_{m\to\infty} F_{f^r x_0, x}(q^{-m}u) = 1$. Hence $\lim_{n\to\infty} F_{f^n x_0, x}(u) = 1$, which means that $\lim_{n\to\infty} f^n x_0 = x$. □

Corollary 3.55 *Let $(S, \mathcal{F}, T)$ be a complete Menger space such that T is a t-norm which is continuous at $(1,1)$ and of H-type. If $f : S \to S$ is a probabilistic (q,n)-locally contraction, then there exists a unique fixed point x of the mapping f and $\lim_{n\to\infty} f^n x_0 = x$ for every $x_0 \in S$.*

Proof. Let $x_0 \in S$, $m \in \mathbb{N}$ and $sn(x_0) < m \le (s+1)n(x_0)$. Then for every $u \in \mathbb{R}$

$$\begin{aligned} F_{f^m x_0, x_0}(u) &\ge T(F_{f^m x_0, f^{n(x_0)} x_0}(qu), F_{f^{n(x_0)} x_0, x_0}(u - qu)) \\ &\ge T(F_{f^{m-n(x_0)} x_0, x_0}(u),\ F_{f^{n(x_0)} x_0, x_0}(u - qu)) \\ &\vdots \\ &\ge (g(u))_T^{(s)}, \end{aligned}$$

where

$$g(u) = \min_{r \in \{1,2,\ldots,n(x_0)\}} \{F_{f^r x_0, x_0}(u - qu)\}.$$

Since $\lim_{u\to\infty} g(u) = 1$ and the family $(u_T^{(s)})_{s\in\mathbb{N}}$ is equicontinuous at the point $u = 1$, the proof is complete. □

If $n(x) = 1$ for every $x \in S$ and $T = T_{\mathbf{M}}$ from Corollary 3.55 the fixed point theorem of Sehgal and Bharucha-Reid follows.

Definition 3.56 *Let $(S, \mathcal{F})$ be a probabilistic metric space and $f : S \to S$. A point $x \in S$ is regular for f if $O(x, f)$ is bounded.*

Definition 3.57 *Let $(S, \mathcal{F})$ be a probabilistic metric space and $f : S \to S$. Two points $x, y \in S$ are asymptotic under f if*

$$\lim_{n\to\infty} F_{f^n(x), f^n(y)}(\varepsilon) = 1 \quad \textit{for every } \varepsilon > 0.$$

If (S,d) is a metric space similar definitions are introduced in [126].

Theorem 3.58 *Let $(S,\mathcal{F},T)$ be a complete Menger space, T a t-norm such that $\sup_{a<1} T(a,a) = 1$, $f : S \to S$ a continuous mapping such that each point of S is regular for f and every pair of points in S is asymptotic under f. If there exists $q \in (0,1)$ such that every $x \in S$ and for every $\varepsilon > 0$*

$$D_{O(f(x),f)}(\varepsilon) \geq D_{O(x,f)}\Big(\frac{\varepsilon}{q}\Big), \tag{3.32}$$

then there exists a unique fixed point z of the mapping f and $z = \lim_{n\to\infty} f^n(x)$ for every $x \in S$.

Proof. From (3.32) it follows that for every $\varepsilon > 0$ and every $n \in \mathbb{N}$

$$D_{O(f^n(x),f)}(\varepsilon) \geq D_{O(f^{n-1}(x),f)}\Big(\frac{\varepsilon}{q}\Big) \geq \ldots \geq D_{O(x,f)}\Big(\frac{\varepsilon}{q^n}\Big). \tag{3.33}$$

We prove that for every $\varepsilon > 0$ and $\lambda \in (0,1)$ there exists $n_0 \in \mathbb{N}$ such that

$$F_{f^m(x),f^s(x)}(\varepsilon) > 1 - \lambda \text{ for every } m, s \geq n_0.$$

Since $\sup_{\varepsilon} D_{O(x,f)}(\varepsilon) = 1$ there exists $t(\lambda) > 0$ such that

$$D_{O(x,f)}(t(\lambda)) > 1 - \frac{\lambda}{2}.$$

If we choose $n_0 \in \mathbb{N}$ such that $\dfrac{\varepsilon}{q^{n_0}} \geq t(\lambda)$, then

$$D_{O(x,f)}\Big(\frac{\varepsilon}{q^{n_0}}\Big) \geq D_{O(x,f)}(t(\lambda)) > 1 - \frac{\lambda}{2},$$

and therefore from (3.33) it follows that

$$D_{O(f^n(x),f)}(\varepsilon) > 1 - \frac{\lambda}{2} \quad \text{for every} \quad n \geq n_0.$$

This means that

$$\sup_{\delta<\varepsilon} \inf_{u,v\in O(f^n(x),f)} F_{u,v}(\delta) > 1 - \frac{\lambda}{2} \quad \text{for every } n \geq n_0. \tag{3.34}$$

Since $F_{x,y}(\varepsilon) \geq F_{x,y}(\delta)$, for every $\delta < \varepsilon$ and every $x, y \in S$, (3.34) implies that

$$\begin{aligned}\inf_{u,v\in O(f^n(x),f)} F_{u,v}(\varepsilon) &\geq \sup_{\delta<\varepsilon} \inf_{u,v\in O(f^n(x),f)} F_{u,v}(\delta)\\ &> 1 - \frac{\lambda}{2}\end{aligned}$$

for every $n \geq n_0$ and therefore

$$F_{u,v}(\varepsilon) > 1 - \frac{\lambda}{2} \quad \text{for every} \quad u, v \in O(f^n(x), f) \text{ and every } n \geq n_0.$$

Hence we have that $F_{f^m(x),f^r(x)}(\varepsilon) > 1 - \lambda/2$ for every $m, r \geq n_0$. This means that $(f^m x)_{m\in\mathbb{N}}$ is a Cauchy sequence and since S is complete there exists $z = \lim\limits_{n\to\infty} f^n(x)$. From the continuity of the mapping f it follows that z is a fixed point of the mapping f. Suppose now that $y \in S$ and $f(y) = y$. Then we have that $F_{y,z}(\varepsilon) = F_{f^n(y),f^n(z)}(\varepsilon)$ and since y and z are asymptotic under f it follows that

$$F_{y,z}(\varepsilon) = \lim_{n\to\infty} F_{f^n(y),f^n(x)}(\varepsilon) = 1 \text{ for every } \varepsilon > 0,$$

which implies that $y = z$. □

Remark 3.59 If $(S, \mathcal{F}, T)$ is a complete Menger space with a continuous t-norm T which is of H-type and $f : S \to S$ is a probabilistic q-contraction it is easy to see that all the conditions of Theorem 3.58 are satisfied.

An interesting fixed point theorem was proved in Hausdorff probabilistic semi-metric space by Seppo Heikkila and Seppo Seikkala in [127].

Let $(S, \mathcal{F})$ be a probabilistic semi-metric space. By $\mathbb{R}_+^{(0,1)}$ we denote the space of all non-negative valued functions on $(0, 1)$ endowed with the topology of pointwise convergence. The partial ordering $\leq$ in $\mathbb{R}_+^{(0,1)}$ is defined by $u \leq v$ if and only if $u(\alpha) \leq v(\alpha)$ for all $\alpha \in (0, 1)$ where $u < v$ means that $u(\alpha) < v(\alpha)$ for all $\alpha \in (0, 1)$. Let $\alpha \in (0, 1)$ and $x, y \in S$. Then by

$$d(x, y)(\alpha) = \sup\{t \mid F_{x,y}(t) \leq 1 - \alpha\}$$

an $\mathbb{R}_+^{(0,1)}$ valued mapping d on $S \times S$ is defined and has the following properties:

1) $d(x, y) = d(y, x)$ for every $(x, y) \in S \times S$;
2) $d(x, y) = \mathbf{0}$ if and only if $x = y$ ($\mathbf{0}$ is the zero function on $(0, 1)$).

A sequence $(x_n)_{n\in\mathbb{N}}$ from S converges to $x \in S$ in the (ε, λ)-topology if and only if $(d(x_n, x))_{n\in\mathbb{N}}$ converges to $\mathbf{0}$ in $\mathbb{R}_+^{(0,1)}$. S is complete if and only if a sequence $(x_n)_{n\in\mathbb{N}}$ converges in S whenever it satisfies

$$d(x_{n+m}, x_n) \leq u_n \quad \text{for every} \quad m, n \in \mathbb{N}$$

for a sequence $(u_n)_{n\in\mathbb{N}}$ in $\mathbb{R}_+^{(0,1)}$ which converges to $\mathbf{0}$.

Let $b > \mathbf{0}$ and $x_0 \in S$. Then

$$[\mathbf{0}, b] = \{u \mid u \in \mathbb{R}_+^{(0,1)}, \ \mathbf{0} \leq u \leq b\}$$

and

$$\begin{aligned} B_b(x_0) &= \{x \mid x \in S,\ d(x, x_0) < b\} \\ &= \{x \mid x \in S,\ F_{x,x_0}(b(\alpha)) > 1 - \alpha \text{ for all } \alpha \in (0,1)\}. \end{aligned}$$

In [127] the following fixed point theorem was proved.

Theorem 3.60 *Let $(S, \mathcal{F})$ be a Hausdorff probabilistic semi-metric space and $f : S \to S$. Suppose that the mapping f satisfies the conditions (H_1) and (H_2) where:*

(H_1) there exists $b > \mathbf{0}$ such that for all $\alpha \in (0,1)$, $u \in [\mathbf{0}, b]$ and $x, y \in S$

$$F_{x,y}(u(\alpha)) > 1 - \alpha \quad \Rightarrow \quad F_{fx,fy}(Qu(\alpha)) > 1 - \alpha$$

where the mapping $Q : [\mathbf{0}, b] \to [\mathbf{0}, b]$ satisfies the condition $\lim\limits_{n\to\infty} Q^n b = \mathbf{0}$;

(H_2) there exists $x_0 \in S$ such that for all $\alpha \in (0,1)$,

$$F_{x_0,x}(b(\alpha)) > 1 - \alpha \quad \Rightarrow \quad F_{x_0,fx}(b(\alpha)) > 1 - \alpha.$$

If S is complete then $x = \lim\limits_{n\to\infty} f^n x_0$ exists and x is the unique fixed point of the mapping f in $B_b(x)$.

Proof. First, let us prove that (H_1) implies the uniqueness of the fixed point of the mapping f in $B_b(x)$ if x is a fixed point of f. From (H_1) it follows that for all $\alpha \in (0,1)$, $u \in [\mathbf{0}, b]$, and $x, y \in S$

$$d(x,y)(\alpha) < u(\alpha) \quad \Rightarrow \quad d(fx, fy)(\alpha) < Qu(\alpha). \tag{3.35}$$

If x is a fixed point of the mapping f and $y \in B_b(x)$ such that $y = fy$, then from the implication (3.35) we obtain

$$d(x,y) = d(f^n x, f^n y) < Q^n b, \quad \text{for every } n \in \mathbb{N},$$

and since $\lim\limits_{n\to\infty} Q^n b = \mathbf{0}$ we have that $d(x,y) = \mathbf{0}$. This implies that $x = y$.

Suppose that S is complete. Condition (H_2) implies that $f(B_b(x_0)) \subseteq B_b(x_0)$ and therefore for every $m \in \mathbb{N}$ we have $d(f^m x_0, x_0) < b$. Hence for every $m, n \in \mathbb{N} \cup \{0\}$

$$d(f^{m+n} x_0, f^n x_0) < Q^n b$$

and since S is complete and $\lim\limits_{n\to\infty} Q^n b = \mathbf{0}$ it follows that there exists $x \in B_b(x_0)$ such that $x = \lim\limits_{n\to\infty} f^n x_0$. It remains to prove that $fx = \lim\limits_{n\to\infty} f^n x_0$, which means that for every $\varepsilon > 0$ and $\alpha \in (0,1)$ there exists $m_0(\alpha, \varepsilon) \in \mathbb{N}$ such that

$$d(fx, f^m x_0)(\alpha) < \varepsilon \text{ for every } m \geq m_0(\alpha, \varepsilon).$$

Choose $\varepsilon > 0$ and $\alpha \in (0,1)$. Then there exists $n_0 \in \mathbb{N}$ such that

$$Q^n b(\alpha) < \varepsilon \text{ for every } n \geq n_0.$$

Further on, $\lim_{n\to\infty} d(x, f^n x_0)(\alpha) = 0$ implies that there exists $m_0(\alpha, \varepsilon) \in \mathbb{N}$ such that

$$d(x, f^{m-1}x_0)(\alpha) < Q^{n_0} b(\alpha) \quad \text{for every } m \geq m_0(\alpha, \varepsilon).$$

By (3.35) we have

$$d(fx, f^m x_0)(\alpha) < Q^{n_0+1} b(\alpha) < \varepsilon \text{ for every } m \geq m_0(\alpha, \varepsilon)$$

and therefore $fx = \lim_{n\to\infty} f^n x_0$. This implies that x is the unique fixed point of the mapping f in $B_b(x)$. □

In [35] a new notion of the contraction in probabilistic metric spaces is introduced and the following fixed point theorem is proved.

Theorem 3.61 *Let $(S, \mathcal{F}, T)$ be a complete Menger space, where $T = T_{\mathbf{M}}$ and $f : S \to S$ be such that the following implication holds for every $x, y \in S$ and every $\alpha, \beta > 0$:*

$$F_{x,y}(\alpha) > 0, F_{x,y}(\beta) < 1 \quad \Rightarrow \quad F_{fx,fy}(t) \geq F_{x,y}\left(\frac{t}{L(\alpha,\beta)}\right) \text{ for every } t > 0, \tag{3.36}$$

where $L : \mathbb{R}^+ \times \mathbb{R}^+ \to (0,1)$. Then f has a unique fixed point.

If $f : S \to S$ satisfies condition (3.36) then we say that f is a *generalized probabilistic B-contraction.*

A generalization of Theorem 3.61 will be proved, using the following theorem.

Theorem 3.62 (Krasnoselski, Zabreiko [173]) *Let (X, d) be a complete metric space and $f : X \to X$ such that there exists a function $L : \mathbb{R}^+ \times \mathbb{R}^+ \to [0,1)$ such that*

$$(\forall a, b > 0)(\forall x, y \in X)(a \leq d(x,y) \leq b \quad \Rightarrow \quad d(fx, fy) \leq L(a,b)d(x,y)). \tag{3.37}$$

Then there exists a unique fixed point of f which is globally attractive.

If f satisfies (3.37) f is the so called *generalized contraction of Krasnoselski's type.*

In the proof of Lemma 3.65 the following Lemma [29] will be used.

Lemma 3.63 *Let $(S, \mathcal{F}, T)$ be a Menger space and T a t-norm such that*

$$\lim_{x\to 1_-} T(x,y) = y \text{ for every } y \in [0,1]. \tag{3.38}$$

If $(p_n)_{n\in\mathbb{N}}$ and $(q_n)_{n\in\mathbb{N}}$ are sequences in S such that

$$\lim_{n\to\infty} p_n = p, \qquad \lim_{n\to\infty} = q$$

then:

a) For every $x \in \mathbb{R}$

$$\liminf_{n\to\infty} F_{p_n,q_n}(x) \geq F_{p,q}(x).$$

b) If x is any point of the continuity of $F_{p,q}$ then

$$\lim_{n\to\infty} F_{p_n,q_n}(x) = F_{p,q}(x).$$

Remark 3.64 If $T \geq T_{\mathbf{L}}$ then the condition $\lim\limits_{x\to 1_-} T(x,y) = y$, for every $y \in [0,1]$ is satisfied since for all $x, y \in [0,1]$

$$y \geq T(x,y) \geq \max(x+y-1,0)$$

and letting $x \to 1_-$ we have that

$$y \geq \lim_{x\to 1_-} T(x,y) \geq y \text{ for every } y \in [0,1].$$

Condition (3.38) implies that $\sup\limits_{a<1} T(a,a) = 1$. Indeed, let $\lambda \in [0,1)$ be given and $\lambda' \in (0,1)$ is such that $1-\lambda' > 1-\lambda$. Since $T(1,1-\lambda') = 1-\lambda'$, by (3.38) we conclude that there exists a $\lambda'' \in (0,1)$ such that

$$T(1-\lambda'', 1-\lambda') > 1-\lambda.$$

If $\lambda_1 = \min(\lambda', \lambda'')$ we have

$$\begin{aligned} T(1-\lambda_1, 1-\lambda_1) &\geq T(1-\lambda'', 1-\lambda') \\ &> 1-\lambda. \end{aligned}$$

Lemma 3.65 (Radu [240]) *Let $\mathcal{F} : S \times S \to \mathcal{D}^+$, F be a fixed element of $\mathcal{D}^+$ and consider the two-place function $d_F : S \times S \to [0,\infty]$, given by*

$$d_F(p,q) = \inf\{a > 0 \mid F_{p,q}(at) \geq F(t) \text{ for all } t > 1\}.$$

If $(S, \mathcal{F}, T_{\mathbf{M}})$ is a Menger space then:

1^0 *d_F is a Luxemburg metric on S* ($\inf \varnothing = +\infty$).
2^0 *(S, d_F) is complete if $(S, \mathcal{F})$ is complete.*
3^0 *The d_F-topology is stronger than the (ε, λ)-topology .*

Proof. 1^0. It is clear that d_F is symmetric and $d_F(p,p) = 0$. If $d_F(p,q) = 0$, then, for each $a > 0$, $F_{p,q}(at) \geq F(t)$ for all $t > 1$. Now if $\varepsilon = at$ is fixed, and $t \to +\infty$, then $F_{p,q}(\varepsilon) = 1$, which implies that $p = q$.

In order to prove the triangle inequality assume that $d_F(p,r) = a < \infty$ and $d_F(r,q) = b < \infty$. Then

$$F_{p,q}((a+b)t) \geq T_{\mathbf{M}}(F_{p,r}(at)\ ,\ F_{r,q}(bt)) \geq F(t)$$

which shows that $d_F(p,q) \leq a+b$, that is $d_F(p,q) \leq d_F(p,r) + d_F(r,q)$. Therefore d_F is a Luxemburg metric .

2^0 Suppose that $(S, \mathcal{F})$ is complete and $(p_n)_{n\in\mathbb{N}}$ is a d_F-Cauchy sequence. If $\varepsilon > 0$ and $\lambda \in (0,1)$ are given, there exists $t_0 > 1$ such that $F(t_0) > 1 - \lambda$. Let $a < \frac{\varepsilon}{t_0}$ and choose n_0 such that $d_F(p_n, p_m) < a$ for all $n, m \geq n_0$. Therefore

$$F_{p_n,p_m}(\varepsilon) \geq F_{p_n,p_m}(at_0) \geq F(t_0) > 1 - \lambda \text{ for all } n, m \geq n_0,$$

which means that $(p_n)_{n\in\mathbb{N}}$ is an $\mathcal{F}$-Cauchy sequence.

Since $(S, \mathcal{F})$ is complete there exists $p \in S$ such that $(p_m)_{m\in\mathbb{N}}$ is $\mathcal{F}$-convergent to p. We have to prove that $\lim\limits_{n\to\infty} d_F(p_n, p) = 0$.

For a given $a > 0$, there exists n_0 such that $F_{p_n,p_m}(at) \geq F(t)$ for all $n, m > n_0$, and each $t > 1$. If at is a point of the continuity of $F_{p_n,p}$ then

$$F_{p_n,p}(at) = \lim_{m\to\infty} F_{p_n,p_m}(at) \geq F(t) \text{ for all } n > n_0.$$

If at is not a point of the continuity of $F_{p_n,p}$, since $F_{p_n,p}$ is a non-decreasing function, there exists a sequence $(at_i^n)_{i\in\mathbb{N}}$ $(t_i^n > 1, i \in \mathbb{N})$ such that $at_1^n < at_2^n \cdots \to at$ and at_i^n is, for every $i \in \mathbb{N}$, the point of the continuity of $F_{p_n,p}$. Then

$$F_{p_n,p}(at_i^n) \geq F(t_i^n) \text{ for all } i \in \mathbb{N}, n > n_0. \tag{3.39}$$

Since $\lim\limits_{i\to\infty} at_i^n = at$ $(a > 0)$ we have that $\lim\limits_{i\to\infty} t_i^n = t$. Functions $F_{p_n,p}$ and F are left continuous and from (3.39) it follows that

$$F_{p_n,p}(at) = \lim_{i\to\infty} F_{p_n,p}(at_i^n) \geq \lim_{i\to\infty} F(t_i^n) = F(t) \text{ for all } t > 1, n > n_0.$$

Hence $d_F(p_n, p) < a$ for all $n > n_0$.

In a similar way the last part of the theorem can be proved. □

Lemma 3.66 *Let $(S, \mathcal{F}, T_M)$ be a Menger space and $f : S \to S$ a generalized probabilistic B-contraction. Then f is a generalized contraction of Krasnoselski's type, relatively to d_F.*

Proof. Let $0 < a < b$ be fixed, and assume that $d_F(x, y) \in [a, b]$. Then

$$\frac{a}{2} < d_F(x, y) \leq b$$

Choose $t_1 > 1$ such that $F(t_1) > 0$. Then $F_{x,y}(bt_1) \geq F(t_1) > 0$ and let $\alpha(b) = bt_1 > 0$. Since $d_F(x, y) > a/2$ there exists $t_{x,y} > 1$ such that $F_{x,y}(at_{x,y}/2) < F(t_{xy})$. Then $F_{x,y}(\frac{a}{2}) \leq F_{xy}(\frac{a}{2}t_{x,y}) < F(t_{x,y})$, and let $\beta(a) = a/2 > 0$.

Since f is a generalized probabilistic B-contraction by (3.36) we obtain

$$F_{fx,fy}(t) \geq F_{x,y}(\frac{t}{L(\alpha(b), \beta(a))}) \geq F(\frac{t}{L(\alpha(b), \beta(a))d_F(x, y)}) \text{ for all } t > 0,$$

which implies that

$$d_F(fx, fy) \leq L(\alpha(b), \beta(a))d_F(x, y).$$

Taking $\overline{L}(a, b) = L(\alpha(b), \beta(a))$ we obtain that

$$d_F(fx, fy) \leq \overline{L}(a, b)d_F(x, y).$$

Thus f is a generalized contraction of Krasnoselski's type relatively to d_F. □

Theorem 3.67 *Let all the conditions of Theorem 3.61 be satisfied. Then f has a unique fixed point which is globally attractive.*

Proof. Let p be an arbitrary element from S and $X = \{x \in S \mid d_F(p, x) < \infty\}$. From the definition of the metric d_F, for $F = F_{p,fp}$, it follows that $d_F(p, fp) \leq 1$, which implies that $fp \in X$. Then (X, d_F) is a complete metric space with $f(X) \subset X$. We can apply Theorem 3.62 and since p is an arbitrary element from S, the theorem is proved. □

In order to generalize Theorem 3.67 to complete Menger spaces $(S, \mathcal{F}, T)$, where T is a continuous t-norm of H-type we prove the following Lemma.

Lemma 3.68 *Let $(S, \mathcal{F}, T)$ be a complete Menger space, T a continuous t-norm of H-type and $(b_n)_{n \in \mathbb{N}}$ a sequence from $(0, 1)$ such that $T(b_n, b_n) = b_n$ $(n \in \mathbb{N})$ and $\lim_{n \to \infty} b_n = 1$. Then $(S, \widetilde{\mathcal{F}}, T_{\mathbf{M}})$ is a complete Menger space with the same (ε, λ)-topology as $(S, \mathcal{F}, T)$, where for every $u \in \mathbb{R}$*

$$\widetilde{F}_{x,y}(u) = \begin{cases} 0 & \text{if } F_{x,y}(u) < b_1, \\ b_n & \text{if } b_n \leq F_{x,y}(u) < b_{n+1}, \\ 1 & \text{if } F_{x,y}(u) = 1. \end{cases}$$

Proof. It is easy to see that $\widetilde{F}_{x,y} \in \mathcal{D}^+$. Let us prove that

$$\widetilde{F}_{x,z}(u+v) \geq T_M(\widetilde{F}_{x,y}(u), \widetilde{F}_{y,z}(v)) \tag{3.40}$$

for every $x, y, z \in S$ and every $u, v > 0$. If $\widetilde{F}_{x,y}(u) = 0$ or $\widetilde{F}_{y,z}(v) = 0$ then (3.40) holds. If $\widetilde{F}_{x,y}(u) = b_n, \widetilde{F}_{y,z}(v) = b_m$ $(n \leq m)$ then $F_{x,y}(u) \geq b_n, F_{y,z}(v) \geq b_m$ and from

$$F_{x,z}(u+v) \geq T(F_{x,y}(u), F_{y,z}(v)) \geq T(b_n, b_n) = b_n$$

it follows that $\widetilde{F}_{x,z}(u+v) \geq b_n = T_{\mathbf{M}}(\widetilde{F}_{x,y}(u), \widetilde{F}_{y,z}(v))$. Since from $b_n > 1-\lambda$ it follows $N(\varepsilon, 1-b_n) \subset \widetilde{N}(\varepsilon, \lambda), \widetilde{N}(\varepsilon, 1-b_n) \subset N(\varepsilon, \lambda)$ the mappings $\mathcal{F}$ and $\widetilde{\mathcal{F}}$ induce the same (ε, λ)-uniformities. □

Theorem 3.69 *Let $(S, \mathcal{F}, T)$ be a complete Menger space, where T is a continuous t-norm of H-type. Then every generalized probabilisric B-contraction $f : S \to S$ has a unique fixed point, and which is globally attractive.*

Proof. We shall prove that (3.36) holds for $\widetilde{\mathcal{F}}$ and f, where $\widetilde{\mathcal{F}}$ is defined in Lemma 3.68. Suppose that, for some $\alpha, \beta > 0$ and $x, y \in S$, we have

$$\widetilde{F}_{x,y}(\alpha) > 0, \qquad \widetilde{F}_{x,y}(\beta) < 1.$$

Then $F_{x,y}(\alpha) > 0$ and $F_{x,y}(\beta) < 1$ and by (3.36)

$$F_{fx,fy}(u) \geq F_{x,y}\Big(\frac{u}{L(\alpha,\beta)}\Big) \text{ for every } u > 0. \tag{3.41}$$

We prove that (3.41) implies

$$\widetilde{F}_{fx,fy}(u) \geq \widetilde{F}_{x,y}\Big(\frac{u}{L(\alpha,\beta)}\Big) \text{ for every } u > 0. \tag{3.42}$$

If $\widetilde{F}_{x,y}\Big(\frac{u}{L(\alpha,\beta)}\Big) = 0$ then (3.42) holds. If $\widetilde{F}_{x,y}\Big(\frac{u}{L(\alpha,\beta)}\Big) = b_n$ then

$$F_{fx,fy}(u) \geq F_{x,y}\Big(\frac{u}{L(\alpha,\beta)}\Big) \geq b_n.$$

Hence

$$\widetilde{F}_{fx,fy}(u) \geq b_n = \widetilde{F}_{x,y}\Big(\frac{u}{L(\alpha,\beta)}\Big).$$

From Theorem 3.67 we obtain the desired conclusion. □

Since any probabilistic q-contraction is a generalized probabilistic B-contraction, using Theorem 3.12 we conclude that the following theorem holds.

Theorem 3.70 *If a continuous t-norm T is such that any generalized probabilistic B-contraction $f : S \to S$ has a fixed point, whenever $(S, \mathcal{F}, T)$ is a complete Menger space, then T is of H-type.*

In [304] Do Hong Tan introduced some new classes of contractive mappings on probabilistic metric spaces and investigated the relation between these classes.

Definition 3.71 *Let $(S, \mathcal{F})$ be a probabilistic metric space. A mapping $f : S \to S$ belongs to the class $[R]$ if there is a non-increasing function $k : (0, \infty) \to (0, 1)$ such that*

$$F_{fx,fy}(t) \geq F_{x,y}\Big(\frac{t}{k(t)}\Big) \text{ for every } x, y \in S \text{ and every } t > 0. \tag{3.43}$$

The metric version of (3.43) is of the form [253]

$$d(fx, fy) \leq k(d(x, y))d(x, y) \text{ for every } x, y \in M \tag{3.44}$$

where (M, d) is a metric space, $f : M \to M$, and k is as in Definition 3.71.

Definition 3.72 *Let $(S, \mathcal{F})$ be a probabilistic metric space. A mapping $f : S \to S$ belongs to the class $[K]$ if there is a function $k : (0, \infty)^2 \to (0, \infty)$ such that for every $\alpha, \beta > 0$, $\alpha \leq \beta$*

$$F_{fx,fy}(t) \geq F_{x,y}\Big(\frac{t}{k(\alpha, \beta)}\Big) \tag{3.45}$$

for every $x, y \in S$ and every $t > 0$ such that $\alpha \leq t \leq \beta$.

Definition 3.73 *Let $(S, \mathcal{F})$ be a probabilistic metric space. A mapping $f : S \to S$ belongs to the class $[S]$ if there is a function $k : (0, \infty) \to (0, 1)$ satisfying the condition $\sup\{k(t) | a \leq t \leq b\} < 1$ for every $a, b > 0$, $a \leq b$, such that (3.43) holds.*

The metric version is inequality (3.44) with k as in Definition 3.73 and it is introduced in [260].

Definition 3.74 *Let $(S, \mathcal{F})$ be a probabilistic metric space. A mapping $f : S \to S$ belongs to the class $[BW]$ if there is a mapping $k : (0, \infty) \to (0, \infty)$, which is upper semicontinuous from the right such that (3.43) holds.*

The metric version is (3.44) with k as in Definition 3.74 and it is introduced in [21].

Definition 3.75 *Let $(S, \mathcal{F})$ be a probabilistic metric space. A mapping $f : S \to S$ belongs to the class $[MK]$ if for each $\varepsilon > 0$ there is a $\delta > 0$ such that*

$$F_{fx,fy}(\varepsilon) \geq F_{x,y}(\varepsilon + \delta) \text{ for every } x, y \in S \tag{3.46}$$

The metric version of (3.46) is introduced in [185] and it is of the form:
For each $\varepsilon > 0$ there is a $\delta > 0$ such that for every $x, y \in M$

$$\varepsilon \leq d(x,y) < \varepsilon + \delta \quad \Rightarrow \quad d(fx, fy) < \varepsilon,$$

where $f : M \to M$ and (M, d) is a metric space.

It is obvious that every probabilistic q-contraction belongs to all introduced classes. Tan proved in [304] the relations

$$[R] \subset [K] \subset [S] \subset [BW] \subset [MK]$$

and in [299] the following theorem is proved.

Theorem 3.76 *Let $(S, \mathcal{F}, T_{\mathbf{M}})$ be a complete Menger space and $f : S \to S$ belongs to the class $[MK]$. Then there exists a unique fixed point $x \in S$ of the mapping f and $x = \lim_{n\to\infty} f^n p$ for every $p \in S$.*

To our knowledge it is an open problem whether Theorem 3.76 holds in a complete Menger space $(S, \mathcal{F}, T)$ for some t-norm $T \neq T_{\mathbf{M}}$, with, possibly, some additional conditions for the growth of the mapping $\mathcal{F}$ (similarly as in Tardiff's fixed point theorem).

3.4 Fixed point theorems of Caristi's type

Banach's contraction principle in metric spaces is a consequence of the well known Caristi fixed point theorem [26], which is one of the most important results in fixed point theory and nonlinear analysis.

Theorem 3.77 (Caristi) *Let (M, d) be a complete metric space and $\varphi : M \to \mathbb{R}$ a lower semicontinuous function with a finite lower bound. Let $f : M \to M$ be any (not necessarily continuous) function such that*

$$d(x, fx) \leq \varphi(x) - \varphi(fx) \text{ for every } x \in M. \tag{3.47}$$

Then f has a fixed point.

Suppose that $f : M \to M$ is a q-contraction on M. Then $d(fx, f^2x) \leq qd(x, fx)$ for every $x \in M$ and so

$$d(x, fx) - qd(x, fx) \leq d(x, fx) - d(fx, f^2x)$$

for every $x \in M$, which implies that

$$d(x, fx) \leq \frac{1}{1-q} d(x, fx) - \frac{1}{1-q} d(fx, f^2x).$$

This means that for $\varphi(x) = \frac{1}{1-q} d(x, fx)$ the inequality (3.47) is satisfied.

A constructive proof of Caristi's theorem is given by F. Browder using the axiom of choice for countable families [23]. This principle can be derived from Zorn's lemma, as was done in a paper by Downing and Kirk [51] and by Cantor intersection theorem [236].

Literature on Caristi's fixed point theory is very large. An interesting paper is also [148] in which connections of Caristi's fixed point theorem and multivalued contractions are investigated.

In this section we prove some fixed point theorems of Caristi's type in Menger spaces and topological spaces of F-type. As a corollary an Ekeland variational principle in Menger spaces is obtained.

If t-norm T is of the H-type, we shall prove the following fixed point theorem of Caristi's type. In Theorem 3.78 $(b_n)_{n\in\mathbb{N}}$ is such a monotone increasing sequence from $(0,1)$ that $x > b_n, y > b_n$ implies $T(x,y) > b_n$ and $b_n \to 1, n \to \infty$.

Theorem 3.78 *Let $(S, \mathcal{F}, T)$ be a complete Menger space such that t-norm T is continuous and of H-type such that $T = (< (\alpha_k, \beta_k), T_k >)_{k\in K}$ and each T_k is strict, $f : S \to S$ a continuous mapping, $\Phi_n : S \to \mathbb{R}^+$ $(n \in \mathbb{N})$, and μ a mapping of $\mathbb{R}^+$ onto $\mathbb{R}^+$, such that μ is non-decreasing and*

$$\mu(a + b) \leq \mu(a) + \mu(b)$$

for every $a, b \in \mathbb{R}^+$. If for every $x \in S$, every $s > 0$ and every $n \in \mathbb{N}$

$$\mu(s) > \Phi_n(x) - \Phi_n(f(x)) \quad \Rightarrow \quad F_{x,f(x)}(s) > b_n, \tag{3.48}$$

then there exists a fixed point $x^* \in S$ of the mapping f and $x^* = \lim_{n\to\infty} f^n(x_0)$ for arbitrary $x_0 \in S$.

Proof. We shall prove that (3.48) implies

$$\mu(d_n(x, f(x))) \leq \Phi_n(x) - \Phi_n(f(x)), \tag{3.49}$$

for every $n \in \mathbb{N}$ and every $x \in S$, where d_n is defined by

$$d_n(x, y) = \sup\{u \mid u \in \mathbb{R}, F_{x,y}(u) \leq b_n\} \quad (n \in \mathbb{N}).$$

In order to prove (3.49) we shall prove the following implication:

$$s > \Phi_n(x) - \Phi_n(f(x)) \quad \Rightarrow \quad \mu(d_n(x, f(x))) \leq s\,. \tag{3.50}$$

Let $s > \Phi_n(x) - \Phi_n(f(x))$. Since μ maps $\mathbb{R}^+$ onto $\mathbb{R}^+$ there exists $s_1 > 0$ such that

$$\mu(s_1) = s > \Phi_n(x) - \Phi_n(f(x))\,.$$

Since from (3.48) it follows that $F_{x,f(x)}(s_1) > b_n$, we obtain that $d_n(x, f(x)) < s_1$ and therefore

$$\mu(d_n(x, f(x))) \leq \mu(s_1) = s\,.$$

Hence (3.50) holds.

Let $x_0 \in S$ and $x_m = f^m(x_0)$ ($m \in \mathbb{N}$). Then for every $m \in \mathbb{N}$

$$\begin{aligned}\mu(d_n(x_{m+1}, x_m)) &= \mu(d_n(f(x_m), x_m)) \\ &\leq \Phi_n(x_m) - \Phi_n(f(x_m)),\end{aligned}$$

which implies that

$$\begin{aligned}\sum_{i=0}^{k} \mu(d_n(x_{i+1}, x_i)) &\leq \Phi_n(x_0) - \Phi_n(x_{k+1}) \\ &\leq \Phi_n(x_0)\,.\end{aligned}$$

By the subadditivity of μ it follows that

$$\mu\left(\sum_{i=0}^{k} d_n(x_{i+1}, x_i)\right) \leq \Phi_n(x_0)\,. \tag{3.51}$$

Relation (3.51) implies that

$$\sum_{i=0}^{k} d_n(x_{i+1}, x_i) \leq \sup\{u \mid u > 0\,,\ \mu(u) = \Phi_n(x_0)\} = M_n,$$

and therefore the series $\sum_{i=0}^{\infty} d_n(x_{i+1}, x_i)$ is convergent.

Since

$$d_n(x_m, x_{m+p}) \leq \sum_{i=m}^{m+p-1} d_n(x_{i+1}, x_i)$$

it follows that $(x_n)_{n\in\mathbb{N}}$ is a Cauchy sequence. If

$$x^* = \lim_{n\to\infty} x_n = \lim_{n\to\infty} f^n(x_0),$$

then from the continuity of f it follows that $x^* = f(x^*)$. □

Corollary 3.79 *Let $(S, \mathcal{F}, T)$ be a complete Menger space such that t-norm T is as in Theorem 3.78, $f : S \to S$ a continuous mapping, $\Phi : S \to [0, \infty)$ and μ a mapping of $\mathbb{R}^+$ onto $\mathbb{R}^+$ such that μ is non-decreasing and*

$$\mu(a + b) \leq \mu(a) + \mu(b)$$

for every $a, b \in \mathbb{R}^+$. *If for every* $s > 0$ *and* $x \in S$

$$F_{x,f(x)}(s) \geq H_0(\mu(s) + \Phi(f(x)) - \Phi(x)), \tag{3.52}$$

then there exists a fixed point $x^* \in S$ *of the mapping* f *and* $x^* = \lim_{n\to\infty} f^n(x_0)$ *for arbitrary* $x_0 \in S$.

Proof. Let $\mu(s) > \Phi(x) - \Phi(f(x))$. Then $H_0(\mu(s) + \Phi(f(x)) - \Phi(x)) = 1$ and (3.52) implies that $F_{x,f(x)}(s) = 1 > b_n$, since $b_n \in (0,1)$. □

Corollary 3.80 *Let* $(S, \mathcal{F}, T)$ *be a complete Menger space such that t-norm* T *is as in Theorem 3.78 and for every* $n \in \mathbb{N}$ *there exists* $q_n \in (0,1)$ *such that the following implication holds for all* $n \in \mathbb{N}$:

$$(\forall s > 0)\ \ F_{x,y}(s) > b_n \quad \Rightarrow \quad F_{f(x),f(y)}(q_n s) > b_n\,. \tag{3.53}$$

Then there exists a fixed point $x^* \in S$ *of the mapping* f *and* $x^* = \lim_{n\to\infty} f^n(x_0)$ *for arbitrary* $x_0 \in S$.

Proof. From (3.53) it follows that

$$d_n(f(x), f(y)) \leq q_n d_n(x,y) \ \text{ for every } x, y \in S\,.$$

Hence

$$d_n(x, f(x)) \leq \frac{d_n(x, f(x))}{1 - q_n} - \frac{d_n(f(x), f^2(x))}{1 - q_n} \ \text{ for every } x \in S$$

and therefore

$$\mu(d_n(x, f(x))) \leq \Phi_n(x) - \Phi_n(f(x))\,,$$

where $\Phi_n(x) = \dfrac{d_n(x, f(x))}{1 - q_n}$ $(n \in \mathbb{N}\,,\ x \in S)$ and $\mu(s) = s$ $(s \in \mathbb{R}^+)$. Since f is continuous, the proof follows by Theorem 3.78. □

The following theorem has been proved by Hicks [129].

Theorem 3.81 *Let* $[U]$ *be a quasi-uniform structure for a set* X, $\preceq$ *a partial order for* X *and* $\phi : X \to [b, \infty]$ $(b \in \mathbb{R})$ *is not identically* ∞. *Suppose that the following conditions hold:*

a) $S(x, \preceq) = \{y \mid y \in X,\ x \preceq y)$ *is complete for every* $x \in X$, *i.e., every non-decreasing Cauchy sequence from* $S(x, \preceq)$ *is converging in* $S(x, \preceq)$;

b) ϕ *is non-increasing;*

c) for each $U \in [U]$, *the exists* $r > 0$ *such that for every* $(x_1, x_2) \in X \times X$

$$(x_1 \preceq x_2) \wedge (\phi(x_1) - \phi(x_2) < r) \quad \Rightarrow \quad (x_1, x_2) \in U.$$

Then for each $x \in X$ *with* $\phi(x) < \infty$, *there exists* $x_0 \in X$ *with* $\phi(x_0) < \infty$ *such that* $x \preceq x_0$ *is maximal element on* $(X, \preceq)$.

Using Theorem 3.81 we prove the following theorem.

Theorem 3.82 *Let X be a complete F-type topological space, $\{d_\lambda \mid \lambda \in D\}$ a generating family of quasi-metrics for the topology of X, $\phi : X \to [b,\infty](b \in \mathbb{R})$ a proper lower semi-continuous mapping and $k : D \to (0,\infty)$ a non-increasing function. Then for each $x \in X$ with $\phi(x) < \infty$ there exists an element $x_0 \in X$ such that:*

(i) $\phi(x_0) < \infty$.
(ii) For every $\lambda \in D$

$$d_\lambda(x, x_0) \leq k(\lambda)(\phi(x) - \phi(x_0)).$$

(iii) For every $y \in X \setminus \{x_0\}, \phi(y) = \infty$ or $\phi(y) < \infty$ and there exists $\lambda \in D$ such that

$$k(\lambda)(\phi(x_0) - \phi(y)) < d_\lambda(x_0, y).$$

Proof. It is easy to prove that $\preceq$ is a partial order on X, where

$$\begin{aligned} x \preceq y \iff (x = y) \ \vee ((\phi(x) < \infty) \wedge (\phi(y) < \infty) \wedge (d_\lambda(x,y) \\ \leq k(\lambda)(\phi(x) - \phi(y))) \text{ for every } \lambda \in D). \end{aligned} \tag{3.54}$$

We shall prove that all the conditions of Theorem 3.81 are satisfied.

We prove that ϕ is non-increasing. Suppose that $x \preceq y$. If $x = y$, then $\phi(x) = \phi(y)$ and if $\phi(x) < \infty$, $\phi(y) < \infty$ and for every $\lambda \in D$

$$0 \leq d_\lambda(x,y) \leq k(\lambda)(\phi(x) - \phi(y))$$

it follows that $\phi(y) \leq \phi(x)$. The condition c) from Theorem 3.81 holds. Indeed, if $U \in [U]$ is of the from

$$U = U(\lambda, t) = \{(x,y) \mid (x,y) \in X \times X,\ d_\lambda(x,y) < t\},$$

then for $r = t/k(\lambda)$ we have the implication

$$(x_1 \preceq x_2) \wedge (\phi(x_1) - \phi(x_2) < r) \quad \Rightarrow \quad d_\lambda(x_1, x_2) < t.$$

It remains to prove that $S(x, \preceq)$ is closed for every $x \in X$. We may assume that $\phi(x) < \infty$. Let $(x_\alpha)_{\alpha \in \mathcal{A}}$ be a generalized sequence from $S(x, \preceq)$ such that $\lim\limits_{\alpha \in \mathcal{A}} x_\alpha = z$. We have to prove that $x \preceq z$, which means that $z \in S(x, \preceq)$. Since $x_\alpha \in S(x, \preceq)$ and $\phi(x) < \infty$ we have that $\phi(x_\alpha) < \infty$, and for every $\alpha \in \mathcal{A}$ and $\lambda \in D$

$$d_\lambda(x_\alpha, x) \leq k(\lambda)(\phi(x) - \phi(x_\alpha)).$$

We prove that for every $\lambda \in D$

$$d_\lambda(x,z) \le k(\lambda)(\phi(x) - \phi(z)).$$

Let $\lambda \in D$ and choose $\mu \in D$ with $\lambda \prec \mu$ such that for all $u, v, w \in X$

$$d_\lambda(u,v) \le d_\mu(u,w) + d_\mu(w,v).$$

Since $\lim\limits_{\alpha \in \mathcal{A}} x_\alpha = z$ there exists, for every $\varepsilon > 0$, $\alpha(\mu,\varepsilon) \in \mathcal{A}$ such that

$$d_\mu(x_\alpha, z) < \varepsilon \text{ for every } \alpha \ge \alpha(\mu,\varepsilon).$$

Then

$$\begin{aligned} d_\alpha(x,z) &\le d_\mu(x,x_\alpha) + d_\mu(x_\alpha, z) \\ &< k(\mu)(\phi(x) - \phi(x_\alpha)) + \varepsilon \text{ for every } \alpha \ge \alpha(\mu,\varepsilon). \end{aligned}$$

Using the lower semi-continuity of ϕ we obtain that

$$\begin{aligned} d_\lambda(x,z) &\le k(\mu)(\phi(x) - \liminf_{\alpha \in \mathcal{A}} \phi(x_\alpha)) + \varepsilon \\ &\le k(\mu)(\phi(x) - \phi(z)) + \varepsilon \le k(\lambda)(\phi(x) - \phi(z)) + \varepsilon. \end{aligned}$$

Since ε is an arbitrary positive number the proof is complete.

Hence all the conditions of Theorem 3.81 are satisfied and there exists $x_0 \in X$ such that (i), (ii) and (iii) hold. □

The following theorem is a generalization of the fixed point theorem 3.1 from [62].

Theorem 3.83 *Let X be a complete F-type topological space, $\{d_\lambda \mid \lambda \in D\}$ a generating family of quasi-metrics for the topology of X, $\phi : X \to [b,\infty] (b \in \mathbb{R})$ a proper lower semi-continuous mapping, $f : X \to X$ and $k : D \to (0,\infty)$ a non-increasing function such that $\phi(x) < \infty$ implies $\phi(fx) < \infty$ and for every $x \in X$ such that $\phi(x) < \infty$*

$$d_\lambda(x, fx) \le k(\lambda)(\phi(x) - \phi(fx)) \quad \textit{for every } \lambda \in D. \tag{3.55}$$

Then, for every $x \in X$ such that $\phi(x) < \infty$ there exists a fixed point $\overline{x} \in X$ of the mapping f such that $\phi(\overline{x}) < \infty$ and for every $\lambda \in D$

$$d_\lambda(x, \overline{x}) \le k(\lambda)(\phi(x) - \phi(\overline{x})). \tag{3.56}$$

If $y \ne \overline{x}$ and $\phi(y) < \infty$ then there exists $\lambda \in D$ such that

$$k(\lambda)(\phi(\overline{x}) - \phi(y)) < d_\lambda(\overline{x}, y).$$

Proof. Let $x \in X$ be such that $\phi(x) < \infty$. From Theorem 3.82 it it follows that there exists $\overline{x} \in X$ such that $\phi(\overline{x}) < \infty$ and that (3.55) holds for every $\lambda \in D$. From Theorem 3.82 and the relation $\phi(f\overline{x}) < \infty$ it follows that $\overline{x} \neq f\overline{x}$ implies the existence of $\lambda \in D$ such that

$$k(\lambda)(\phi(\overline{x}) - \phi(f\overline{x})) < d_\lambda(\overline{x}, f\overline{x}).$$

This contradicts to (3.56). Hence $\overline{x} = f\overline{x}$. □

Corollary 3.84 *Let (X, d, L, R) be a complete fuzzy metric space such that $\lim_{a\to 0_+} R(a,a) = 0$ and for every $x, y \in X$, $\lim_{t\to\infty} d(x,y)(t) = 0$.*

Let $\phi : X \to [b, \infty](b \in \mathbb{R})$ be a proper lower semi-continuous mapping, $f : X \to X$ and $h : (0,1] \to (0,\infty)$ a non-decreasing mapping such that $\phi(x) < \infty$ implies $\phi(fx) < \infty$ and for every $x \in X$ such that $\phi(x) < \infty$

$$\rho_\alpha(x, fx) \leq h(\alpha)(\phi(x) - \phi(fx)) \quad \text{for every } \alpha \in (0,1].$$

Then, for every $x \in X$ such that $\phi(x) < \infty$ there exists a fixed point $\overline{x} \in X$ of the mapping f such that $\phi(\overline{x}) < \infty$ and for every $\alpha \in (0,1]$

$$\rho_\alpha(x, \overline{x}) \leq h(\alpha)(\phi(x) - \phi(\overline{x})).$$

If $y \neq \overline{x}$ and $\phi(y) < \infty$ then there exists $\alpha \in (0,1]$ such that

$$h(\alpha)(\phi(\overline{x}) - \phi(y)) < \rho_\alpha(\overline{x}, y).$$

Proof. If $\rho'_\alpha = \rho_{1-\alpha}$ and $k(\alpha) = h(1-\alpha)$, $\alpha \in [0,1)$, then we can apply Theorem 3.83 for $D = [0,1)$, $d_\alpha = \rho'_\alpha (\alpha \in D)$. □

If $h(\alpha) = 1$, $\alpha \in (0,1]$ from Corollary we obtain Caristi's fixed point theorem from [112].

Caristi's fixed point theorem [26] and Ekeland's variational principle [14], [57] are very important parts of nonlinear functional analysis. Ekeland's variational principle has interesting applications in control theory, theory of differential equations and minimization problems. Many authors generalized Caristi's fixed point theorem and Ekeland's variational principle [229, 232, 281].

Corollary 3.85 (ε-variational principle) *Let (X, d, L, R) be a complete fuzzy metric space such that*

$$\lim_{a\to 0_+} R(a,a) = 0 \quad \text{and} \quad \lim_{t\to\infty} d(x,y)(t) = 0 \quad \text{for every } (x,y) \in X \times X.$$

Let $\Phi : X \to [b,\infty]$ $(b \in \mathbb{R})$ be a proper lower semi-continuous mapping, $\varepsilon > 0$ and $x_\varepsilon \in X$ be given such that $\Phi(x_\varepsilon) \leq \varepsilon + \inf_{y\in X} \Phi(y)$. Then for all $r > 0$ there exists $y_\varepsilon \in X$ such that $\Phi(y_\varepsilon) \leq \Phi(x_\varepsilon)$, $\rho_\alpha(x_\varepsilon, y_\varepsilon) \leq \frac{1}{r}$, for every $\alpha \in (0,1]$, and for every $x \neq y_\varepsilon$ there exists $\alpha \in (0,1]$ such that

$$\Phi(x) > \Phi(y_\varepsilon) - r\varepsilon\rho_\alpha(x, y_\varepsilon). \tag{11.8}$$

Proof. We can use the function ϕ defined by

$$\phi(x) = \frac{\Phi(x) - \inf_{z \in X} \Phi(z)}{r\varepsilon}.$$

Then the conditions of Theorem 3.82 are satisfied with $x = x_\varepsilon$, $k(\alpha) = 1$ for every $\alpha \in (0, 1)$, and ϕ. We take for y_ε the element x_0 from Theorem 3.82 and hence the Corollary is proved. □

We shall prove Ekeland's variational principle and a fixed point theorem of Caristi's type in Menger spaces.

It is easily checked that the relation $\preceq$ defined by

$$x \preceq y \quad \Leftrightarrow \quad F_{x,y}(s) \geq H_0(s - (\Phi(x) - \Phi(y))) \text{ for every } s \geq 0 \tag{3.57}$$

is equivalent with the previously defined relation $\preceq$ in the fuzzy metric space associated with the Menger space. Also the statement $\sup\limits_{a<1} T(a, a) = 1$ for the t-norm T is equivalent to the statement $\lim\limits_{a \to 0_+} R(a, a) = 0$ for the function R in the associated fuzzy metric space.

Hence we obtain the following two theorems.

Theorem 3.86 *Let $(S, \mathcal{F}, T)$ be a complete Menger space such that t-norm T satisfies $\sup\limits_{a<1} T(a, a) = 1$. Let $\Phi : S \to [b, \infty]$ $(b \in \mathbb{R})$ be a proper lower semi-continuous mapping. Let $\varepsilon > 0$ be given and x_ε satisfies*

$$\Phi(x_\varepsilon) < \inf_{z \in S} \Phi(z) + \varepsilon.$$

Then, for every $k > 0$ there exists $y_\varepsilon \in S$ such that

$$\Phi(y_\varepsilon) \leq \Phi(x_\varepsilon),$$

$$F_{x_\varepsilon, y_\varepsilon}(s) \geq H_0\left(s - \frac{1}{k}\right) \quad \textit{for every } s \geq 0,$$

and for every $x \neq y_\varepsilon$ there exists $s_0 > 0$ such that

$$F_{x, y_\varepsilon}(s_0) < H_0\left(s_0 - \frac{1}{k\varepsilon}(\Phi(y_\varepsilon) - \Phi(x))\right).$$

Theorem 3.87 *Let $(S, \mathcal{F}, T)$ be a complete Menger space such that the t-norm T satisfies $\sup\limits_{a<1} T(a, a) = 1$. Let $\Phi : S \to [b, \infty]$, $(b \in \mathbb{R})$ be a proper lower semicontinuous mapping. Let $f : S \to S$ satisfies for all $x \in S, x \preceq fx$. Then for each $x \in S$ with $\Phi(x) < \infty$, there exists an element $x_0 \in S$ which is a fixed point of f, and is such that $\Phi(x_0) < \infty$ and $x \preceq x_0$.*

Further interesting results about fixed point theorems of Caristi's type in Menger spaces are obtained in [281, 330].

3.5 Common fixed point theorems

In this section we shall present two common fixed point theorems. First, we shall give a common fixed point theorem for a sequence of single-valued mappings, proved by Shih-sen Chang in [28] and then a common fixed point theorem for quasi-uniformisable spaces. Since every Menger space $(S, \mathcal{F}, T)$, where $\sup_{a<1} T(a, a) = 1$, is a quasi uniformisable space, some corollaries on common fixed points in Menger spaces are obtained.

In the next theorem we shall assume that

$\mathcal{Q} = \{\Phi \mid \Phi : \mathbb{R}^+ \to \mathbb{R}^+,\ \Phi \text{ is strictly increasing},\ \Phi(0) = 0,\ \Phi^n(t) \to \infty,\ n \to \infty,$ for every $t > 0\}$.

By Φ^n the n-th iteration $\underbrace{\Phi \circ \Phi \circ \cdots \circ \Phi}_{n-\text{times}}$ is denoted. In the next theorem $m \mid n$ $(m, n \in \mathbb{N})$ means that there exists $k \in \mathbb{N}$ such that $km = n$.

Theorem 3.88 *Let $(S, \mathcal{F}, T)$ be a complete Menger space, T a t-norm such that $\lim_{x \to 1} T(x, y) = y$, for every $y \in (0, 1]$ and $L_n : S \to S$ $(n \in \mathbb{N})$. Suppose that there exists a sequence $m_n : S \to \mathbb{N}$ $(n \in \mathbb{N})$ such that for each $n \in \mathbb{N}$ and each $x \in S$, $m_n(x) | m_n(L_n x)$ and that there exists $\Phi \in \mathcal{Q}$ such that for every $s > 0$*

$$F_{L_i^{m_i(x)}x, L_j^{m_j(y)}y}(s) \geq \min_{p,q \in \{x, y, L_i^{m_i(x)}x, L_j^{m_j(y)}y\}} F_{p,q}(\Phi(s)) \tag{3.58}$$

for every $i, j \in \mathbb{N}$, $i \neq j$, and every $x, y \in S$.

If there exists $x_0 \in S$ such that for the sequence $(x_n)_{n \in \mathbb{N}}$ defined by

$$x_n = L_n^{m_n(x_{n-1})} x_{n-1} \quad (n \in \mathbb{N}),$$

the relation (3.59) holds

$$\sup_{s>0} \inf_{n,m \in \mathbb{N}} F_{x_n, x_m}(s) = 1, \tag{3.59}$$

then there exists a unique $x^* \in S$ such that $x^* = L_n x^*$ for every $n \in \mathbb{N}$ and $x^* = \lim_{n \to \infty} x_n$.

Proof. We shall prove that the sequence $(x_n)_{n \in \mathbb{N}}$ defined by $x_n = L_n^{m_n(x_{n-1})} x_{n-1}$ $(n \in \mathbb{N})$ is a Cauchy sequence. For any $i, j \in \mathbb{N}$, $i \neq j$, by (3.58) we obtain that

$$\begin{aligned} F_{x_i, x_j}(s) &= F_{L_i^{m_i(x_{i-1})} x_{i-1}, L_j^{m_j(x_{j-1})} x_{j-1}}(s) \\ &\geq \min_{p,q \in \{x_{i-1}, x_{j-1}, x_i, x_j\}} F_{p,q}(\Phi(s)). \end{aligned} \tag{3.60}$$

Hence for any $m, n \in \mathbb{N}$ $(m < n)$ (3.60) implies that

$$\inf_{m \leq i,j \leq n} F_{x_i, x_j}(s) \geq \inf_{m-1 \leq i,j \leq n} F_{x_i, x_j}(\Phi(s))$$

and therefore

$$\begin{aligned}\inf_{i,j\geq m} F_{x_i,x_j}(s) &\geq \inf_{i,j\geq m-1} F_{x_i,x_j}(\Phi(s))\\ &\geq \sup_{u<\Phi(s)} \inf_{i,j\geq m-1} F_{x_i,x_j}(u).\end{aligned}$$

It can be proved easily that for every $m \in \mathbb{N}$ and every $s \geq 0$

$$\begin{aligned}\inf_{i,j\geq m} F_{x_i,x_j}(s) &\geq \inf_{i,j\geq 0} F_{x_i,x_j}(\Phi^m(s))\\ &\geq \sup_{u<\Phi^m(s)} \inf_{i,j\geq 0} F_{x_i,x_j}(u). \qquad (3.61)\end{aligned}$$

Since $\Phi \in \mathcal{Q}$ and (3.59), (3.61) hold, we have that for every $s > 0$

$$\begin{aligned}\lim_{m\to\infty} \inf_{i,j\geq m} F_{x_i,x_j}(s) &\geq \lim_{m\to\infty}\left(\sup_{u<\Phi^m(s)} \inf_{i,j\geq 0} F_{x_i,x_j}(u)\right)\\ &= \begin{cases} 0 & \text{if } s=0,\\ \sup_{u>0}\inf_{i,j\geq 0} F_{x_i,x_j}(u) & \text{if } s>0\end{cases}\\ &= \begin{cases} 0 & \text{if } s=0,\\ 1 & \text{if } s>0\end{cases}\\ &= H_0(s).\end{aligned}$$

Hence for any $\varepsilon > 0$ and $\lambda \in (0,1)$ there exists $n_0(\varepsilon,\lambda) \in \mathbb{N}$ such that

$$m \geq n_0(\varepsilon,\lambda) \quad \Rightarrow \quad \inf_{i,j\geq m} F_{x_i,x_j}(\varepsilon) > 1-\lambda$$

and therefore

$$i,j \geq n_0(\varepsilon,\lambda) \Rightarrow F_{x_i,x_j}(\varepsilon) > 1-\lambda.$$

This means that $(x_n)_{n\in\mathbb{N}}$ is a Cauchy sequence, and since S is complete there exists $x^* = \lim_{n\to\infty} x_n$. We shall prove that

$$x^* = L_n^{m_n(x^*)}x^* \quad \text{for every} \quad n \in \mathbb{N}. \qquad (3.62)$$

From (3.58) we have that for every $i \in \mathbb{N}$, every $s \geq 0$ and every $n > i$

$$\begin{aligned}F_{x_n,L_i^{m_i(x^*)}x^*}(s) &= F_{L_n^{m_n(x_{n-1})}x_{n-1},L_i^{m_i(x^*)}x^*}(s) \qquad (3.63)\\ &\geq \min_{p,q\in\{x_{n-1},x^*,x_n,L_i^{m_i(x^*)}x^*\}} F_{p,q}(\Phi(s)). \qquad (3.64)\end{aligned}$$

Let G_0 be the set of all discontinuity points of $F_{x^*, L_i^{m_i(x^*)}x^*}(s)$. Since Φ^m is strictly increasing, $\Phi^{-m}(G_0)$ is the set of all discontinuity points of $F_{x^*, L_i^{m_i(x^*)}x^*}(\Phi^m(s))$, $m \in \mathbb{N}$. The sets $G_0, \Phi^{-m}(G_0)$ $(m \in \mathbb{N})$ are all countable sets, and therefore

$$G = G_0 \cup \left(\bigcup_{m=1}^{\infty} \Phi^{-m}(G_0) \right)$$

is countable. Let $\widetilde{G} = \mathbb{R}^+ \setminus G$. If $s = 0$ or $s \in \widetilde{G}$ (s is a common continuity point of $F_{x^*, L_i^{m_i(x^*)}x^*}(\cdot)$ and $F_{x^*, L_i^{m_i(x^*)}x^*}(\Phi^m(\cdot))$, $m \in \mathbb{N}$) from (3.64) and Lemma 3.63, it follows that

$$\begin{aligned} F_{x^*, L_i^{m_i(x^*)}x^*}(s) &= \lim_{n\to\infty} F_{x_n, L_i^{m_i(x^*)}x^*}(s) \\ &\geq F_{x^*, L_i^{m_i(x^*)}x^*}(\Phi(s)), \end{aligned}$$

and therefore

$$F_{x^*, L_i^{m_i(x^*)}x^*}(s) \geq F_{x^*, L_i^{m_i(x^*)}x^*}(\Phi^n(s)).$$

If $n \to \infty$ it follows from the condition $\Phi \in \mathcal{Q}$, that

$$F_{x^*, L_i^{m_i(x^*)}x^*}(s) = H_0(s) \text{ for every } s \in \widetilde{G} \text{ or } s = 0.$$

If $s \in G$, $s > 0$, there exist $s_1, s_2 \in \widetilde{G}$ such that $0 < s_1 < s < s_2$ and

$$\begin{aligned} 1 &= H_0(s_1) \\ &= F_{x^*, L_i^{m_i(x^*)}x^*}(s_1) \\ &\leq F_{x^*, L_i^{m_i(x^*)}x^*}(s) \\ &\leq F_{x^*, L_i^{m_i(x^*)}x^*}(s_2) \\ &= 1. \end{aligned}$$

Hence $F_{x^*, L_i^{m_i(x^*)}x^*}(s) = H_0(s)$ for every $s > 0$, and therefore (3.62) holds.

Let $y^* \in S$ be such that $y^* = L_j^{m_j(y^*)}y^*$ for some $j \in \mathbb{N}$. Then for any $i \in \mathbb{N}$ such that $j \neq i$

$$\begin{aligned} F_{x^*, y^*}(s) &= F_{L_i^{m_i(x^*)}x^*, L_j^{m_j(y^*)}y^*}(s) \\ &\geq \min_{p,q \in \{x^*, y^*\}} F_{p,q}(\Phi(s)) \\ &= F_{x^*, y^*}(\Phi(s)), \end{aligned}$$

and therefore $F_{x^*, y^*}(s) \geq F_{x^*, y^*}(\Phi^n(s))$ for every $n \in \mathbb{N}$ and every $s \geq 0$. Since $\Phi \in \mathcal{Q}$ we obtain that $F_{x^*, y^*}(s) = H_0(s)$ for every $s \geq 0$, which means that $x^* = y^*$.

From the condition that $m_n(x)|m_n(L_nx)$, for all $n \in \mathbb{N}$ and all $x \in S$ it follows that there exists $k_i \in \mathbb{N}$, such that

$$m_i(L_ix^*) = k_i m_i(x^*). \tag{3.65}$$

From (3.65) we obtain that for any $i \in \mathbb{N}$

$$\begin{aligned} L_ix^* &= L_iL_i^{m_i(x^*)}x^* \\ &= L_iL_i^{2m_i(x^*)}x^* \\ &= L_iL_i^{k_im_i(x^*)}x^* \\ &= L_iL_i^{m_i(L_ix^*)}x^* \\ &= L_i^{m_i(L_ix^*)}L_ix^* \end{aligned}$$

which implies $x^* = L_ix^*$ for every $i \in \mathbb{N}$. □

Corollary 3.89 *Let $(S, \mathcal{F}, T)$, $(L_n)_{n\in\mathbb{N}}$ and $(m_n(x))_{n\in\mathbb{N}}$ be as in Theorem 3.88 and there exists $h \in (0,1)$ such that for every $s \geq 0$*

$$F_{L_i^{m_i(x)}x, L_j^{m_j(y)}y}(s) \geq \min_{p,q\in\{x,y,L_i^{m_i(x)}x,L_j^{m_j(y)}y\}} F_{p,q}\left(\frac{s}{h}\right), \tag{6.9}$$

for every $i,j \in \mathbb{N}$, $i \neq j$ and every $x, y \in S$. If there exists x_0 such that (3.59) holds, then there exists a unique x^* such that $x^* = L_nx^*$ for every $n \in \mathbb{N}$.

Remark 3.90 If $m_n(x)$ does not depend on $x \in S$, then $m_i(x)|m_i(Tx)$ holds and therefore if for every $s \geq 0$

$$F_{L_i^{m_i}x, L_j^{m_j}y}(s) \geq \min_{p,q\in\{x,y,L_i^{m_i}x,L_j^{m_j}y\}} F_{p,q}\left(\frac{s}{h}\right) \tag{3.66}$$

for every $i, j \in \mathbb{N}$, $i \neq j$ and every $x, y \in S$, then there exists a unique common fixed point for the sequence of mappings $(L_n)_{n\in\mathbb{N}}$.

D.H. Tan introduced the following definition.

Definition 3.91 *Let S and I be arbitrary sets, $g : I \to I$ and for every $i \in I$, $d_i : S \times S \to \mathbb{R}^+$. The triplet $(S, (d_i)_{i\in I}, g)$ is said to be a quasi-uniformizable space if for every $x, y, z \in S$ and $i \in I$ the following hold:*

a) $d_i(x, y) \geq 0$, $d_i(x, x) = 0$;

b) $d_i(x, y) = d_i(y, x)$;

c) $d_i(x, y) \leq d_{g(i)}(x, z) + d_{g(i)}(z, y)$.

A quasi-uniformizable space $(S,(d_i)_{i\in I},g)$ is Hausdorff if the relation $d_i(x,y)=0$, for every $i\in I$, implies that $x=y$. A Hausdorff quasi-uniformizable space $(S,(d_i)_{i\in I},g)$ becomes a Hausdorff topological space if the fundamental system of neighbourhoods of $x\in S$ is given by the family $\mathcal{V}_x=(B(x;\varepsilon,i))_{\substack{\varepsilon>0\\ i\in I}}$, where

$$B(x;\varepsilon,i)=\{y\mid y\in S,\ d_i(x,y)<\varepsilon\}\quad(\varepsilon>0,\ i\in I).$$

A Menger space $(S,\mathcal{F},T)$, such that $\sup\limits_{a<1}T(a,a)=1$, is a quasi-uniformizable space $(S,(d_\lambda)_{\lambda\in J},g)$ where $J=(0,1)$,

$$d_\lambda(x,y)=\sup\{s\mid F_{x-y}(s)\le 1-\lambda\},\ \lambda\in J,\ x,y\in S$$

and $g:J\to J$ is defined in the following way.

From $\sup\limits_{a<1}T(a,a)=1$ it follows that for every $\lambda\in(0,1)$ there exists $\delta_\lambda\in(0,1)$ such that

$$T(1-\delta_\lambda,1-\delta_\lambda)\ge 1-\lambda.$$

Then we can take that $g(\lambda)=\delta_\lambda,\lambda\in(0,1)$. We shall prove only c). Let r_1 and r_2 are arbitrary numbers such that $d_{g(\lambda)}(x,z)<r_1$ and $d_{g(\lambda)}(z,y)<r_2$. Then

$$F_{x,z}(r_1)>1-g(\lambda)=1-\delta_\lambda,\quad F_{z,y}(r_2)>1-g(\lambda)=1-\delta_\lambda.$$

This implies that

$$\begin{aligned}F_{x,y}(r_1+r_2)&\ge T(F_{x,z}(r_1),F_{z,y}(r_2))\\&\ge T(1-\delta_\lambda,1-\delta_\lambda)\\&> 1-\lambda\end{aligned}$$

and therefore $d_\lambda(x,y)<r_1+r_2$, which means that c) is satisfied.

Every locally convex space $(X,(p_i)_{i\in I})$, where $(p_i)_{i\in I}$ is the family of seminorms on X, is a quasi-uniformizable space, where $g:I\to I$ is defined by $g(i)=i$, for every $i\in I$. In the papers [9, 117] some fixed point theorems for mappings $L:X\to X$ were proved, where L satisfies the condition

$$p_i(Lx-Ly)\le q_i p_{f(i)}(x-y)\text{ for every } i\in I \text{ and every } x,y\in X,$$

where f is a mapping of I into I. An application to differential equations in the field of Mikusiński's operators is given in [117] and some interesting applications are given by Angelov [9].

Theorem 3.92 *Let $(S,(d_i)_{i\in I},g)$ be a sequentially complete Hausdorff quasi-uniformizable space, $f:I\to I$, L_1 and L_2 continuous mappings from S into S, $L:S\to L_1S\cap L_2S$ a continuous mapping which commutes with L_1 and L_2 and the following conditions are satisfied:*

1) *for every* $i \in I$, *there exists* $q_i : \mathbb{R}^+ \to [0, 1)$, *which is a non-decreasing function, for which* $\overline{\lim_{n\to\infty}} q_{f^n(i)}(t) < 1$ *for every* $t \in \mathbb{R}^+$ *and every* $i \in I$ *and*

$$d_i(Lx, Ly) \le q_i(d_{f(i)}(L_1x, L_2y)) \cdot d_{f(i)}(L_1x, L_2y) \text{ for every } x, y \in S;$$

2) *there exists* $x_0 \in S$ *such that for every* $i \in I$

$$\sup_{j\in O(i,f), p\in\mathbb{N}} d_j(Lx_0, Lx_p) \le K_i \quad (K_i \in \mathbb{R}^+)$$

where the sequence $(x_p)_{p\in\mathbb{N}}$ *is defined by*

$$L_1x_{2n-1} = Lx_{2n-2}, \quad L_2x_{2n} = Lx_{2n-1} \ (n \in \mathbb{N}).$$

Then there exists $z \in S$ *such that* $Lz = L_1z = L_2z$. *If, in addition, for every* $i \in I$

$$\sup_{j\in O(i,f)} d_j(L^3x_1, L^2x_0) \le M_i \quad (M_i \in \mathbb{R}^+)$$

then Lz *is a common fixed point for* L, L_1 *and* L_2, *and which is the unique common fixed point for* L, L_1 *and* L_2 *in the set*

$$\left\{u \mid u \in S, (\forall i \in I)(\exists V_i \in \mathbb{R}^+)\left(\sup_{j\in O(i,f)} d_j(Lz, u) \le V_i\right)\right\}.$$

Proof. Let $(x_n)_{n\in\mathbb{N}}$ be such a sequence from S that

$$L_2x_{2k} = Lx_{2k-1} \text{ and } L_1x_{2k-1} = Lx_{2k-2} \text{ for every } k \in \mathbb{N},$$

and 2) holds. Then for every $i \in I$ and every $k \in \mathbb{N}$ we have

$$\begin{aligned}
& d_i(Lx_{2k}, Lx_{2k-1}) \\
\le\ & q_i(d_{f(i)}(L_2x_{2k}, L_1x_{2k-1})) \cdot d_{f(i)}(L_2x_{2k}, L_1x_{2k-1}) \\
=\ & q_i(d_{f(i)}(Lx_{2k-1}, Lx_{2k-2})) \cdot d_{f(i)}(Lx_{2k-1}, Lx_{2k-2}) \\
\le\ & q_i(d_{f(i)}(Lx_{2k-1}, Lx_{2k-2})) \cdot q_{f(i)}(d_{f^2(i)}(L_1x_{2k-1}, L_2x_{2k-2})) \cdot d_{f^2(i)}(L_1x_{2k-1}, L_2x_{2k-2}) \\
=\ & q_i(d_{f(i)}(Lx_{2k-1}, Lx_{2k-2})) \cdot q_{f(i)}(d_{f^2(i)}(Lx_{2k-2}, Lx_{2k-3})) \cdot d_{f^2(i)}(Lx_{2k-2}, Lx_{2k-3}) \\
\vdots\ & \\
\le\ & q_i(d_{f(i)}(Lx_{2k-1}, Lx_{2k-2})) \cdots q_{f^{2k-2}(i)}(d_{f^{2k-1}(i)}(Lx_1, Lx_0)) \cdot d_{f^{2k-1}(i)}(Lx_1, Lx_0)
\end{aligned}$$

and similarly

$$\begin{aligned}
& d_i(Lx_{2k+1}, Lx_{2k}) \\
\le\ & q_i(d_{f(i)}(Lx_{2k}, Lx_{2k-1})) \cdot \prod_{s=0}^{2k-2} q_{f^{s+1}(i)}(d_{f^{s+2}(i)}(Lx_{2k-s-1}, Lx_{2k-s-2})) \cdot d_{f^{2k}(i)}(Lx_1, Lx_0).
\end{aligned}$$

Since $q_i(t) < 1$ for every $i \in I$ and every $t \in \mathbb{R}^+$, it follows for every $i \in I$ and every $n \in \mathbb{N}$ that $d_j(Lx_n, Lx_{n-1})) \leq K_i$ for every $j \in O(i, f)$, and therefore

$$d_i(Lx_{2k}, Lx_{2k-1}) \leq \prod_{s=0}^{2k-2} q_{f^s(i)}(K_i)K_i,$$

$$d_i(Lx_{2k+1}, Lx_{2k}) \leq \prod_{s=0}^{2k-1} q_{f^s(i)}(K_i)K_i,$$

where $f^0(i) = i$ for every $i \in I$. Using the condition $\overline{\lim_{n\to\infty}} q_{f^n(i)}(K_i) \leq Q_i < 1$ for every $i \in I$, we conclude that there exists $n_i \in \mathbb{N}$ such that

$$q_{f^n(i)}(K_i) \leq Q_i \quad \text{for every} \quad n \geq n_i$$

which implies that

$$d_i(Lx_n, Lx_{n-1}) \leq S_i Q_i^n \quad \text{for every} \quad i \in I,$$

for some $S_i \in \mathbb{R}^+$ $(i \in I)$.

Next, we prove that $(Lx_n)_{n\in\mathbb{N}}$ is a Cauchy sequence, which means that for every $i \in I$ and every $\varepsilon > 0$ there exists $n(i, \varepsilon) \in \mathbb{N}$ such that

$$d_i(Lx_n, Lx_{n+p}) < \varepsilon \text{ for every } n \geq n(i, \varepsilon) \text{ and every } p \in \mathbb{N}.$$

Let $m \geq k$. From the definition of the sequence $(x_n)_{n\in\mathbb{N}}$ it follows that

$$\begin{aligned} d_i(Lx_{2k}, Lx_{2m+1}) \quad &\leq \quad q_i(d_{f(i)}(Lx_{2m}, Lx_{2k-1}))\ldots q_{f^{2k-1}(i)}(d_{f^{2k}(i)}(Lx_0, Lx_{2m+1-2k})) \\ &\cdot d_{f^{2k}(i)}(Lx_0, Lx_{2m+1-2k}) \end{aligned}$$

and similarly for $2k > 2m + 1$

$$\begin{aligned} d_i(Lx_{2k}, Lx_{2m+1}) \quad &\leq \quad q_i(d_{f(i)}(Lx_{2m}, Lx_{2k-1}))\cdots q_{f^{2m}(i)}(d_{f^{2m+1}(i)}(Lx_0, Lx_{2k-2m-1})) \\ &\cdot d_{f^{2m+1}(i)}(Lx_0, Lx_{2k-2m-1}). \end{aligned}$$

Using the condition 2), since $q_i(t) < 1$, $i \in I$, $t \in \mathbb{R}^+$, we obtain

$$d_i(Lx_{2k}, Lx_{2m+1}) \leq q_i(K_i)\cdots q_{f^{2k-1}(i)}(K_i)K_i \quad (m \geq k)$$

and

$$d_i(Lx_{2k}, Lx_{2m+1}) \leq q_i(K_i)\cdots q_{f^{2m}(i)}(K_i)K_i \quad (2k > 2m + 1).$$

This implies that

$$d_i(Lx_n, Lx_{2m+1}) \leq q_i(K_i)\cdots q_{f^{n-1}(i)}(K_i)K_i$$

for every $i \in I$ and for $n = 2k$, $p = 2m+1$, or $n = 2k+1$, $p = 2m+1$.

Let $p = 2m$ and $n = 2k$ or $n = 2k+1$. Then

$$\begin{aligned} d_i(Lx_n, Lx_{n+p}) &\leq d_{g(i)}(Lx_n, Lx_{n+1}) + d_{g(i)}(Lx_{n+1}, Lx_{n+1+p-1}) \\ &\leq S_{g(i)}Q_{g(i)}^{n+1} + \prod_{s=0}^{n} q_{f^s(g(i))}(K_{g(i)})K_{g(i)}. \end{aligned}$$

Since $\overline{\lim_{s\to\infty}} q_{f^s(g(i))}(K_{g(i)}) < 1$ for every $i \in I$, it follows that $(Lx_n)_{n\in\mathbb{N}}$ is a Cauchy sequence. Let

$$z = \lim_{n\to\infty} Lx_n = \lim_{n\to\infty} L_1x_{2n-1} = \lim_{n\to\infty} L_2x_{2n}.$$

By the continuity of L and L_1 we have that

$$L_1z = \lim_{n\to\infty} L_1Lx_{2n+1} = \lim_{n\to\infty} LL_1x_{2n+1} = Lz.$$

Similarly we have $Lz = L_2z$. Now, we prove that Lz is a common fixed point for L, L_1 and L_2 if $\sup_{j\in O(i,f)} d_j(L^3x_1, L^2x_0) \leq M_i$ for every $i \in I$.

First, we prove that for every $i \in I$

$$\sup_{j\in O(i,f)} d_j(L^2z, Lz) \leq M_i.$$

Since $d_j(L^2z, Lz) = \lim_{n\to\infty} d_j(L^2Lx_{2n+1}, LLx_{2n})$ it is enough to prove that for every $i \in I$

$$d_j(L^3x_{2n+1}, L^2x_{2n}) \leq M_i \text{ for every } n \in \mathbb{N} \text{ and } j \in O(i,f) \quad (i \in I).$$

The pairs L, L_1 and L, L_2 commute and we obtain that

$$\begin{aligned} d_j(L^3x_{2n+1}, L^2x_{2n}) &\leq q_j(d_{f(j)}(L_2(Lx_{2n}), L_1(L^2x_{2n+1})) \cdot d_{f(j)}(L_2(Lx_{2n}), L_1(L^2x_{2n+1})) \\ &= q_j(d_{f(j)}(L(L_2x_{2n}), L^2(L_1x_{2n+1})) \cdot d_{f(j)}(L(L_2x_{2n}), L^2(L_1x_{2n+1})) \\ &= q_j(d_{f(j)}(LLx_{2n-1}, L^2Lx_{2n})) \cdot d_{f(j)}(LLx_{2n-1}, L^2Lx_{2n})) \\ &= q_j(d_{f(j)}(L^2x_{2n-1}, L^3x_{2n})) \cdot d_{f(j)}(L^2x_{2n-1}, L^3x_{2n})) \\ &\leq d_{f(j)}(L^2x_{2n-1}, L^3x_{2n})) \\ &\vdots \\ &\leq d_{f^{2n}(j)}(L^3x_1, L^2x_0) \\ &\leq M_i \end{aligned}$$

for every $n \in \mathbb{N}$ since $f^{2n}(j) \in O(i,f)$.

We are going to prove that $d_i(L^2z, Lz) = 0$ for every $i \in I$. This follows from the inequalities

$$\begin{aligned}
d_i(L^2z, Lz) &\leq q_i(d_{f(i)}(L_1Lz, L_2z)) \cdot d_{f(i)}(L_1Lz, L_2z) \\
&= q_i(d_{f(i)}(L^2z, Lz)) \cdot d_{f(i)}(L^2z, Lz)) \\
&\vdots \\
&\leq q_i(d_{f(i)}(L^2z, Lz)) \cdot q_{f(i)}(d_{f^2(i)}(L^2z, Lz)) \cdots q_{f^n(i)}(d_{f^{n+1}(i)}(L^2z, Lz)) \\
&\quad \cdot d_{f^{n+1}(i)}(L^2z, Lz)) \\
&\leq q_i(M_i)q_{f(i)}(M_i) \cdots q_{f^n(i)}(M_i)M_i
\end{aligned}$$

for every $i \in I$. Since $\overline{\lim_{n\to\infty}} q_{f^n(i)}(M_i) < 1$ it follows that $d_i(L^2z, Lz) = 0$ for every $i \in I$, which implies that $L^2z = Lz$. Hence from $Lz = L_1z = L_2z$ it follows that

$$L^2z = L_1Lz = L_2Lz = Lz$$

which means that Lz is a common fixed point for mappings L, L_1 and L_2.

Suppose that $y = Ly = L_1y = L_2y$ and for every $i \in I$

$$\sup_{j\in O(i,f)} d_j(Lz, y) \leq V_i \qquad (V_i \in \mathbb{R}^+)$$

We prove that $y = Lz$. For every $i \in I$ we have

$$\begin{aligned}
d_i(Lz, y) &= d_i(L(Lz), Ly) \\
&\leq q_i(d_{f(i)}(L_1(Lz), L_2y) \cdot d_{f(i)}(L_1(Lz), L_2y) \\
&= q_i(d_{f(i)}(Lz, y)) \cdot d_{f(i)}(Lz, y) \\
&\vdots \\
&\leq q_i(d_{f(i)}(Lz, y)) \cdots q_{f^n(i)}(d_{f^{n+1}(i)}(Lz, y))d_{f^{n+1}(i)}(Lz, y)
\end{aligned}$$

and therefore

$$d_i(Lz, y) \leq q_i(V_i)q_{f(i)}(V_i) \cdots q_{f^n(i)}(V_i)V_i.$$

Since $\overline{\lim_{n\to\infty}} q_{f^n(i)}(V_i) < 1$ it follows that $d_i(Lz, y) = 0$ for every $i \in I$. Hence $Lz = y$. □

Corollary 3.93 *Let $(S, (d_i)_{i\in I})$ be a sequentially complete Hausdorff uniformizable space, $f : I \to I$, L_1 and L_2 continuous mappings from S into S, $L : S \to L_1S \cap L_2S$ continuous such that condition 1) of Theorem 3.92 is satisfied and there exist $x_0 \in S$ and $x_1 \in S$ such that $L_1x_1 = Lx_0$ and for every $i \in I$*

$$\sup_{n\in\mathbb{N}} d_{f^n(i)}(Lx_0, Lx_1) \leq K_i \qquad (K_i \in \mathbb{R}^+).$$

Then there exists $z \in S$ such that $Lz = L_1 z = L_2 z$. and if for every $i \in I$

$$\sup_{n\in\mathbb{N}} d_{f^n(i)}(L^3 x_1, L^2 x_0) \le M_i \qquad (M_i \in \mathbb{R}^+)$$

then Lz is a common fixed point for L, L_1 and L_2. Furthermore, if for every $i \in I$

$$\sup_{n\in\mathbb{N}} d_{f^n(i)}(L^2 x_1, L^2 x_0) \le V_i \qquad (V_i \in \mathbb{R}^+) \tag{3.67}$$

then there exists one and only one element $y \in S$ such that $y = Ly = L_1 y = L_2 y$ and for every $i \in I$

$$\sup_{n\in\mathbb{N}} d_{f^n(i)}(y, L^2 x_0) \le N_i(y) \qquad (N_i(y) \in \mathbb{R}^+). \tag{3.68}$$

Proof. Every uniformizable space $(S, (d_i)_{i\in I})$ is a quasi-uniformizable space, where $g(i) = i$ for every $i \in I$. We have that for every $i \in I$ and every $p \ge 2$

$$d_i(Lx_p, Lx_0) \le d_i(Lx_p, Lx_{p-1}) + d_i(Lx_{p-1}, Lx_{p-2}) + \cdots + d_i(Lx_1, Lx_0),$$

and since for every $i \in I$, every $j \in O(i, f)$, and every $p \in \mathbb{N}$

$$d_j(Lx_p, Lx_{p-1}) \le \prod_{s=0}^{p-2} q_{f^s(j)}(K_i) K_i,$$

it follows that

$$d_j(Lx_p, Lx_0) \le \sum_{r=1}^{p} \Big(\prod_{s=0}^{r-2} q_{f^s(j)}(K_i)\Big) K_i. \tag{3.69}$$

Further on, from $\overline{\lim_{n\to\infty}} q_{f^n(i)}(K_i) < 1$ it follows that (3.69) implies condition 2) from Theorem 3.92. This implies that there exists $z \in S$ such that Lz is a common fixed point for L, L_1 and L_2.

We shall prove that (3.67) implies the existence of one and only one element $y \in S$ such that $y = Ly = L_1 y = L_2 y$ and that (3.68) holds.

Let $u = Lu = L_1 u = L_2 u$, $v = Lv = L_1 v = L_2 v$ and suppose that (3.68) hold for $y = u$ and $y = v$ respectively. Then we have

$$\begin{aligned}
d_i(u, v) &= d_i(Lu, Lv) \\
&\le q_i(d_{f(i)}(L_1 u, L_2 v)) \cdot d_{f(i)}(L_1 u, L_2 v) \\
&= q_i(d_{f(i)}(u, v)) \cdot d_{f(i)}(u, v) \\
&\ \ \vdots \\
&\le q_i(d_{f(i)}(u, v)) \cdot q_{f(i)}(d_{f^2(i)}(u, v)) \cdots q_{f^n(i)}(d_{f^{n+1}(i)}(u, v)) \cdot d_{f^{n+1}(i)}(u, v) \\
&\le d_{f^{n+1}(i)}(u, v).
\end{aligned}$$

Since for every $i \in I$ and every $j \in O(i,f)$

$$d_j(u,v) \le N_i(u) + N_i(v),$$

it follows that

$$d_i(u,v) \le \prod_{s=0}^{n} q_{f^s(i)}\left(N_i(u) + N_i(v)\right)\left(N_i(u) + N_i(v)\right)$$

and therefore $d_i(u,v) = 0$ for every $i \in I$, which implies $u = v$. We prove that $y = Lz$ satisfies (3.68), where $z = \lim_{n\to\infty} Lx_{2n}$. First we show that there exists $N_i \in \mathbb{R}^+$, for every $i \in I$, such that for every $n \in \mathbb{N}$ and every $j \in O(i,f)$

$$d_j(LLx_{2n}, L^2x_0) \le N_i. \tag{3.70}$$

From (3.70) and $d_i(Lz, L^2x_0) = \lim_{n\to\infty} d_i(LLx_{2n}, L^2x_0)$ it follows (3.68).

We have

$$\begin{aligned} d_i(LLx_{2n}, LLx_{2n-1}) \le\ & q_i(d_{f(i)}(L^2x_{2n-1}, L^2x_{2n-2})) \cdot q_{f(i)}(d_{f^2(i)}(L^2x_{2n-2}, L^2x_{2n-3})) \\ & \cdots q_{f^{2n-2}(i)}(d_{f^{2n-1}(i)}(L^2x_1, Lx^2x_0)) \cdot d_{f^{2n-1}(i)}(L^2x_1, L^2x_0), \end{aligned}$$

and since for every $j \in O(i,f)$ and every $i \in I$ $d_j(L^2x_1, L^2x_0) \le K_i$, it follows that

$$d_j(LLx_{2n}, LLx_{2n-1}) \le q_j(V_i) \cdots q_{f^{2n-2}(j)}(V_i)V_i,$$

and similarly

$$d_j(LLx_{2n-1}, LLx_{2n-2}) \le q_j(V_i) \cdots q_{f^{2n-3}(j)}(V_i)V_i$$

for every $j \in O(i,f)$ and every $i \in I$. Therefore

$$\begin{aligned} d_j(LLx_{2n}, LLx_0) &\le d_j(LLx_{2n}, LLx_{2n-1}) + \cdots + d_j(L^2x_1, L^2x_0) \\ &\le V_i(1 + q_j(V_i) + \cdots + q_j(V_i) \cdots q_{f^{2n-2}(j)}(V_i)), \end{aligned}$$

and from $\overline{\lim_{n\to\infty}} q_{f^n(i)}(V_i) < 1$ it follows that for every $i \in I$ there exists $N_i \in \mathbb{R}^+$ such that (3.70) holds. □

Corollary 3.94 *Let $(S, (d_i)_{i\in I})$ be a sequentially complete Hausdorff uniformizable space, $f : I \to I$ and L a mapping from S into S satisfying the following conditions:*

a) for each $i \in I$ there exists a non-decreasing function $q_i : \mathbb{R}^+ \to [0,1]$ such that

$$d_i(Lx, Ly) \le q_i(d_{f(i)}(x,y)) \cdot d_{f(i)}(x,y) \text{ for every } x, y \in S;$$

b) for each $i \in I$ and $t \in \mathbb{R}^+$

$$\overline{\lim_{n\to\infty}} q_{f^n(i)}(t) < 1;$$

c) there is $x_0 \in S$ such that for each $i \in I$

$$\sup_{n\in\mathbb{N}} d_{f^n(i)}(x_0, Lx_0) \leq K_i \quad (K_i \in \mathbb{R}^+).$$

Then there exists a unique $x \in S$ such that $x = Lx$ and for each $i \in I$

$$\sup_{n\in\mathbb{N}} d_{f^n(i)}(x_0, x) \leq S_i \quad (S_i \in \mathbb{R}^+).$$

Proof. It is easy to see that all the conditions of the preceding corollary are satisfied for $L_1 = L_2 = \mathrm{Id}$, $\mathrm{Id}\ x = x$ for every $x \in S$. □

Corollary 3.95 *Let $(S, \mathcal{F}, T)$ be a complete Menger space such that $\sup_{a<1} T(a,a) = 1$, $f : (0,1) \to (0,1)$, L_1 and L_2 continuous mappings from S into S, $L : S \to L_1S \cap L_2S$ a continuous mapping such that L commutes with L_1 and L_2 and the following conditions are satisfied:*

1) for every $\alpha \in (0,1)$ there exists a right continuous, non-decreasing mapping

$$q_\alpha : \mathbb{R}^+ \to [0,1) \text{ for which } \overline{\lim_n}\, q_{f^n(\alpha)}(t) < 1 \text{ for every } t \in \mathbb{R}^+$$

and for every $\alpha \in (0,1)$, every $r > 0$ and every $x, y \in S$

$$F_{L_1x,L_2y}(r) > 1 - f(\alpha) \quad \Rightarrow \quad F_{Lx,Ly}(q_\alpha(r)r) > 1 - \alpha;$$

2) there exists $x_0 \in S$ such that for every $\alpha \in (0,1)$ there exists $K_\alpha \in \mathbb{R}^+$ such that for every $p \in \mathbb{N}$ and every $n \in \mathbb{N}$

$$F_{Lx_0,Lx_p}(K_\alpha) > 1 - f^n(\alpha),$$

where the sequence $(x_n)_{n\in\mathbb{N}}$ is defined by

$$L_1x_{2n-1} = L_{x_{2n-2}} \text{ and } L_2x_{2n} = Lx_{2n-1} \qquad (n \in \mathbb{N}).$$

Then there exists $z \in S$ such that

$$Lz = L_1z = L_2z.$$

If, in addition, for every $\alpha \in (0,1)$ there exists $M_\alpha \in \mathbb{R}^+$ such that for every $n \in \mathbb{N}$

$$F_{L^3x_1,L^2x_0}(M_\alpha) > 1 - f^n(\alpha),$$

then Lz is a common fixed point for mapping L, L_1, and L_2, which is the unique common fixed point in the set

$$\{u \mid u \in S, \forall\alpha \in (0,1), \exists P_\alpha \in \mathbb{R}^+ \text{ such that } F_{Lz,u}(P_\alpha) > 1 - f^n(\alpha), \forall n \in \mathbb{N}\}.$$

Proof. We have only to prove that 1) implies that for every $\lambda \in (0,1)$ and every $x, y \in S$

$$d_\lambda(Lx, Ly) \le q_\lambda(d_\lambda(L_1x, L_2y)) \cdot d_{f(\lambda)}(L_1x, L_2y). \tag{3.71}$$

Suppose that $d_{f(\lambda)}(L_1x, L_2y) < r$. Then $F_{L_1x,L_2y}(r) > 1 - f(\lambda)$ and by 1) we obtain that $F_{Lx,Ly}(q_\lambda(r)r) > 1 - \lambda$. Hence $d_\lambda(Lx, Ly) < q_\lambda(r)r$ and since q_λ is right-continuous we obtain that (3.71) holds. □

Corollary 3.96 *Let $(S, \mathcal{F}, T)$ be a complete Menger space and T a t-norm of H-type such that $\sup\limits_{a<1} T(a,a) = 1$, $L_1, L_2 : S \to S$ continuous mappings and $L : S \to L_1S \cap L_2S$ a continuous mapping which commutes with L_1 and L_2, such that for some $q \in (0,1)$*

$$F_{Lx,Ly}(u) \ge F_{L_1x,L_2y}\left(\frac{u}{q}\right) \quad \textit{for every } x, y \in S \textit{ and every } u > 0. \tag{3.72}$$

Then L, L_1 and L_2 have a unique common fixed point.

Proof. Since in this case $f(\lambda) = \lambda$ for every $\lambda \in (0,1)$, it remains only to prove that

$$\sup_{\varepsilon} \inf_{p\in\mathbb{N}} F_{Lx_0,Lx_p}(\varepsilon) = 1, \tag{3.73}$$

where $(x_p)_{p\in\mathbb{N}}$ is defined for an arbitrary $x_0 \in S$ in the following way:

$$Lx_{2n-2} = L_1x_{2n-1}, \qquad Lx_{2n-1} = L_2x_{2n} \qquad (n \in \mathbb{N}).$$

For every $n, p \in \mathbb{N}$ we have that

$$\begin{aligned}
F_{Lx_0,Lx_p}\left(\frac{\varepsilon}{q^n}\right) \ \ge\ & \underbrace{T\Big(T\Big(\cdots\Big(T}_{\text{(p-1)-times}}\left(F_{Lx_p,Lx_{p-1}}\Big(\frac{\varepsilon}{q^n}\cdot q^{p-1}\Big),\right. \\
& F_{Lx_{p-1},Lx_{p-2}}\Big(\frac{\varepsilon(1-q)}{q^n}\cdot q^{p-2}\Big)\Big)\Big), \dots, F_{Lx_1,Lx_0}\Big(\frac{\varepsilon(1-q)}{q^n}\Big)\Big) \\
\ge\ & \underbrace{T\Big(T\Big(\cdots\Big(T}_{\text{(p-1)-times}}\left(F_{Lx_1,Lx_0}\Big(\frac{\varepsilon}{q^n}\Big),\right. \\
& F_{Lx_1,Lx_0}\Big(\frac{\varepsilon(1-q)}{q^n}\Big)\Big)\Big), \dots, F_{Lx_1,Lx_0}\Big(\frac{\varepsilon(1-q)}{q^n}\Big)\Big) \\
\ge\ & \left(F_{Lx_1,Lx_0}\left(\frac{\varepsilon(1-q)}{q^n}\right)\right)_T^{(p-1)},
\end{aligned}$$

i.e.,

$$F_{Lx_0,Lx_p}\left(\frac{\varepsilon}{q^n}\right) \ge \left(F_{Lx_1,Lx_0}\left(\frac{\varepsilon(1-q)}{q^n}\right)\right)_T^{(p-1)}. \tag{3.74}$$

From (3.74) it follows that for every $n, p \in \mathbb{N}$

$$F_{Lx_0,Lx_p}\left(\frac{\varepsilon}{q^n}\right) \geq \mathop{\mathrm{T}}_{i=n}^{\infty} F_{Lx_1,Lx_0}\left(\frac{\varepsilon(1-q)}{q^n}\right), \tag{3.75}$$

and (3.75) implies that the set $\{Lx_p \mid p \in \mathbb{Z}_+\}$ is probabilistic bounded. □

Chapter 4

Probabilistic B-contraction principles for multi-valued mappings

The inequality $F_{fx,fy}(qs) \geq F_{x,y}(s)$ $(s \geq 0)$, where $q \in (0,1)$, is generalized for multi-valued mappings in many directions. In this chapter we consider three generalizations of the above inequality for multi-valued mappings, and for such a kind of mappings some fixed point theorems are proved. In section 4.1 a fixed point theorem is proved for multi-valued mappings which satisfy a multi-valued version of the strict probabilistic (b_n)-contraction condition introduced in section 3.3. We introduce in section 4.2 the notion of a multi-valued probabilistic Ψ-contraction, and by using the notion of the function of non-compactness a fixed point theorem is proved. Using Hausdorff distance S.B. Nadler obtained in [205] a generalization of the Banach contraction principle in metric spaces, and in section 4.3 a probabilistic version of Nadler's fixed point theorem is proved. As a corollary a multi-valued version of Tardiff's fixed point theorem is obtained. In section 4.4 a probabilistic version of Itoh's fixed point theorem from [146] is given, and section 4.5 contains a fixed point result for probabilistic non-expansive multi-valued mappings of Nadler's type, defined on probabilistic metric spaces with convex structures.

We assume in the whole chapter for the probabilistic metric spaces $(S, \mathcal{F}, \tau)$ that Range$(\mathcal{F}) \subset \mathcal{D}^+$.

4.1 Multi-valued contractions of Miheţ's type

Let $(S, \mathcal{F}, T)$ be a Menger space. By $CB(S)$ we shall denote the family of all non-empty closed (in the (ε, λ)-topology) and probabilistic bounded subsets of S.

The probabilistic distance between A and B from $CB(S)$, $\widetilde{F}_{A,B} : \mathbb{R} \to [0,1]$ is defined by

$$\widetilde{F}_{A,B}(u) = \sup_{s<u} T\left(\inf_{p\in A}\sup_{q\in B} F_{p,q}(s),\ \inf_{q\in B}\sup_{p\in A} F_{p,q}(s)\right), \quad u \in \mathbb{R}.$$

The proof of Theorem 4.2 is based on the following lemma from [216].

Lemma 4.1 *Let X be a non-empty compact uniform space, $(d_i)_{i\in I}$ a family of pseudo-metrics which generates the uniformity of X and $f : X \to CB(X)$ (the family of all non-empty, closed and bounded subsets of X) has the following property:*

For every $i \in I$ there exists a constant $k_i \in (0,1)$ such that

$$D_i(fx, fy) \le k_i d_i(x,y) \ \text{ for every } \ x, y \in X,$$

where for every $i \in I, A, B \in CB(X)$

$$D_i(A,B) = \max\left\{\sup_{a\in A}\inf_{b\in B} d_i(a,b),\ \sup_{b\in B}\inf_{a\in A} d_i(a,b)\right\}.$$

Then there exists $x \in X$ such that $x \in fx$.

Let T be a t-norm. We recall that $T \in \mathcal{H}$ if and only if there exists a non-decreasing sequence $(b_n)_{n\in\mathbb{N}}$ from $(0,1)$ such that $\lim_{n\to\infty} b_n = 1$ and the following implication holds:

For every $n \in \mathbb{N}$

$$1 \ge x > b_n, \quad 1 \ge y > b_n \quad \Rightarrow \quad T(x,y) > b_n.$$

The following theorem is proved by D. Miheţ in [194].

Theorem 4.2 *Let $(S, \mathcal{F}, T)$ be a compact Menger space, $T \in \mathcal{H}$ and $f : S \to CB(S)$ a multi-valued mapping. If for every $n \in \mathbb{N}$ there exists a $k_n \in (0,1)$ such that for every $p, q \in S$, and every $s > 0$,*

$$F_{p,q}(s) > b_n \Rightarrow \widetilde{F}_{fp,fq}(k_n s) > b_n, \tag{4.1}$$

then there exists $x \in S$ such that $x \in fx$.

Proof. If $T \in \mathcal{H}$ then the family of pseudo-metrics $(d_n)_{n\in\mathbb{N}}$, defined by

$$d_n(p,q) = \sup\{t \mid F_{p,q}(t) \le b_n\} \ \ (p, q \in S),$$

generates the (ε, λ)-uniformity.

The proof is similar to the proof given in [95]. We prove that (4.1) implies (4.2), where

$$D_n(fp, fq) \leq k_n d_n(p, q), \tag{4.2}$$

for every $n \in \mathbb{N}$ and every $p, q \in S$.

If (4.2) does not hold, there exist $n \in \mathbb{N}$ and $p, q \in S$ such that

$$D_n(fp, fq) > k_n d_n(p, q). \tag{4.3}$$

Let $s = \frac{1}{k_n} D_n(fp, fq)$. Then (4.3) implies that $s > d_n(p, q)$ and therefore $F_{p,q}(s) > b_n$. Using (4.1) we conclude that $\widetilde{F}_{fp,fq}(k_n s) > b_n$, i.e., $\widetilde{F}_{fp,fq}(D_n(fp, fq)) > b_n$. We shall prove that for all $A, B \in CB(S)$, $\widetilde{F}_{A,B}(D_n(A, B)) \leq b_n$ by showing that

$$D_n(A, B) \leq \sup\{s \mid \widetilde{F}_{A,B}(s) \leq b_n\}. \tag{4.4}$$

In order to prove (4.4), we shall prove that

$$\sup\{s \mid \widetilde{F}_{A,B}(s) \leq b_n\} = \sup\left\{s \mid T\left(\inf_{p\in A} \sup_{q\in B} F_{p,q}(s),\ \inf_{q\in B} \sup_{p\in A} F_{p,q}(s)\right) < b_n\right\}, \tag{4.5}$$

and

$$\sup\left\{s \mid T\left(\inf_{p\in A} \sup_{q\in B} F_{p,q}(s),\ \inf_{q\in B} \sup_{p\in A} F_{p,q}(s)\right) \leq b_n\right\}$$

$$= \max\left\{\sup\left\{s \mid \inf_{p\in A} \sup_{q\in B} F_{p,q}(s) \leq b_n\right\},\ \sup\left\{s \mid \inf_{q\in B} \sup_{p\in A} F_{p,q}(s) \leq b_n\right\}\right\}. \tag{4.6}$$

Let

$$G(s) = T(\inf_{p\in A} \sup_{q\in B} F_{p,q}(s),\ \inf_{q\in B} \sup_{p\in A} F_{p,q}(s)).$$

It is easy to see that

$$\sup\{s \mid G(s) \leq b_n\} = \sup\left\{s \mid \sup_{u<s} G(u) \leq b_n\right\}.$$

Let

$$P(s) = \inf_{p\in A} \sup_{q\in B} F_{p,q}(s), R(s) = \inf_{q\in B} \sup_{p\in A} F_{p,q}(s).$$

Since $T \leq T_{\mathbf{M}}$ we have that

$$\{s \mid P(s) \leq b_n\} \subset \{s \mid T(P(s), R(s)) \leq b_n\}$$

and

$$\{s \mid R(s) \leq b_n\} \subset \{s \mid T(P(s), R(s)) \leq b_n\},$$

which implies that

$$\max\{\sup\{s \mid P(s) \le b_n\},\ \sup\{s \mid R(s) \le b_n\}\} \le \sup\{s \mid T(P(s), R(s)) \le b_n\}.$$

In order to prove (4.6) we shall suppose that there exists $\delta > 0$ such that

$$\max\{\sup\{s \mid P(s) \le b_n\},\ \sup\{s \mid R(s) \le b_n\}\} < \delta < \sup\{s \mid T(P(s), R(s)) \le b_n\}.$$

Then

$$P(\delta) > b_n, \quad R(\delta) > b_n, \quad T(P(\delta), R(\delta)) \le b_n,$$

which is a contradiction since $T \in \mathcal{H}$. Hence

$$\sup\{s \mid \widetilde{F}_{A,B}(s) \le b_n\} = \max\{\sup\{s \mid P(s) \le b_n\},\ \sup\{s \mid R(s) \le b_n\}\}.$$

From

$$\sup_{q \in B} \inf_{p \in A} d_n(p, q) \le \sup\{s \mid P(s) \le b_n\}$$

$$\sup_{p \in A} \inf_{q \in B} d_n(p, q) \le \sup\{s \mid R(s) \le b_n\},$$

we obtain that

$$D_n(A, B) \le \sup\{s \mid \widetilde{F}_{A,B}(s) \le b_n\}.$$

Since (4.2) holds the theorem follows from Lemma 4.1. □

4.2 Multi-valued probabilistic Ψ-contractions

The following definition is introduced in [107].

Definition 4.3 *Let $(S, \mathcal{F})$ be a probabilistic metric space and $\Psi : [0, \infty) \to [0, \infty)$. We say that a mapping $f : S \to 2^S$ (2^S is the family of all non-empty subsets of the set S) is a multi-valued probabilistic (Ψ)-contraction if for every $x, y \in S$ and every $p \in fx$ there exists $q \in fy$ such that*

$$F_{p,q}(\Psi(\varepsilon)) \ge F_{x,y}(\varepsilon), \ \textit{for every } \varepsilon > 0.$$

If $f : S \to S$ is a single-valued mapping which is a probabilistic Ψ-contraction for the function $\Psi(u) = qu$ $(u > 0), q \in (0, 1)$, then f is a probabilistic q-contraction.

Definition 4.4 *Let $\gamma \in \{\alpha, \beta\}$, $g : [0, \infty) \to [0, \infty)$ and $f : K \to 2^S$, where K is a probabilistic bounded subset of S. If for every $u > 0$*

$$\gamma_{f(B)}(g(u)) \ge \gamma_B(u),$$

for every $B \subset K$, then f is a (γ, g)-condensing mapping.

Definition 4.5 *A mapping $f : K \to 2^S$ ($K \subset S$) is densifying on K with respect to the function $\gamma \in \{\alpha, \beta\}$ if $f(K)$ is probabilistic bounded subset of S and for every $B \subset K$:*

$$\gamma_{f(B)}(u) \leq \gamma_B(u), \text{ for every } u > 0 \Rightarrow B \text{ is precompact.} \tag{4.7}$$

Let $\mathcal{P}$ be the class of all mappings Ψ which maps $[0, \infty)$ onto $[0, \infty)$ which are strictly increasing and $\lim_{n\to\infty} \Psi^n(x) = 0$, for every $x \in [0, \infty)$. If $\Psi \in \mathcal{P}$ let $\Psi^{-1} = \bar{\Psi}$.

By Com(M) and $C(M)$ we shall denote the family of all nonempty, compact and closed subsets of M, respectively.

The next theorem is proved in [107].

Theorem 4.6 *Let $(S, \mathcal{F}, T)$ be a complete Menger space with a continuous t-norm T, M a non-empty, closed and probabilistic bounded subset of S and $f : M \to$ Com(M) a multi-valued probabilistic Ψ-contraction. If $\Psi \in \mathcal{P}$ then there exists $x \in M$ such that $x \in fx$.*

Proof. Let $x_0 \in M$ and $x_1 \in fx_0$. Since f is a multi-valued probabilistic Ψ-contraction, there exists $x_2 \in fx_1$ such that for every $s > 0$

$$F_{x_2,x_1}(s) \geq F_{x_1,x_0}(\bar{\Psi}(s)).$$

Hence we can define a sequence $(x_n)_{n\in\mathbb{N}}$ from M such that $x_{n+1} \in fx_n$, for every $n \in \mathbb{N}$ and

$$F_{x_{n+1},x_n}(s) \geq F_{x_1,x_0}(\bar{\Psi}^n(s)) \quad \text{for every } n \in \mathbb{N} \text{ and every } s > 0.$$

First, we prove that the set $\overline{\{x_n | \ n \in \mathbb{N}\}}$ is compact, which implies that there exists a convergent subsequence $(x_{n_k})_{k\in\mathbb{N}}$ of the sequence $(x_n)_{n\in\mathbb{N}}$.

Denote by β^M the function β in the induced probabilistic metric space $(M, \mathcal{F}, T)$. We shall prove that for every $s > 0$ and every $A \subset M$

$$\beta^M_{f(A)}(\Psi(s)) \geq \beta^M_A(s), \tag{4.8}$$

i.e., f is a (Ψ, β^M)-condensing mapping.

Since β is a left-continuous function it is enough to prove that for every $u \in (0, s)$

$$\beta^M_A(s-u) \leq \beta^M_{f(A)}(\Psi(s)). \tag{4.9}$$

If $\beta^M_A(s-u) = 0$ then (4.9) holds. Suppose that $\beta^M_A(s-u) > 0$ and prove (4.9), i.e., the following implication

$$(\forall r > 0) \left(r < \beta^M_A(s-u) \quad \Rightarrow \quad r \leq \beta^M_{f(A)}(\Psi(s))\right).$$

Let $0 < r < \beta^M_A(s-u)$. From the definition of $\beta^M_A(\cdot)$ it follows that there exists a finite subset $A_0 \subset M$ such that $\mathbf{F}_{A,A_0}(s-u) > r$ and so

$$\inf_{z\in A} \max_{w\in A_0} F_{z,w}(s-u) > r. \tag{4.10}$$

Using (4.10) we conclude that for every $z \in A$ there exists $w(z) \in A_0$ such that $F_{z,w(z)}(s-u) > r$. Hence for every $y \in f(A)$, $y \in fz$, $z \in A$, there exists $x \in f(A_0), x \in fw(z), w(z) \in A_0$ such that

$$F_{y,x}(\Psi(s-u)) \geq F_{z,w(z)}(s-u) > r.$$

From the continuity of T and relation $T(r,1) = r$, we obtain that for every $\delta \in (0,r)$ there exists $\lambda(\delta) \in (0,1)$ such that

$$1 \geq h > 1 - \lambda(\delta) \Rightarrow T(r,h) > r - \delta.$$

Let $A_0 = \{x_1, x_2, \ldots, x_n\}$. From the compactness of fx_i $(i \in \{1,2,\ldots,n\})$ there follows the existence of a finite set $\{z_1^i, z_2^i, \ldots, z_{n(i)}^i\} \subset M$ $(i \in \{1,2,\ldots,n\})$ such that

$$fx_i \subset \bigcup_{j=1}^{n(i)} N_{z_j^i}\Big(\frac{\Psi(s)-\Psi(s-u)}{2}, \lambda(\delta)\Big),$$

where $N_v(\varepsilon, \lambda) = \{z \,|\, F_{z,v}(\varepsilon) > 1 - \lambda\}$. We shall prove that

$$\mathbf{F}_{f(A), \bigcup\limits_{i=1}^{n}\bigcup\limits_{j=1}^{n(i)}\{z_j^i\}}(\Psi(s)) > r - \delta$$

which means that

$$\sup_{a<\Psi(s)} \inf_{y \in f(A)} \max_{w \in B_0} F_{y,w}(a) > r - \delta, \tag{4.11}$$

where $B_0 = \bigcup\limits_{i=1}^{n}\bigcup\limits_{j=1}^{n(i)}\{z_j^i\}$. Let $y \in f(A)$ and $x \in f(A_0)$ such that

$$F_{y,x}(\Psi(s-u)) > r.$$

If $x \in fx_i$ for some $i \in \{1,2,\ldots,n\}$ then there exists z_j^i $(j \in \{1,2,\ldots,n(i)\})$ such that $x \in N_{z_j^i}((\Psi(s)-\Psi(s-u))/2,\ \lambda(\delta))$ which means that

$$F_{x,z_j^i}\left(\frac{\Psi(s)-\Psi(s-u)}{2}\right) > 1 - \lambda(\delta). \tag{4.12}$$

From (4.12) we have that

$$\begin{aligned} F_{y,z_j^i}\left(\frac{\Psi(s)+\Psi(s-u)}{2}\right) &\geq T\Big(F_{y,x}(\Psi(s-u), F_{x,z_j^i}\left(\frac{\Psi(s)-\Psi(s-u)}{2}\right)\Big) \\ &> r - \delta \end{aligned}$$

and this implies, since Ψ is strictly increasing, that

$$\begin{aligned} \sup_{a<\Psi(s)} \inf_{y \in f(A)} \max_{w \in B_0} F_{y,w}(a) &\geq \inf_{y \in f(A)} \max_{w \in B_0} F_{y,w}\left(\frac{\Psi(s)+\Psi(s-u)}{2}\right) \\ &> r - \delta. \end{aligned}$$

From (4.11) we obtain

$$r-\delta\le\beta^M_{f(A)}(\Psi(s))$$

for every $\delta\in(0,r)$, and since δ is an arbitrary element from $(0,r)$ we have that $r\le\beta^M_{f(A)}(\Psi(s))$. Hence for every $s>0$ and every $A\subset M$

$$\beta^M_{f(A)}(\Psi(s))\ge\beta^M_A(s), \tag{4.13}$$

i.e., f is (Ψ,β^M)-condensing on M. Relation (4.13) implies that f is densifying on M in respect to β^M. Indeed, suppose that $\beta^M_{f(A)}(s)\le\beta^M_A(s)$ for every $s>0$. Then from (4.13) we obtain that for every $n\in\mathbb{N}$ and $s>0$

$$\beta^M_A(s)\ge\beta^M_{f(A)}(s)\ge\beta^M_A(\bar\Psi(s))\ge\cdots\ge\beta^M_A(\bar\Psi^n(s)). \tag{4.14}$$

Since $\lim_{n\to\infty}\bar\Psi^n(s)=+\infty$ from (4.14), we obtain that $\beta^M_A(s)=1$ for every $s>0$. Hence $\beta^M_A=H_0$ which implies that $\bar A$ is compact.

Take $A=\{x_n|\, n\in\mathbb{N}\}$. Since $x_{n+1}\in fx_n$ $(n\in\mathbb{N})$ we have

$$\begin{aligned}\beta^M_{f(\{x_n|n\ge1\})}(u)&\le\beta^M_{\{x_n|n\ge2\}}(u)\\&=\beta^M_{\{x_n|n\ge1\}}(u)\end{aligned}$$

and therefore $\beta^M_{f(A)}(u)\le\beta^M_A(u)$. Since f is densifying on M in respect to β^M, it follows that $\overline{\{x_n|\, n\in\mathbb{N}\}}$ is compact. Let $\lim_{k\to\infty}x_{n_k}=x\in M$. From the inequality

$$F_{x_{n_k+1},x_{n_k}}(s)\ge F_{x_1,x_0}(\bar\Psi^{n_k}(s))\text{ for every }k\in\mathbb{N}\text{ and every }s>0,$$

and the relation $\lim_{k\to\infty}\bar\Psi^{n_k}(s)=\infty$ it follows that $\lim_{k\to\infty}x_{n_k+1}=x$. We shall prove that $x\in fx$. Since $x_{n_k+1}\in fx_{n_k}$ $(k\in\mathbb{N})$, it remains to prove that the mapping f is closed.

Let $(v_n)_{n\in\mathbb{N}}$ and $(w_n)_{n\in\mathbb{N}}$ be two sequences from M such that $w_n\in fv_n$ $(n\in\mathbb{N})$ and $\lim_{n\to\infty}w_n=w$, $\lim_{n\to\infty}v_n=v$.

If we prove that $w\in\overline{f(v)}$, since $\overline{f(v)}=f(v)$, this means that $w\in fv$.

Let $\varepsilon>0$ and $\delta\in(0,1)$. Since f is a multi-valued probabilistic Ψ-contraction, for every $n\in\mathbb{N}$ there exists $z_n\in fv$ such that

$$F_{z_n,w_n}(s)\ge F_{v_n,v}(\bar\Psi(s))\quad\text{for every }s>0.$$

Then

$$\begin{aligned}F_{w,z_n}(\varepsilon)&\ge T(F_{w,w_n}(\varepsilon/2),F_{w_n,z_n}(\varepsilon/2))\\&\ge T(F_{w,w_n}(\varepsilon/2),F_{v_n,v}(\bar\Psi(\varepsilon/2)).\end{aligned}$$

From the continuity of the mapping T and the relation $T(1,1) = 1$, it follows that there exists $u \in (0,1)$ such that $T(s,s) > 1 - \delta$ for every $s \in (u,1)$.

If for $n = n_0(\varepsilon, s) \in \mathbb{N}$ and $s \in (u,1)$

$$F_{w,w_n}\left(\frac{\varepsilon}{2}\right) > s \quad \text{and} \quad F_{v_n,v}\left(\bar{\Psi}\left(\frac{\varepsilon}{2}\right)\right) > s$$

then

$$F_{w,z_{n_0}}(\varepsilon) \geq T(s,s) > 1 - \delta,$$

which means that $z_{n_0} \in N_w(\varepsilon, \delta)$. Hence $w \in \overline{fv}$, which completes the proof. □

Remark 4.7 If a t-norm T is of H-type then it is enough to assume in Theorem 4.6 only that fx is closed for every $x \in M$, see [61].

From Theorem 4.6 we obtain the following corollary for fuzzy metric spaces.

We suppose that R is associative, continuous, and that $R(a,0) = a$ for every $a \in [0,1]$.

Corollary 4.8 *Let (X, d, L, R) be a complete fuzzy metric space such that*

$$\lim_{u\to\infty} d(x,y)(u) = 0 \quad \textit{for every } (x,y) \in X \times X$$

and that for every $\alpha \in (0,1)$ there exists M_α such that $\sup\limits_{x,y\in X} \rho_\alpha(x,y) < M_\alpha$. Let $f : X \to \mathrm{Com}(X)$ and $\Psi \in \mathcal{P}$ be such that the following condition is satisfied: For every $x, y \in X$ and every $p \in fx$ there exists $q \in fy$ such that

$$s \geq \lambda_1(x,y) \quad \Rightarrow \quad d(x,y)(s) \geq 1 - F_{p,q}(\Psi(s)).$$

Then there exists $x \in X$ such that $x \in fx$.

4.3 Probabilistic Nadler q-contraction

S.B. Nadler proved in [205] a generalization of the Banach contraction principle for multi-valued mappings $f : X \to CB(X)$, where (X,d) is a metric space, by introducing the condition

$$D(fx, fy) \leq q d(x,y),$$

where D is the Hausdorff metric and $q \in (0,1)$. We shall give a probabilistic version of Nadler's fixed point theorem.

In [83] the following definition is given.

Definition 4.9 *Let $(S, \mathcal{F})$ be a probabilistic metric space, M a non-empty subset of S and f a mapping from M into the family of all non-empty subsets of S. The* *mapping f is said to be a probabilistic Nadler q-contraction, where $q \in (0,1]$, if the following condition is satisfied:*

For every $u, v \in M$, every $x \in fu$, and every $\delta > 0$ there exists $y \in fv$ such that for every $\varepsilon > 0$

$$F_{x,y}(\varepsilon) \geq F_{u,v}\Big(\frac{\varepsilon - \delta}{q}\Big). \tag{4.15}$$

If f is a single-valued mapping, then the notion of a probabilistic Nadler q-contraction coincides with the notion of a probabilistic q-contraction introduced by Sehgal and Bharucha-Reid, since the function $F_{u,v}(\cdot)$ is left-continuous. If (4.15) is satisfied for $q = 1$ we say that f is multi-valued probabilistic non-expansive mapping of Nadler's type. If $f : S \to S$ is a single-valued mapping, since $F_{u,v}$ is left-continuous, (4.15) reduces to

$$F_{fu,fv}(\varepsilon) \geq F_{u,v}(\varepsilon) \text{ for every } u, v \in M \text{ and every } \varepsilon > 0.$$

It is easy to prove that if (S, d) is a metric space and f is a mapping from a nonempty subset M of S into the family $CB(S)$, which is a q-contraction in the sense of Nadler [205], then f is also a probabilistic Nadler q-contraction in the associated Menger space $(S, \mathcal{F}, T_{\mathbf{M}})$.

The following example of a probabilistic Nadler q -contraction type mapping is connected with random operators.

Example 4.10 Suppose that M is a separable metric space, $(\Omega, \mathcal{A}, P)$ a probability space and $(S, \mathcal{F}, T_{\mathbf{L}})$ the Menger space of all classes of equivalence of measurable mappings from Ω into M. Further, let $f : \Omega \times M \to CB(M)$ be a random operator, which means that for every $x \in M$ the mapping $f(\cdot, x) : \Omega \to CB(M)$ is measurable, i.e., for every open subset G of M

$$f^{-1}(G) = \{\omega \mid \omega \in \Omega, f(\omega) \cap G \neq \varnothing\} \in \mathcal{A}.$$

Suppose that for every $x, y \in M$ and every $\omega \in \Omega$

$$D(f(\omega, x), f(\omega, y)) \leq q d(x, y) \; (q \in (0, 1)).$$

For every measurable mapping $X : \Omega \to M$ ($X \in S$), the mapping $\omega \mapsto f(\omega, X(\omega))$ is measurable and therefore (see [136]) there exists a countable family $(u_n)_{n \in \mathbb{N}}$ of measurable selectors of this mapping, such that

$$f(\omega, X(\omega)) = \overline{\{u_n(\omega) \mid n \in \mathbb{N}\}}.$$

Let the mapping $\overline{f}$ be defined in the following way:

$$\overline{f}X = \{U \mid U \in S, U(\omega) \in f(\omega, X(\omega)) \text{ for every } \omega \in \Omega\}.$$

Then the mapping $\overline{f}$ is a probabilistic Nadler q-contraction in the space $(S, \mathcal{F}, T_{\mathbf{L}})$. Indeed, let $X, Y \in S$, $V \in \overline{f}X$ and $\delta > 0$. This means that for every $\omega \in \Omega$, $V(\omega) \in f(\omega, X(\omega))$. Since for every $\omega \in \Omega$

$$D(f(\omega, X(\omega)), f(\omega, Y(\omega)) \leq qd(X(\omega), Y(\omega)),$$

it follows (Proposition 4, [146]) that there exists $W \in \overline{f}Y$ such that

$$d(V(\omega), W(\omega)) < D(f(\omega, X(\omega)), f(\omega, Y(\omega))) + \delta \text{ for every } \omega \in \Omega.$$

Then

$$d(V(\omega), W(\omega)) < qd(X(\omega), Y(\omega)) + \delta \text{ for every } \omega \in \Omega.$$

Hence if $\omega \in \Omega$ is such that $d(X(\omega), Y(\omega)) < \dfrac{\varepsilon - \delta}{k}$ then $d(V(\omega), W(\omega)) < \varepsilon$ and therefore

$$P\left(\left\{\omega \mid \omega \in \Omega, d(X(\omega), Y(\omega)) < \frac{\varepsilon - \delta}{k}\right\}\right) \leq P(\{\omega \mid \omega \in \Omega, d(V(\omega), W(\omega)) < \varepsilon\}).$$

This means that

$$F_{V,W}(\varepsilon) \geq F_{X,Y}\left(\frac{\varepsilon - \delta}{k}\right) \quad \text{for every } \varepsilon > 0.$$

Definition 4.11 *Let $(S, \mathcal{F})$ be a probabilistic metric space, M a non-empty subset of S and $f : M \to 2^S \setminus \{\emptyset\}$ a mapping from M into the family of all non-empty subsets of S. The mapping f is weakly demicompact if for every sequence $(x_n)_{n \in \mathbb{N}}$ from M such that $x_{n+1} \in fx_n$, for every $n \in \mathbb{N}$ and $\lim_{n \to \infty} F_{x_{n+1}, x_n}(\varepsilon) = 1$, for every $\varepsilon > 0$, there exists a convergent subsequence $(x_{n_k})_{k \in \mathbb{N}}$.*

Remark 4.12 Every densifying mapping on a complete space S is a weakly demicompact mapping. Indeed, let $f : M \to 2^S$ be a densifying mapping. If $(x_n)_{n \in \mathbb{N}}$ is a sequence from M such that $x_{n+1} \in fx_n$ $(n \in \mathbb{N})$ then for $\gamma \in \{\alpha, \beta\}$ we have that

$$\gamma_{\{x_n | n \geq 2\}}(u) = \gamma_{\{x_n | n \geq 1\}}(u), \text{ for every } u > 0$$

and therefore from $\{x_{n+1} \mid n \geq 1\} \subseteq f(\{x_n \mid n \geq 1\})$ we have that

$$\gamma_{f(\{x_n | n \geq 1\})}(u) \leq \gamma_{\{x_{n+1} | n \geq 1\}}(u) = \gamma_{\{x_n | n \geq 1\}}(u).$$

Since f is densifying with respect to γ we have, using (4.7) for $B = \{x_n \mid n \geq 1\}$, that B is a precompact set. This means, if S is complete, that there exists a convergent sequence $(x_{n_k})_{k \in \mathbb{N}}$.

Theorem 4.13 *Let $(S, \mathcal{F}, T)$ be a complete Menger space, T a t-norm such that $\sup_{a<1} T(a, a) = 1$, M a non-empty and closed subset of S, $f : M \to C(M)$ a probabilistic Nadler q-contraction, $q \in (0, 1)$, such that at least one of the following two conditions are satisfied:*

(i) f is weakly demicompact.

(ii) There exists $x_0 \in M$, $x_1 \in fx_0$ and $\mu \in (q,1)$ such that

$$\lim_{n\to\infty} \mathop{\mathrm{T}}_{i=n}^{\infty} F_{x_0,x_1}\left(\frac{1}{\mu^i}\right) = 1.$$

Then there exists $x \in M$ such that $x \in fx$.

Proof. Since f is a probabilistic Nadler q-contraction we can construct a sequence $(p_n)_{n\in\mathbb{N}}$ from M, such that $p_1 = x_1 \in fx_0$, $p_{n+1} \in fp_n$ and

$$F_{p_{n+1},p_n}(\varepsilon) \geq F_{p_n,p_{n-1}}\left(\frac{\varepsilon - q^n}{q}\right), \text{ for every } \varepsilon > 0 \text{ and every } n \in \mathbb{N}. \tag{4.16}$$

From (4.16) it follows that for every $\varepsilon > 0$ and every $n \in \mathbb{N}$

$$F_{p_{n+1},p_n}(\varepsilon) \geq F_{x_1,x_0}\left(\frac{\varepsilon - nq^n}{q^n}\right).$$

For every $\varepsilon > 0$, $\lim_{n\to\infty} F_{x_1,x_0}\left(\frac{\varepsilon - nq^n}{q^n}\right) = 1$, which implies that

$$\lim_{n\to\infty} F_{p_{n+1},p_n}(\varepsilon) = 1, \text{ for every } \varepsilon > 0. \tag{4.17}$$

Suppose that (i) holds. Then (4.17) implies the existence of a convergent subsequence $(p_{n_k})_{k\in\mathbb{N}}$ of the sequence $(p_n)_{n\in\mathbb{N}}$.

Suppose now that (ii) holds. Let $\sigma = \frac{q}{\mu}$. Since $0 < \sigma < 1$ the series $\sum_{i=1}^{\infty} \sigma^i$ is convergent and there exists $m_0 \in \mathbb{N}$ such that $\sum_{i=m_0}^{\infty} \sigma^i < 1$. Hence for every $m > m_0$ and every $s \in \mathbb{N}$

$$\varepsilon > \varepsilon \sum_{i=m_0}^{\infty} \sigma^i > \varepsilon \sum_{i=m}^{m+s} \sigma^i,$$

which implies that

$$\begin{aligned}
F_{p_{m+s+1},p_m}(\varepsilon) &\geq F_{p_{m+s+1},p_m}\Big(\varepsilon\sum_{i=m}^{m+s}\sigma^i\Big)\\
&\geq \underbrace{T\circ T\circ\cdots\circ T}_{s-\text{times}}(F_{p_{m+s+1},p_{m+s}}(\varepsilon\cdot\sigma^{m+s}),\\
&\qquad F_{p_{m+s},p_{m+s-1}}(\varepsilon\cdot\sigma^{m+s-1})),\ldots,F_{p_{m+1},p_m}(\varepsilon\cdot\sigma^m))\\
&\geq \underbrace{T\circ T\circ\cdots\circ T}_{s-\text{times}}\Big(F_{x_1,x_0}\Big(\frac{\varepsilon\sigma^{m+s}-(m+s)q^{m+s}}{q^{m+s}}\Big),\\
&\qquad F_{x_1,x_0}\Big(\frac{\varepsilon\sigma^{m+s-1}-(m+s-1)q^{m+s-1}}{q^{m+s-1}}\Big)\Big),\ldots,\\
&\qquad F_{x_1,x_0}\Big(\frac{\varepsilon\sigma^m-mq^m}{q^m}\Big)\Big)\\
&= \underbrace{T\circ T\circ\cdots\circ T}_{s-\text{times}}\Big(F_{x_1,x_0}\Big(\frac{\varepsilon}{(\frac{q}{\sigma})^{m+s}}-(m+s)\Big),\\
&\qquad F_{x_1,x_0}\Big(\frac{\varepsilon}{(\frac{q}{\sigma})^{m+s-1}}-(m+s-1)\Big)\Big),\ldots,\\
&\qquad F_{x_1,x_0}\Big(\frac{\varepsilon}{(\frac{q}{\sigma})^{m}}-m\Big)\Big)\\
&= \mathop{\mathsf{T}}_{i=m}^{m+s} F_{x_1,x_0}\Big(\frac{\varepsilon}{\mu^i}-i\Big)
\end{aligned}$$

Since $\mu\in(0,1)$ there exists $m_1(\varepsilon)>m_0$ such that $\varepsilon/\mu^m-m>\varepsilon/2\mu^m$ for every $m>m_1(\varepsilon)$, and therefore for every $m>m_1(\varepsilon)$ and every $s\in\mathbb{N}$

$$\begin{aligned}
F_{p_{m+s+1},p_m}(\varepsilon) &\geq \mathop{\mathsf{T}}_{i=m}^{m+s} F_{x_1,x_0}\left(\frac{\varepsilon}{2\mu^i}\right)\\
&\geq \mathop{\mathsf{T}}_{i=m}^{\infty} F_{x_1,x_0}\left(\frac{\varepsilon}{2\mu^i}\right).
\end{aligned}$$

It is obvious that $\lim\limits_{m\to\infty}\mathsf{T}_{i=m}^{\infty}F_{x_1,x_0}\left(\frac{1}{\mu^i}\right)=1$ implies $\lim\limits_{m\to\infty}\mathsf{T}_{i=m}^{\infty}F_{x_1,x_0}\left(\frac{\varepsilon}{2\mu^i}\right)=1$ for every $\varepsilon>0$. Since $\lim\limits_{m\to\infty}\mathop{\mathsf{T}}\limits_{i=m}^{\infty}F_{x_1,x_0}(\frac{\varepsilon}{2\mu^i})=1$ for every $\varepsilon>0$ it follows that for every $\lambda\in(0,1)$ there exists $m_2(\varepsilon,\lambda)>m_1(\varepsilon)$ such that

$$F_{p_{m+s+1},p_m}(\varepsilon)>1-\lambda\ \text{ for every } m\geq m_2(\varepsilon,\lambda) \text{ and every } s\in\mathbb{N}.$$

This means that $(p_n)_{n\in\mathbb{N}}$ is a Cauchy sequence, and since S is complete there exists $\lim\limits_{n\to\infty}p_n$. Hence in both cases there exists a subsequence $(p_{n_k})_{k\in\mathbb{N}}$ such that

$\lim_{k\to\infty} p_{n_k}$ exists. We shall prove that $x = \lim_{k\to\infty} p_{n_k}$ is a fixed point of f. Since fx is closed $fx = \overline{fx}$, and therefore it remains to prove that $x \in \overline{fx}$, i.e., that for every $\varepsilon > 0$ and $\lambda \in (0,1)$ there exists $b(\varepsilon,\lambda) \in fx$ such that $F_{x,b(\varepsilon,\lambda)}(\varepsilon) > 1-\lambda$. From the condition $\sup_{a<1} T(a,a) = 1$ it follows that there exists $\eta(\lambda) \in (0,1)$ such that

$$u > 1-\eta(\lambda) \quad \Rightarrow \quad T(u,u) > 1-\lambda.$$

Since f is a Nadler q-contraction, for every $k \in \mathbb{N}$ there exists $b_k(\varepsilon) \in f(x)$ such that

$$F_{p_{n_k+1},b_k(\varepsilon)}\left(\frac{\varepsilon}{2}\right) \geq F_{p_{n_k},x}\left(\frac{\varepsilon}{4q}\right). \tag{4.18}$$

Let $k_1(\varepsilon,\lambda) \in \mathbb{N}$ be such that

$$F_{p_{n_k},x}\left(\frac{\varepsilon}{4q}\right) > 1-\frac{\eta(\lambda)}{2} \quad \text{for every } k \geq k_1(\varepsilon,\lambda).$$

Since $\lim_{k\to\infty} p_{n_k} = x$ such a number $k_1(\varepsilon,\lambda)$ exists. Then (4.18) implies that

$$F_{p_{n_k+1},b_k(\varepsilon)}\left(\frac{\varepsilon}{2}\right) > 1-\frac{\eta(\lambda)}{2}$$

for every $k \geq k_1(\varepsilon,\lambda)$.

From (4.17) it follows that $\lim_{k\to\infty} p_{n_k+1} = x$ and therefore there exists $k_2(\varepsilon,\lambda) \in \mathbb{N}$ such that

$$F_{x,p_{n_k+1}}\left(\frac{\varepsilon}{2}\right) > 1-\frac{\eta(\lambda)}{2}, \text{ for every } k \geq k_2(\varepsilon,\lambda).$$

Let $k_3(\varepsilon,\lambda) = \max\{k_1(\varepsilon,\lambda),\, k_2(\varepsilon,\lambda)\}$. Then for $k > k_3(\varepsilon,\lambda)$ we have

$$F_{x,b_k(\varepsilon)}(\varepsilon) \geq T\left(F_{x,p_{n_k+1}}\left(\frac{\varepsilon}{2}\right),\, F_{p_{n_k+1},b_k(\varepsilon)}\left(\frac{\varepsilon}{2}\right)\right) > 1-\lambda.$$

Hence if $k > k_3(\varepsilon,\lambda)$ then we can choose $b(\varepsilon,\lambda) = b_k(\varepsilon) \in fx$. □

Corollary 4.14 *Let $(S,\mathcal{F},T)$ be a complete Menger space, T a t-norm of H-type, M a non-empty and closed subset of S and $f : M \to C(M)$ a probabilistic Nadler q-contraction, $q \in (0,1)$. Then there exists $x \in M$ such that $x \in fx$.*

Proof. We shall prove that (ii) from Theorem 4.13 holds for every $\mu \in (0,1)$. Since for every $n \in \mathbb{N}$ and $s \in \mathbb{N}$

$$\mathop{\mathrm{T}}_{i=n}^{n+s} F_{x_0,x_1}\left(\frac{1}{\mu^i}\right) \geq \mathop{\mathrm{T}}_{i=n}^{n+s} F_{x_0,x_1}\left(\frac{1}{\mu^n}\right) \quad \text{and} \quad \lim_{n\to\infty}\frac{1}{\mu^n} = \infty$$

there exists $n_1(\lambda) \in \mathbb{N}$ such that

$$F_{x_0,x_1}\left(\frac{1}{\mu^n}\right) > 1-\eta(\lambda) \text{ for every } n \geq n_1(\lambda),$$

where $\eta(\lambda) \in (0,1)$ is such that

$$\mathop{\mathsf{T}}_{i=n}^{n+s}(1-\eta(\lambda)) > 1-\lambda \text{ for every } n \in \mathbb{N},\ s \in \mathbb{N}.$$

Since T if of H-type such an element $\eta(\lambda)$ exists.

Hence

$$\mathop{\mathsf{T}}_{i=n}^{\infty} F_{x_0,x_1}(\frac{1}{\mu^n}) \geq 1-\lambda \text{ for every } n \geq n_1(\lambda).$$

This implies that (ii) from Theorem 4.13 holds. □

Corollary 4.15 *Let $(S,\mathcal{F},T)$ be a complete Menger space, T a t-norm such that $T \geq T_{\mathbf{L}}$, M a nonempty and closed subset of S and $f: M \to C(M)$ a probabilistic Nadler q-contraction, $q \in (0,1)$. If there exist $x_0 \in M$ and $x_1 \in fx_0$ such that*

$$\int_1^\infty \ln u \, dF_{x_1,x_0}(u) < \infty, \tag{4.19}$$

then there exists $x \in M$ such that $x \in fx$.

Proof. Tardiff proved that (4.19) implies (ii) from Theorem 4.13. □

4.4 A fixed point theorem of Itoh's type

Let us recall that a multi-valued mapping f is closed on S if for every sequence $(x_n)_{n\in\mathbb{N}}$ from S and every sequence $(y_n)_{n\in\mathbb{N}}$ such that $y_n \in fx_n$, for every $n \in \mathbb{N}$, the following implication holds:

$$x_n \to x, y_n \to y \quad \Rightarrow \quad y \in fx.$$

Itoh proved in [147] the following fixed point theorem for multi-valued mappings.

Theorem 4.16 (Itoh) *Let (X,d) be a complete metric space, $f: X \to CB(X)$ an upper semi-continuous mapping, K a non-empty, bounded subset of X such that $f(K)$ is bounded and the following conditions are satisfied:*

1. $\inf\limits_{x\in K} d(x,fx) = 0$;
2. *for every $A \subset K$ the following implication holds:*

$$\alpha(A) > 0 \quad \Rightarrow \quad \alpha(f(A)) < \alpha(A).$$

Then there exists at least one element $x \in \overline{K}$ such that $x \in fx$.

The following fixed point theorem, which is a generalization of Theorem 4.16 is proved in [105].

Theorem 4.17 *Let $(S, \mathcal{F}, T)$ be a complete Menger space with a continuous t-norm $T, f : S \to CB(S)$ a closed mapping and there exists a non-empty probabilistic bounded subset K of S such that $f(K)$ is probabilistic bounded and the following conditions are satisfied:*

(i) There exists a sequence $(x_n)_{n\in\mathbb{N}}$ from K and a sequence $(y_n)_{n\in\mathbb{N}}$ such that for every $n \in \mathbb{N}$, $y_n \in fx_n$ and

$$\lim_{n\to\infty} F_{x_n,y_n}(\varepsilon) = 1 \quad \text{for every} \quad \varepsilon > 0;$$

(ii) The mapping f is densifying on K in respect to the function γ, where $\gamma \in \{\alpha, \beta\}$.

Then there exists at least one element $x \in \overline{K}$ such that $x \in fx$.

The proof of Theorem 4.17 is based on the following lemma.

Lemma 4.18 *Let $(S, \mathcal{F}, T)$ be a Menger space, T a continuous t-norm and $(x_n)_{n\in\mathbb{N}}$, $(y_n)_{n\in\mathbb{N}}$ be two sequences from S such that for every $u > 0$*

$$\lim_{n\to\infty} F_{x_n,y_n}(u) = 1.$$

If $\gamma = \alpha$ or $\gamma = \beta$, and the sets $\{x_n \mid n \in \mathbb{N}\}$, $\{y_n \mid n \in \mathbb{N}\}$ are probabilistic bounded then for every $u > 0$

$$\gamma_{\{x_n|n\in\mathbb{N}\}}(u) = \gamma_{\{y_n|n\in\mathbb{N}\}}(u). \tag{4.20}$$

Proof. First, we shall prove that for every $u > 0$ and every $\varepsilon \in (0, u)$

$$\beta_{\{y_n|n\in\mathbb{N}\}}(u - \varepsilon) \leq \beta_{\{x_n|n\in\mathbb{N}\}}(u). \tag{4.21}$$

If (4.21) holds then $\beta_{\{y_n|n\in\mathbb{N}\}}(u) \leq \beta_{\{x_n|n\in\mathbb{N}\}}(u)$ for every $u > 0$ since the function $\beta_{\{y_n|n\in\mathbb{N}\}}$ is left-continuous.

Suppose that for some $u > 0$ and $\varepsilon > 0$ we have that

$$\beta_{\{y_n|n\in\mathbb{N}\}}(u - \varepsilon) > 0. \tag{4.22}$$

Namely, if (4.22) is not satisfied then (4.21) holds. Let $\rho > 0$ be such that $\rho < \beta_{\{y_n|n\in\mathbb{N}\}}(u - \varepsilon)$. From the definition of the function β_A ($A \subset S$), it follows that there exists a finite set $A_f \subset S$ such that

$$\mathbf{F}_{\{y_n|n\in\mathbb{N}\},A_f}(u - \varepsilon) \geq \rho,$$

which means that $\sup\limits_{s<u-\varepsilon} \inf\limits_{n\in\mathbb{N}} \max_{z\in A_f} F_{y_n,z}(s) \geq \rho$. This implies that

$$\inf_{n\in\mathbb{N}} \max_{z\in A_f} F_{y_n,z}(u - \varepsilon) \geq \rho,$$

since $F_{y_n,z}$ is a monotone non-decreasing function. Hence for every $n \in \mathbb{N}$

$$\max_{z\in A_f} F_{y_n,z}(u-\varepsilon) \geq \rho$$

which implies that for every $n \in \mathbb{N}$ there exists $z(n) \in A_f$ such that $F_{y_n,z(n)}(u-\varepsilon) \geq \rho$. The t-norm T is continuous, and since $T(1,\rho) = \rho$ it follows that for every $\delta > 0$ there exists a number $\widetilde{\delta} \in (0,1)$ such that

$$1 \geq s > 1-\widetilde{\delta} \quad \Rightarrow \quad T(s,\rho) > \rho - \delta.$$

Let $n_0(\varepsilon,\widetilde{\delta}) \in \mathbb{N}$ such that $F_{x_n,y_n}(\varepsilon/2) > 1-\widetilde{\delta}$ for every $n \geq n_0(\varepsilon,\widetilde{\delta})$. From the inequality

$$\begin{aligned} F_{x_n,z(n)}(u-\frac{\varepsilon}{2}) &\geq T\left(F_{x_n,y_n}\left(\frac{\varepsilon}{2}\right), F_{y_n,z(n)}\left(u-\varepsilon\right)\right) \\ &\geq T\left(F_{x_n,y_n}\left(\frac{\varepsilon}{2}\right), \rho\right), \end{aligned}$$

we obtain for every $n \geq n_0(\varepsilon,\widetilde{\delta})$ that $F_{x_n,z(n)}(u-\varepsilon/2) > \rho - \delta$.

This implies that

$$\sup_{s<u} \inf_{n\geq n_0(\varepsilon,\widetilde{\delta})} \max_{z\in A_f} F_{x_n,z}(s) > \rho - \delta$$

and therefore

$$\rho - \delta \leq \beta_{\{x_n | n > n_0(\varepsilon,\widetilde{\delta})\}}(u).$$

Since the set $\{x_n \mid n < n_0(\varepsilon,\widetilde{\delta})\}$ is precompact we have that $\beta_{\{x_n|n<n_0(\varepsilon,\widetilde{\delta})\}}(u) = 1$ for every $u > 0$. Using the property of the function β that for every $A, C \subset S$ which are probabilistic bounded $\beta_{A\cup C}(u) = \min\{\beta_A(u), \beta_C(u)\}$ we obtain

$$\beta_{\{x_n|n\in\mathbb{N}\}}(u) = \min\{\beta_{\{x_n|n\geq n_0(\varepsilon,\widetilde{\delta})\}}(u), 1\} = \beta_{\{x_n|n\geq n_0(\varepsilon,\widetilde{\delta})\}}(u).$$

Hence for arbitrary $\delta > 0$ we have that

$$\beta_{\{x_n|n\in\mathbb{N}\}}(u) \geq \rho - \delta,$$

which implies that for every $u > 0$, $\beta_B(u) \geq \rho$ and therefore (4.21) is proved. Analogously $\beta_{\{y_n|n\in\mathbb{N}\}}(u) \geq \beta_{\{x_n|n\in\mathbb{N}\}}(u)$ for every $u > 0$. Hence (4.20) holds for $\gamma = \beta$.

Suppose now that the function γ is equal to α. Let us prove that for every $u > 0$ and every $\varepsilon > 0$

$$\alpha_{\{y_n|n\in\mathbb{N}\}}(u-\varepsilon) \leq \alpha_{\{x_n|n\in\mathbb{N}\}}(u).$$

Let $\alpha_{\{y_n|n\in\mathbb{N}\}}(u-\varepsilon) > 0$ and $\rho < \alpha_{\{y_n|n\in\mathbb{N}\}}(u-\varepsilon)$. From the definition of the function $\alpha_{\{y_n|n\in\mathbb{N}\}}(u-\varepsilon)$ it follows that there exist $A_1, A_2, \ldots, A_n \subset S$ such that

$$\{y_n \mid n \in \mathbb{N}\} = \bigcup_{j=1}^{n} A_j \quad \text{and} \quad D_{A_j}(u-\varepsilon) \geq \rho \quad \text{for every } j \in \{1, 2, \ldots, n\}.$$

This means that for every $j \in \{1, 2, \ldots, n\}$

$$\sup_{s<u-\varepsilon} \inf_{x,y\in A_j} F_{x,y}(s) \geq \rho$$

and therefore $\inf_{x,y\in A_j} F_{x,y}(u-\varepsilon) \geq \rho$, which implies that for every $x, y \in A_j$, $F_{x,y}(u-\varepsilon) \geq \rho$. Let $\delta > 0$ be an arbitrary positive number. Since the function $(u, w) \mapsto T(u, T(\rho, w))$ is continuous on $[0,1]^2$ and $T(1, T(\rho, 1)) = \rho$ there exists a $\widetilde{\delta} \in (0,1)$ such that

$$u, w \in (1-\widetilde{\delta}, 1] \quad \Rightarrow \quad T(u, T(\rho, w)) > \rho - \delta.$$

Let $B_j = \left\{z \mid F_{z,y}(\varepsilon/4) > 1-\widetilde{\delta} \text{ for some } y \in A_j\right\}$ $(j \in \{1, 2, \ldots, n\})$. Since $\lim\limits_{n\to\infty} F_{x_n,y_n}(\varepsilon/4) = 1$ there exists $n_1(\varepsilon, \widetilde{\delta}) \in \mathbb{N}$ such that $F_{x_n,y_n}(\varepsilon/4) > 1-\widetilde{\delta}$, for every $n \geq n_1(\varepsilon, \widetilde{\delta})$. From the relation $\{y_n \mid n \in \mathbb{N}\} = \bigcup_{j=1}^{n} A_j$ it follows that for every $n \in \mathbb{N}$ there exists $j(n) \in \{1, 2, \ldots, n\}$ such that $y_n \in A_{j(n)}$, which implies that for every $n \geq n_1(\varepsilon, \widetilde{\delta})$, $x_n \in B_{j(n)}$. This means that

$$\{x_n \mid n > n_1(\varepsilon, \widetilde{\delta})\} \subset \bigcup_{j=1}^{n} B_j.$$

We prove that for every $j \in \{1, 2, \ldots, n\}$, $D_{B_j}(u) \geq \rho - \delta$.

If $x \in B_j$ and $y \in B_j$, then there exists $\widetilde{x} \in A_j$ and $\widetilde{y} \in A_j$ such that

$$F_{x,\widetilde{x}}(\varepsilon/4) > 1-\widetilde{\delta}, \quad F_{y,\widetilde{y}}(\varepsilon/4) > 1-\widetilde{\delta}$$

and since $F_{\widetilde{x},\widetilde{y}}(u-\varepsilon) \geq \rho$, we have that

$$\begin{aligned} F_{x,y}(u-\varepsilon/2) &\geq T\left(F_{x,\widetilde{x}}(\varepsilon/4), T\left(F_{\widetilde{x},\widetilde{y}}(u-\varepsilon), F_{\widetilde{y},y}(\varepsilon/4)\right)\right) \\ &\geq T\left(F_{x,\widetilde{x}}(\varepsilon/4), T\left(\rho, F_{\widetilde{y},y}(\varepsilon/4)\right)\right) \\ &> \rho - \delta. \end{aligned}$$

Hence $\inf\limits_{x,y\in B_j} F_{x,y}(u-\varepsilon/2) \geq \rho - \delta$ which implies that

$$\sup_{s<u} \inf_{x,y\in B_j} F_{x,y}(s) \geq \rho - \delta$$

and therefore

$$\alpha_{\{x_n|n\geq n_1(\varepsilon,\widetilde{\delta})\}}(u) \geq \rho - \delta.$$

Using the relation $\alpha_{\{x_n|n\in\mathbb{N}\}}(u) = \alpha_{\{x_n|n\geq n_1(\varepsilon,\tilde{\delta})\}}(u)$ we obtain $\alpha_{\{x_n|n\in\mathbb{N}\}}(u) \geq \rho - \delta$ for every $\delta > 0$, and therefore $\alpha_{\{x_n|n\in\mathbb{N}\}}(u) \geq \rho$. Hence $\alpha_{\{y_n|n\in\mathbb{N}\}}(u) \leq \alpha_{\{x_n|n\in\mathbb{N}\}}(u)$ for every $u > 0$ and analogously $\alpha_{\{x_n|n\in\mathbb{N}\}}(u) \leq \alpha_{\{y_n|n\in\mathbb{N}\}}(u)$, for every $u > 0$. This means that (4.20) holds for $\gamma = \alpha$. □

Proof of Theorem 4.17. Let $B = \{x_n \mid n \in \mathbb{N}\}$, where the sequence $(x_n)_{n\in\mathbb{N}}$ is from condition (i), and suppose that the function γ is β. We shall prove that for every $u > 0$

$$\beta_{f(B)}(u) \leq \beta_B(u). \tag{4.23}$$

From (i) and Lemma 4.18 we have that $\beta_B(u) = \beta_{\{y_n|n\in\mathbb{N}\}}(u)$, $u > 0$.
Since $y_n \in fx_n$ $(n \in \mathbb{N})$ we have

$$\beta_{f(B)}(u) \leq \beta_{\{y_n|n\in\mathbb{N}\}}(u) = \beta_B(u) \text{ for every } u > 0.$$

Since the mapping f is densifying in respect to the function β, there exists a convergent subsequence $(x_{n_k})_{k\in\mathbb{N}}$, and suppose that $\lim\limits_{k\to\infty} x_{n_k} = x \in \bar{K}$. Then from condition (i) it follows that for every $\varepsilon > 0$, $\lim\limits_{k\to\infty} F_{x_{n_k},y_{n_k}}(\varepsilon) = 1$, and therefore $y_{n_k} \to x$ $(k \to \infty)$. The mapping f is closed and from the relation $y_{n_k} \in fx_{n_k}$ $(k \in \mathbb{N})$ it follows that $x \in fx$.

Analogously as in the case when $\gamma = \beta$ we obtain that $\alpha_{f(B)}(u) \leq \alpha_B(u)$, for every $u > 0$. This implies that $\overline{B}$ is compact and therefore there exists $x \in \overline{K}$ such that $x \in fx$. □

Corollary 4.19 *Let $(S, \mathcal{F}, T)$ be a complete Menger space with a continuous t-norm $T, f : S \to CB(S)$ a closed mapping and there exists a nonempty probabilistic bounded subset K of S such that $f(K)$ is probabilistic bounded and the following conditions are satisfied:*

(i)

$$\inf_{x\in K} \inf_{y\in fx} \sup\{u \mid u > 0,\ F_{x,y}(u) \leq 1 - u\} = 0;$$

(ii) the mapping f is densifying on K in respect to the function γ, where $\gamma \in \{\alpha, \beta\}$.

Then there exists at least one element $x \in \overline{K}$ such that $x \in fx$.

Proof. It is obvious that (i) implies that for every $n \in \mathbb{N}$ there exists $x_n \in K$ and $y_n \in fx_n$ such that

$$\sup\{u \mid u > 0,\ F_{x_n,y_n}(u) \leq 1 - u\} < 2^{-n}.$$

Hence for every $n \in \mathbb{N}$ we have that $F_{x_n,y_n}(2^{-n}) > 1 - 2^{-n}$. Then for every $\varepsilon > 0$ we have $\lim\limits_{n\to\infty} F_{x_n,y_n}(\varepsilon) = 1$, and the condition (i) from Theorem 4.17 is satisfied.

Indeed, suppose that $\varepsilon > 0$ and $\delta \in (0,1)$. Then there exists $n_0(\delta) \in \mathbb{N}$ such that $2^{-n} < \delta$, for every $n > n_0(\delta)$ and let $n_1(\varepsilon) \in \mathbb{N}$ be such that $2^{-n} < \varepsilon$, for every $n > n_1(\varepsilon)$. If $n(\varepsilon,\delta) = \max\{n_0(\delta), n_1(\varepsilon)\}$, then for every $n > n(\varepsilon,\delta)$ we have that

$$F_{x_n,y_n}(\varepsilon) \geq F_{x_n,y_n}(2^{-n}) > 1 - 2^{-n} > 1 - \delta,$$

which implies that $\lim_{n\to\infty} F_{x_n,y_n}(\varepsilon) = 1$. □

From Theorem 4.17 we obtain Theorem 4.16 as a corollary.

We shall recall the notion of the upper semi-continuous mapping. Let X and Y be two Hausdorff topological spaces. A set valued (multi-valued) mapping $F : X \to 2^Y$ is said to be *upper semi-continuous at a given point* $x_0 \in X$ if and only if for each open subset W of Y such that $F(x_0) \subseteq W$ there exists a neighbourhood V of x_0 such that $F(V) \subseteq W$.

Corollary 4.20 *Let (X,d) be a complete metric space, $f : X \to CB(X)$ an upper semi-continuous mapping, K a non-empty bounded subset of X such that $f(K)$ is bounded, and the following conditions are satisfied:*

1. $\inf_{x\in K} d(x, fx) = 0$;
2. *for every $A \subset K$ the following implication holds:*

$$\alpha(A) > 0 \quad \Rightarrow \quad \alpha(f(A)) < \alpha(A).$$

Then there exists at least one element $x \in \overline{K}$ such that $x \in fx$.

Proof. We shall prove that all the conditions of Theorem 4.17 are satisfied. We know that (X,d) may be considered as the Menger space $(X,\mathcal{F},T_{\mathbf{M}})$, where the mapping $\mathcal{F}$ is defined by

$$F_{x,y}(u) = \begin{cases} 0, & d(x,y) \geq u \\ 1, & d(x,y) < u, \end{cases} \qquad u \geq 0,\ x,y \in X,$$

and that the (ε,λ)-topology is the same as the topology induced by the metric d. It is easy to see that condition 1 implies the existence of two sequences $(x_n)_{n\in\mathbb{N}}$ ($x_n \in K$, $n \in \mathbb{N}$) and $(y_n)_{n\in\mathbb{N}}$ such that $y_n \in fx_n$ ($n \in \mathbb{N}$) and $\lim_{n\to\infty} d(x_n,y_n) = 0$. Hence from the definition of the mapping $F_{x_n,y_n}(\cdot)$ it follows that for every $u > 0$ there exists $n_0(u) \in \mathbb{N}$ such that $F_{x_n,y_n}(u) = 1$ for every $n > n_0(u)$, and therefore $\lim_{n\to\infty} F_{x_n,y_n}(u) = 1$.

It remains to prove that the mapping f is closed and densifying with respect to the Kuratowski function α. It is easy to prove that if (X,d) is a metric space and $(X,\mathcal{F},T_{\mathbf{M}})$ is the associated probabilistic metric space, for every bounded set $A \subset X$ there holds

$$\alpha_A(t) = H_0(t - \alpha(A)) \ (t > 0).$$

Suppose that for every $u > 0$ and some $A \subset K$

$$\alpha_{f(A)}(u) \leq \alpha_A(u)$$

and that $\overline{A}$ is not compact. Since X is complete it follows that $\alpha(A) > 0$ and therefore from the condition 2 it follows that $\alpha(f(A)) < \alpha(A)$. This implies that for every $u \in \mathbb{R}$

$$u - \alpha(f(A)) > u - \alpha(A)$$

and therefore

$$H_0(u - \alpha(f(A))) \geq H_0(u - \alpha(A)).$$

Since for $u = \alpha(A)$ we have that

$$H_0(u - \alpha(f(A)) = 1 > H_0(0) = 0$$

we conclude that $\alpha_{f(A)}(u) > \alpha_A(u)$ which is a contradiction. Since the mapping f is upper semi-continuous and $f(x)$ is closed for every $x \in X$, it follows that the mapping f is closed. □

4.5 Fixed point theorems in probabilistic metric spaces with a convex structure

W. Takahashi introduced in [297] the notion of a metric space with a convex structure. This class of metric spaces includes normed linear spaces and metric spaces of the hyperbolic type (see books [282] and [285]). Iterative processes on metric spaces of the hyperbolic type are investigated by W.A. Kirk and K. Goebel ([282, 285]). Some fixed point theorems in such spaces are proved in [147, 206, 254, 282, 285, 297, 298].

In this section we give a generalization of the Takahashi definition of metric spaces with a convex structure in [297] to the case of Menger spaces. A fixed point theorem for a probabilistic non-expansive multi-valued mappings of Nadler's type in such a kind of probabilistic metric spaces is proved.

Let us recall that a metric space (S, d) is with a convex structure in the sense of Takahashi, if there exists a mapping $W : S \times S \times [0, 1] \to S$ such that for every $(u, x, y, \delta) \in S \times S \times S \times [0, 1]$

$$d(u, W(x, y, \delta)) \leq \delta d(u, x) + (1 - \delta)d(u, y).$$

Definition 4.21 *Let $(S, \mathcal{F}, T)$ be a Menger space. A mapping $W : S \times S \times [0, 1] \to S$ is said to be a convex structure on S if for every $(x, y) \in S \times S$*

$$W(x, y, 0) = y, W(x, y, 1) = x \text{ and for every } \delta \in (0, 1),\ u \in S,\ \varepsilon > 0$$

$$F_{u,W(x,y,\delta)}(2\varepsilon) \geq T\left(F_{u,x}\left(\frac{\varepsilon}{\delta}\right), F_{u,y}\left(\frac{\varepsilon}{1-\delta}\right)\right).$$

It is easy to see that every metric space (S, d) with a convex structure W can be considered as a Menger space $(S, \mathcal{F}, T_{\mathbf{M}})$ (the associated Menger space) with the same function W.

A random normed space is a Menger space with the convex structure $W(x, y, \delta) = \delta x + (1-\delta)y$ ($x, y \in S$ and $\delta \in [0,1]$).

A nontrivial example of a Menger space with a convex structure is the following one.

Example 4.22 Let (M, d) be a separable metric space with a convex structure W which has the property that for every $\delta \in [0,1]$ the mapping $(x, y) \mapsto W(x, y, \delta)$ is continuous on $M \times M$. Let $(\Omega, \mathcal{A}, P)$ be a probability space. We shall prove that the Menger space $(S, \mathcal{F}, T_{\mathbf{L}})$ has a convex structure, where S is the space of all measurable mappings from Ω into M (the space of equivalence classes) and for every $X, Y \in S$

$$F_{X,Y}(u) = P\Big(\{\omega \mid \omega \in \Omega, d(X(\omega), Y(\omega)) < u\}\Big), \quad u \in \mathbb{R}.$$

Let $\overline{W} : S \times S \times [0,1] \to S$ be defined by the relation

$$\overline{W}(X, Y, \delta)(\omega) = W(X(\omega), Y(\omega), \delta) \quad \text{for every } \omega \in \Omega \text{ and every } X, Y \in S$$

and every $\delta \in [0,1]$.

Since X and Y are measurable mappings and the mapping $(x, y) \mapsto W(x, y, \delta)$ is continuous for every fixed $\delta \in [0,1]$, it follows that for every $X, Y \in S$, $\overline{W}(X, Y, \delta) \in S$.

Using the definition of the mapping $F_{X,Y}(\cdot)$ and the property of the probability P that for every $A, B \in \mathcal{A}$

$$P(A \cap B) = P(A) + P(B) - P(A \cup B)$$

it is easy to prove that

$$F_{U,\overline{W}(X,Y,\delta)}(2\varepsilon) \geq F_{U,X}\left(\frac{\varepsilon}{\delta}\right) + F_{U,Y}\left(\frac{\varepsilon}{1-\delta}\right) - 1 \quad (X, Y, U \in S, \delta \in (0,1), \varepsilon > 0).$$

Since

$$\overline{W}(X, Y, 0)(\omega) = W(X(\omega), Y(\omega), 0) = Y(\omega)$$

and

$$\overline{W}(X, Y, 1)(\omega) = W(X(\omega), Y(\omega), 1) = X(\omega),$$

for every $\omega \in \Omega$, it follows that the mapping $\overline{W}$ is a convex structure on the probabilistic metric space $(S, \mathcal{F}, T_{\mathbf{L}})$.

In the following we assume that a convex structure W on a Menger space $(S, \mathcal{F}, T)$ satisfies the condition

$$F_{W(x,z,\delta),W(y,z,\delta)}(\varepsilon\delta) \geq F_{x,y}(\varepsilon) \quad \text{for every } (x,y,z) \in S \times S \times S, \tag{4.24}$$

for every $\varepsilon > 0$ and $\delta \in (0,1)$. A similar condition for metric spaces with a convex structure was introduced in Itoh's paper [147]. In a random normed space this condition (4.24) is satisfied.

The notion of a W-convex subset of a Menger space with the convex structure W can be introduced analogously, as in a normed space.

Definition 4.23 *Let $(S, \mathcal{F}, T)$ be a Menger space with a convex structure W. We say that a subset M of S is W-convex if for every $x, y \in M$ and every $\delta \in (0,1)$: $W(x,y,\delta) \in M$.*

Lemma 4.24 *Let $(S, \mathcal{F}, T)$ be a Menger space with a convex structure W, M a nonempty, probabilistic bounded and W-convex subset of S and $\delta \in (0,1)$. Then, for every $B \subset M$, $u > 0$, and $\gamma \in \{\alpha, \beta\}$*

$$\gamma_{W(B,x,\delta)}(u) \geq \gamma_B\left(\frac{u}{\delta}\right),$$

where x is an arbitrary element from the set M.

Proof. Let $\gamma = \alpha$, $u > 0, \delta \in (0,1)$, $B \subset M$ and $x \in M$. Suppose that $\alpha_B(u/\delta) > 0$ and $0 < \rho < \alpha_B(u/\delta)$. From the definition of the function α it follows that there exist $A_1, A_2, \ldots, A_n \subset S$ such that

$$B = \bigcup_{i=1}^{n} A_i, \quad D_{A_i}\left(\frac{u}{\delta}\right) \geq \rho \text{ for every } i \in \{1, 2, \ldots, n\}.$$

Since $W(B, x, \delta) = \bigcup_{i=1}^{n} C_i$ where $C_i = W(A_i, x, \delta)$ for every $i \in \{1, 2, \ldots, n\}$, it remains to prove that $D_{C_i}(u) \geq \rho$ for every $i \in \{1, 2, \ldots, n\}$. We have that for every $i \in \{1, 2, \ldots, n\}$

$$\begin{aligned} D_{C_i}(u) &= D_{W(A_i,x,\delta)}(u) \\ &= \sup_{s<u} \inf_{v,w \in W(A_i,x,\delta)} F_{v,w}(s) \\ &= \sup_{s<u} \inf_{\bar{v},\bar{w} \in A_i} F_{W(\bar{v},x,\delta),W(\bar{w},x,\delta)}\left(\frac{s\delta}{\delta}\right). \end{aligned}$$

Using relation (4.24) we obtain

$$D_{C_i}(u) \geq \sup_{s<u} \inf_{v,w \in A_i} F_{v,w}\left(\frac{s}{\delta}\right) = D_{A_i}\left(\frac{u}{\delta}\right) \geq \rho.$$

This implies that

$$\Big\{\rho \mid \rho > 0, \exists A_1, A_2, \dots, A_n \subset S, B = \bigcup_{i=1}^{n} A_i, D_{A_i}\Big(\frac{u}{\delta}\Big) \geq \rho, \forall i \in \{1, 2, \dots, n\}\Big\}$$

is a subset of the set

$$\Big\{\rho \mid \rho > 0, \exists C_1, C_2, \dots, C_n \subset S, W(B, x, \delta) = \bigcup_{i=1}^{n} C_i, D_{C_i}(u) \geq \rho, \forall i \in \{1, 2, \dots, n\}\Big\}$$

and therefore for every $u > 0$, $\alpha_B\Big(\frac{u}{\delta}\Big) \leq \alpha_{W(B,x,\delta)}(u)$.

Suppose now that $\gamma = \beta$. Let $\rho > 0$ be such that $\rho < \beta_B\Big(\frac{u}{\delta}\Big)$. Then there exists a finite subset A_f of S such that

$$\mathbf{F}_{B,A_f}\Big(\frac{u}{\delta}\Big) \geq \rho \quad \text{, i.e.,} \quad \sup_{s<u/\delta} \inf_{v \in B} \sup_{w \in A_f} F_{v,w}(s) \geq \rho.$$

Furthermore, from (4.24) we have

$$\begin{aligned}
\mathbf{F}_{W(B,x,\delta),W(A_f,x,\delta)}(u) &= \sup_{s<u} \inf_{v \in W(B,x,\delta)} \sup_{w \in W(A_f,x,\delta)} F_{v,w}(s) \\
&= \sup_{s<u} \inf_{\bar{v} \in B} \sup_{\bar{w} \in A_f} F_{W(\bar{v},x,\delta),W(\bar{w},x,\delta)}\Big(\frac{s\delta}{\delta}\Big) \\
&\geq \sup_{s<u} \inf_{v \in B} \sup_{w \in A_f} F_{v,w}\Big(\frac{s}{\delta}\Big) \\
&= \mathbf{F}_{B,A_f}\Big(\frac{u}{\delta}\Big) \\
&\geq \rho
\end{aligned}$$

and therefore $\beta_B\Big(\frac{u}{\delta}\Big) \leq \beta_{W(B,x,\delta)}(u)$, for every $u > 0$. □

Theorem 4.25 is a generalization of Theorem 1 from [147].

Theorem 4.25 *Let $(S, \mathcal{F}, T)$ be a complete Menger space with a convex structure W and a continuous t-norm T, M a non-empty closed probabilistic bounded and W-convex subset of S, and $f : M \to \mathrm{Com}(M)$ a probabilistic multi-valued non-expansive mapping of Nadler's type and such that for at least one function $\gamma \in \{\alpha, \beta\}$ the following implication holds:*

For every $B \subset M$

$$\gamma_{f(B)}(u) \leq \gamma_B(u) \quad \text{for every } u > 0 \quad \Rightarrow \quad \overline{B} \text{ is compact.} \tag{4.25}$$

Then there exists $x \in M$ such that $x \in fx$.

Proof. Let $(k_n)_{n\in\mathbb{N}}$ be a sequence from the interval $(0,1)$ such that $\lim_{n\to\infty} k_n = 1$. Let x_0 be an element from M. For every $n \in \mathbb{N}$ and every $x \in M$ let

$$f_n x = W(fx, x_0, k_n).$$

Since M is a W-convex subset of S it follows that $f_n x \subset M$ $(x \in M)$. We shall prove that the mapping f_n satisfies all the conditions of Theorem 4.13. This will imply that for every $n \in \mathbb{N}$ there exists $x_n \in M$ such that $x_n \in f_n x_n$. Since W is continuous in respect to the first variable it follows from the definition of $f_n x$ that $f_n x \in C(M)$. Let us prove that f_n is a probabilistic Nadler k_n-contraction. Let $u, v \in M$, $x \in f_n u$, and $\delta > 0$. Since $x \in f_n u$ it follows that there exists $z \in fu$ such that $x = W(z, x_0, k_n)$. Since f is a probabilistic multi-valued non-expansive mapping of Nadler's type there exists $y' \in fv$ such that for every $\varepsilon > 0$

$$F_{y',z}(\varepsilon) \geq F_{v,u}\Big(\varepsilon - \frac{\delta}{k_n}\Big)$$

and let $y = W(y', x_0, k_n) \in f_n v$. Then from (4.24) we have that

$$\begin{aligned} F_{y,x}(\varepsilon) &= F_{W(y',x_0,k_n),W(z,x_0,k_n)}\Big(k_n \frac{\varepsilon}{k_n}\Big) \\ &\geq F_{y',z}\Big(\frac{\varepsilon}{k_n}\Big) \\ &\geq F_{v,u}\Big(\frac{\varepsilon-\delta}{k_n}\Big) \quad \text{for every } \varepsilon > 0, \end{aligned}$$

which means that f_n is a probabilistic Nadler k_n-contraction. We prove that for every $n \in \mathbb{N}$ the mapping f_n is weakly demicompact. Suppose that $(z_m)_{m\in\mathbb{N}}$ is a sequence from M such that $z_{m+1} \in f_n z_m$ $(m \in \mathbb{N})$ and that $\lim_{m\to\infty} F_{z_m,z_{m+1}}(\varepsilon) = 1$ for every $\varepsilon > 0$. If we prove that f_n is densifying in respect to γ then, as in Theorem 4.17, we shall obtain that the set $\{z_m \mid m \in \mathbb{N}\}$ is compact, which means that f_n is weakly demicompact. Suppose that for some $B \subset M : \gamma_{f_n(B)}(u) < \gamma_B(u)$, for every $u > 0$. Since $f_n B = W(f(B), x_0, k_n)$ this means that

$$\gamma_{W(f(B),x_0,k_n)}(u) < \gamma_B(u) \quad \text{for every } u > 0.$$

From Lemma 4.24 it follows that $\gamma_{W(f(B),x_0,k_n)}(u) \geq \gamma_{f(B)}\Big(\dfrac{u}{k_n}\Big)$ and since $k_n \in (0,1)$ from $\dfrac{u}{k_n} > u$ we obtain that

$$\gamma_B(u) > \gamma_{W(f(B),x_0,k_n)}(u) \geq \gamma_{f(B)}(u).$$

Since f is densifying on M with respect to γ, it follows that $\overline{B}$ is compact and hence f_n is weakly demicompact. From Theorem 4.13 it follows that there exists

$x_n \in M$ such that $x_n \in f_n x_n$ $(n \in \mathbb{N})$. In order to apply Theorem 4.17 we shall prove that there exists a sequence $(y_n)_{n\in\mathbb{N}}$ from the set M such that $y_n \in f x_n$ $(n \in \mathbb{N})$ and $\lim_{n\to\infty} F_{x_n,y_n}(\varepsilon) = 1$ for every $\varepsilon > 0$. From $x_n \in f_n x_n$ for every $n \in \mathbb{N}$, it follows that there exists $y_n \in f x_n$ $(n \in \mathbb{N})$ such that $x_n = W(y_n, x_0, k_n)$ $(n \in \mathbb{N})$ and therefore for every $\varepsilon > 0$

$$\begin{aligned} F_{x_n,y_n}(\varepsilon) &= F_{W(y_n,x_0,k_n),y_n}(\varepsilon) \\ &\geq T\Big(F_{y_n,y_n}\Big(\frac{\varepsilon}{2k_n}\Big), F_{y_n,x_0}\Big(\frac{\varepsilon}{2(1-k_n)}\Big)\Big) \\ &= T\Big(1, F_{y_n,x_0}\Big(\frac{\varepsilon}{2(1-k_n)}\Big)\Big) \\ &= F_{y_n,x_0}\Big(\frac{\varepsilon}{2(1-k_n)}\Big). \end{aligned}$$

Since the set M is probabilistic bounded $\sup_{x\in\mathbb{R}} D_M(x) = 1$, and so $y_n, x_0 \in M$ implies that $\lim_{n\to\infty} F_{x_n,y_n}(\varepsilon) = 1$. If we prove that the mapping f is closed, from Theorem 4.17 it follows that there exists $x \in M$ such that $x \in fx$.

Suppose that $(x_n)_{n\in\mathbb{N}}$ and $(y_n)_{n\in\mathbb{N}}$ are sequences from M such that $y_n \in f x_n$ $(n \in \mathbb{N})$ and $\lim_{n\to\infty} x_n = x$, $\lim_{n\to\infty} y_n = y$. We shall prove that $y \in fx$. Since fx is closed, it is enough to prove that $y \in \overline{fx}$. Let $\varepsilon > 0$ and $\lambda \in (0,1)$. We prove that there exists $b \in fx$ such that $b \in N_y(\varepsilon, \lambda)$. Let $u = x_n, v = x$ and $\delta = \varepsilon/4$. Since $y_n \in f x_n$ and the mapping f is a multi-valued probabilistic non-expansive mapping of Nadler's type, there exists $b_n \in fx$ such that

$$F_{y_n,b_n}\Big(\frac{\varepsilon}{2}\Big) \geq F_{x_n,x}\Big(\frac{\varepsilon}{4}\Big) \quad \text{for every } n \in \mathbb{N}.$$

We have that

$$\begin{aligned} F_{y,b_n}(\varepsilon) &\geq T\Big(F_{y,y_n}\Big(\frac{\varepsilon}{2}\Big), F_{y_n,b_n}\Big(\frac{\varepsilon}{2}\Big)\Big) \\ &\geq T\Big(F_{y,y_n}\Big(\frac{\varepsilon}{2}\Big), F_{x_n,x}\Big(\frac{\varepsilon}{4}\Big)\Big). \end{aligned}$$

Since the mapping T is continuous it is easy to see that there exists $n_0(\varepsilon, \lambda) \in \mathbb{N}$ such that $F_{y,b_{n_0}}(\varepsilon) > 1 - \lambda$ and therefore $b_{n_0} \in N_y(\varepsilon, \lambda)$. □

Remark 4.26 It is obvious from the proof that Theorem 4.25 holds, if we suppose that $(S, \mathcal{F}, T)$ is a Menger space with a continuous t-norm T and M is a complete, probabilistic bounded and nonempty subset of S which satisfies an additional condition: M is a W-star-shaped subset of S which satisfies probabilistic condition I. This means that there exists an element x_0 from M, the so called star center of M, and a mapping $W : M \times \{x_0\} \times [0,1] \to M$ such that for every $x, y \in M$, $\delta \in (0,1), \varepsilon > 0$

$$F_{y,W(x,x_0,\delta)}(2\varepsilon) \geq T\left(F_{y,x}\left(\frac{\varepsilon}{\delta}\right), F_{y,x_0}\left(\frac{\varepsilon}{1-\delta}\right)\right),$$

and

$$F_{W(x,x_0,\delta),W(y,x_0,\delta)}(\varepsilon\delta) \geq F_{x,y}(\varepsilon).$$

Similar conditions in metric spaces were introduced by Itoh in [147].

Corollary 4.27 *Let (X, d) be a bounded, complete, W-star-shaped metric space satisfying condition I. Let f be a non-expansive, condensing mapping of X into* Com(X). *Then there exists $x \in X$ such that $x \in fx$.*

Proof. Since every multi-valued non-expansive mapping is a multi-valued probabilistic non-expansive mapping of Nadler's type in the associated Menger space $(S, \mathcal{F}, T_{\mathbf{M}})$, the proof follows immediately. □

Corollary 4.28 *Let $(S, \mathcal{F}, T)$ be a complete random normed space with a continuous t-norm T, M a nonempty, closed, convex and probabilistic bounded subset of S, and f as in Theorem 4.25. Then there exists $x \in M$ such that $x \in fx$.*

Proof. Every random normed space is a Menger space with the convex structure W defined by $W(x, y, \delta) = \delta x + (1 - \delta)y \ \ (x, y, \in S, \ \delta \in [0, 1])$. Hence M is convex if and only if it is W-convex. □

If f is a single-valued mapping the following fixed point theorem for non-expansive mapping is proved in [99].

Theorem 4.29 *Let $(S, \mathcal{F}, T)$ be a complete Menger space with a convex structure W and a continuous t-norm T, M a closed and W-star-shaped subset of S and $f : M \to M$ a non-expansive mapping such that $f(M)$ is probabilistic bounded. If there exists $m \in \mathbb{N}$ such that f^m is densifying on the set $W(f(M), x_0, (0, 1))$ with respect to $\gamma \in \{\alpha, \beta\}$, where x_0 is the star center of M, then there exists $x \in M$ such that $x = fx$.*

Proof. Let $(k_n)_{n\in\mathbb{N}}$ be a sequence of numbers from the interval $(0, 1)$ such that $\lim\limits_{n\to\infty} k_n = 1$ and for every $n \in \mathbb{N}$ and $x \in M$

$$f_n x = W(fx, x_0, k_n).$$

It is easy to see that $f_n : M \to M$ is probabilistic k_n-contraction and that $f_n(M)$ is a probabilistic bounded subset of M, for every $n \in \mathbb{N}$. This implies that Fix $(f_n) \neq \emptyset$, for every $n \in \mathbb{N}$ and let $x_n = f_n(x_n) = W(f(x_n), x_0, k_n)$, for every $n \in \mathbb{N}$. As in Theorem 4.25, for every $\varepsilon > 0$

$$\lim_{n\to\infty} F_{x_n,f(x_n)}(\varepsilon) = 1. \tag{4.26}$$

It is obvious that $x_n \in W(f(M), x_0, (0, 1))$, for every $n \in \mathbb{N}$. Since

$$F_{x_n,f^m(x_n)}(\varepsilon) \geq \underbrace{T\Big(\ldots\Big(T}_{(\mathrm{m}-2)-\mathrm{times}}\Big(F_{x_n,f(x_n)}\Big(\frac{\varepsilon}{2}\Big), F_{x_n,f(x_n)}\Big(\frac{\varepsilon}{2^2}\Big)\Big), \ldots, F_{x_n,f(x_n)}\Big(\frac{\varepsilon}{2^{m-1}}\Big)\Big)$$

by the continuity of T and (4.26) we conclude that for every $\varepsilon > 0$

$$\lim_{n\to\infty} F_{x_n,f^m(x_n)}(\varepsilon) = 1. \tag{4.27}$$

By Lemma 4.18

$$\gamma_{\{x_n|n\in\mathbb{N}\}}(\varepsilon) = \gamma_{\{f^m(x_n)|n\in\mathbb{N}\}}(\varepsilon) \text{ for every } \varepsilon > 0,$$

and since f^m is γ-densifying on $W(f(M), x_0, (0,1))$ we have that $\overline{\{x_n \mid n \in \mathbb{N}\}}$ is a compact subset of M. Hence there exists a convergent subsequence $(x_{n_k})_{k\in\mathbb{N}}$. If $\lim_{k\to\infty} x_{n_k} = x$, the continuity of f and (4.26) imply that $x = fx$. □

As a consequence we obtain the non-probabilistic result from [75].

Corollary 4.30 *Let $(X, \|\cdot\|)$ be a Banach space, M a closed and star-shaped subset of X, $f : M \to M$ a non-expansive mapping such that $f(M)$ is bounded and for some $m \in \mathbb{N}$, $\overline{f^m(M)}$ is compact. Then there exists $x \in M$ such that $x = fx$.*

4.6 A common fixed point theorem for sequence of mappings

Definition 4.31 *Let $(S, \mathcal{F})$ be a probabilistic metric space and $L_n : S \to 2^S, n \in \mathbb{N}$. The family $(L_n)_{n\in\mathbb{N}}$ is demicompact if for every sequence $(x_n)_{n\in\mathbb{N}}$ in S, such that $x_{n+1} \in L_n x_n (n \in \mathbb{N})$, the following implication holds:*

$$\lim_{n\to\infty} F_{x_{n+1},x_n}(\varepsilon) = 1, \forall \varepsilon > 0 \Rightarrow \textit{ there exists a convergent subsequence } (x_{n_k})_{k\in\mathbb{N}}.$$

If A is a closed and probabilistic bounded subset of S, i.e.,

$$\sup_{s>0} (\sup_{u<s} \inf_{p,q\in A} F_{p,q}(u)) = 1,$$

then *the probabilistic distance between A and $x \in S$* is the function $F_{x,A}$ defined by

$$F_{x,A}(s) = \sup_{u<s} \sup_{y\in A} F_{x,y}(u) \text{ for every } s \geq 0.$$

It is easy to prove that in a Menger space $(S, \mathcal{F}, T)$ with a continuous t-norm T the following inequality holds for every $u_1, u_2 > 0$ and every $y \in S$

$$F_{x,A}(u_1 + u_2) \geq T(F_{x,y}(u_1), F_{y,A}(u_2)),$$

and that $F_{x,A}(s) = 1$ for every $s > 0$ if and only if $x \in A$. For every $A, B \in CB(S)$ and $x \in A$, $F_{x,B}(s) \geq \widetilde{F}_{A,B}(s)$ for every $s \geq 0$.

Theorem 4.32 *Let $(S, \mathcal{F}, T)$ be a complete Menger space such that t-norm T is continuous, $L_i : S \to CB(S)$ for every $i \in \mathbb{N}$, and the following condition is satisfied:*

(i) there exists $0 < k < 1$ such that for every $i, j \in \mathbb{N}$, $i \neq j$, and every $x, y \in S$

$$\widetilde{F}_{L_i x, L_j y}(ks) \geq F_{x,y}(s) \quad \text{for every} \quad s \geq 0,$$

and for every $x \in S$, every $n \in \mathbb{N}$, every $a \in L_n x$, and every $\delta > 0$, there exists $b \in L_{n+1} a$ such that

$$F_{a,b}(s) \geq \widetilde{F}_{L_n x, L_{n+1} a}(s - \delta) \text{ for every } s \geq 0.$$

If (ii) or (iii) is satisfied then there exists $x^* \in S$ such that $x^* \in \bigcap_{n \in \mathbb{N}} L_n x^*$, where

(ii) The family $(L_n)_{n \in \mathbb{N}}$ is demicompact.

(iii) There exists $x_0 \in S$, $x_1 \in L_1 x_0$ and $\mu \in (k, 1)$ such that

$$\lim_{n \to \infty} \mathop{\mathrm{T}}_{i=n}^{\infty} F_{x_0, x_1}\left(\frac{1}{\mu^i}\right) = 1.$$

Proof. Let $x_0 \in S$ and $x_1 \in L_1 x_0$. If $\delta = k$ there exists $x_2 \in L_2 x_1$ such that

$$F_{x_1, x_2}(s) \geq \widetilde{F}_{L_1 x_0, L_2 x_1}(s - k) \geq F_{x_0, x_1}\left(\frac{s - k}{k}\right) \quad \text{for every } s \geq 0.$$

Analogously there exists $x_3 \in L_3 x_2$ such that for every $s \geq 0$

$$\begin{aligned}
F_{x_2, x_3}(s) &\geq \widetilde{F}_{L_2 x_1, L_3 x_2}(s - k^2) \\
&\geq F_{x_1, x_2}\left(\frac{s - k^2}{k}\right) \\
&\geq F_{x_0, x_1}\left(\frac{\frac{s - k^2}{k} - k}{k}\right) \\
&= F_{x_0, x_1}\left(\frac{s - 2k^2}{k^2}\right).
\end{aligned}$$

Continuing in this way we obtain a sequence $(x_n)_{n \in \mathbb{N}}$ in S such that:

a) $x_n \in L_n x_{n-1}$ for every $n \in \mathbb{N}$;

b) $F_{x_n, x_{n+1}}(s) \geq F_{x_{n-1}, x_n}\left(\dfrac{s - k^n}{k^n}\right)$ for every $s \geq 0$ and every $n \in \mathbb{N}$.

From b) it follows that

$$F_{x_n,x_{n+1}}(s) \geq F_{x_0,x_1}\left(\frac{s-nk^n}{k^n}\right) \text{ for every } s \geq 0, \tag{4.28}$$

and therefore for every $s > 0$

$$\lim_{n\to\infty} F_{x_n,x_{n+1}}(s) = 1. \tag{4.29}$$

If the family of functions $(L_n)_{n\in\mathbb{N}}$ is demicompact, since $x_{n+1} \in L_{n+1}x_n$ $(n \in \mathbb{N})$, from (4.29) we conclude that there exists a convergent subsequence $(x_{n_k})_{k\in\mathbb{N}}$.

Suppose that (iii) is satisfied. Similarly as in Theorem 4.13 it can be proved that $(x_n)_{n\in\mathbb{N}}$ is a Cauchy sequence, i.e., that for every $\varepsilon > 0$ and $\lambda \in (0,1)$ there exists $n(\varepsilon,\lambda) \in \mathbb{N}$ such that

$$F_{x_{n+p},x_n}(\varepsilon) > 1-\lambda \text{ for every } n \geq n(\varepsilon,\lambda) \text{ and every } p \in \mathbb{N}.$$

Since S is complete we conclude that in both cases (ii) and (iii) there exists $x^* = \lim\limits_{n\to\infty} x_n$. We shall prove that $x^* \in L_i x^*$, for every $i \in \mathbb{N}$. Let $\alpha > 1$ and $i \neq n_k + 1$. Then

$$\begin{aligned} F_{x^*,L_ix^*}(\varepsilon) &\geq T\left(F_{x^*,x_{n_k+1}}\left(\left(1-\frac{1}{\alpha}\right)\varepsilon\right), F_{x_{n_k+1},L_ix^*}\left(\frac{\varepsilon}{\alpha}\right)\right) \\ &\geq T\left(F_{x^*,x_{n_k+1}}\left(\left(1-\frac{1}{\alpha}\right)\varepsilon\right), \widetilde{F}_{L_{n_k+1}x_{n_k},L_ix^*}\left(\frac{\varepsilon}{\alpha}\right)\right) \\ &\geq T\left(F_{x^*,x_{n_k+1}}\left(\left(1-\frac{1}{\alpha}\right)\varepsilon\right), F_{x_{n_k},x^*}\left(\frac{\varepsilon}{k\cdot\alpha}\right)\right). \end{aligned}$$

Since $\lim\limits_{k\to\infty} x_{n_k} = x^*$ (4.29) implies $\lim\limits_{k\to\infty} x_{n_k+1} = x^*$, and from the continuity of T we obtain that

$$\begin{aligned} F_{x^*,L_ix^*}(\varepsilon) &\geq \lim_{k\to\infty} T\left(F_{x^*,x_{n_k+1}}\left(\left(1-\frac{1}{\alpha}\right)\varepsilon\right), F_{x_{n_k},x^*}\left(\frac{\varepsilon}{k\alpha}\right)\right) \\ &= T(1,1) = 1. \end{aligned}$$

Hence $x^* \in L_i x^*$ for every $i \in \mathbb{N}$. □

Corollary 4.33 *Let $(S,\mathcal{F},T)$ be a complete Menger space such that t-norm T is continuous and $L_i : S \to CB(S)$ for every $i \in \mathbb{N}$, and the following conditions are satisfied:*

(i) there exists $0 < k < 1$ such that for any $i, j \in \mathbb{N}$, $i \neq j$, and any $x, y \in S$

$$\widetilde{F}_{L_ix,L_jy}(ks) \geq F_{x,y}(s) \quad \textit{for every} \quad s \geq 0,$$

and for any $x \in S$, any $n \in \mathbb{N}$, any $a \in L_n x$ and any $\delta > 0$, there exists $b \in L_{n+1}a$ such that

$$F_{a,b}(s) \geq \widetilde{F}_{L_n x, L_{n+1}a}(s - \delta) \ \ \text{for every} \ \ s \geq 0;$$

(ii) T is of H-type.

Then there exists $x^* \in S$ such that $x^* \in \bigcap_{n \in \mathbb{N}} L_n x^*$.

Corollary 4.34 *Let $(S, \mathcal{F}, T)$ be a complete Menger space such that the t-norm T is continuous, $L_i : S \to CB(S)$ for every $i \in \mathbb{N}$ and the following conditions are satisfied:*

(i) there exists $0 < k < 1$ such that for any $i, j \in \mathbb{N}$, $i \neq j$, and any $x, y \in S$

$$\widetilde{F}_{L_i x, L_j y}(ks) \geq F_{x,y}(s) \quad \text{for every} \quad s \geq 0,$$

and for any $x \in S$, any $n \in \mathbb{N}$, any $a \in L_n x$ and any $\delta > 0$, there exists $b \in L_{n+1}a$ such that

$$F_{a,b}(s) \geq \widetilde{F}_{L_n x, L_{n+1}a}(s - \delta) \ \ \text{for every} \ \ s \geq 0;$$

(ii) there exists $x_0 \in S$ and $x_1 \in L_1 x_0$ such that

$$\int_1^\infty \ln u \, dF_{x_1, x_0}(u) < \infty.$$

Then there exists $x^* \in S$ such that $x^* \in \bigcap_{n \in \mathbb{N}} L_n x^*$.

Chapter 5

Hicks' contraction principle

T. Hicks in [128] considered another notion of probabilistic contraction mapping than probabilistic q-contraction, which is incomparable with probabilistic q-contraction, [262]. In section 5.1 two types of generalizations of Hick's notion of C-contraction is introduced and some fixed point theorems for these new classes of mappings are proved. A multi-valued generalization of the notion of C-contraction is given in section 5.2 and two fixed point theorems for multi-valued mappings are proved, where some results on infinitary operations from section 1.8 are used.

We assume in the whole chapter that Range$(\mathcal{F}) \subset \mathcal{D}^+$ for the probabilistic metric spaces $(S, \mathcal{F}, \tau)$.

5.1 Hicks' contraction principle for single-valued mappings and its generalizations

Definition 5.1 *Let $(S, \mathcal{F})$ be a probabilistic metric space and $f : S \to S$. The mapping f is a C-contraction if there exists $k \in (0, 1)$ such that for every $p, q \in S$ and for every $t > 0$*

$$F_{p,q}(t) > 1 - t \quad \Rightarrow \quad F_{fp,fq}(kt) > 1 - kt. \tag{5.1}$$

If $f : S \to S$ is a C-contraction and $(S, \mathcal{F}, T_{\mathbf{M}})$ is a complete Menger space Hicks proved that f has a unique fixed point. V. Radu proved in [245] that this result holds in a complete Menger space $(S, \mathcal{F}, T)$, where T is such that

$$\sup_{a<1} T(a, a) = 1.$$

In [246] the following generalization of the notion of C-contraction is given, where the family of functions $\mathcal{M}$ is defined in section 2.4.

Definition 5.2 *Let $(S, \mathcal{F})$ be a probabilistic metric space and $f : S \to S$. The mapping f is a generalized C-contraction if there exist a continuous, decreasing* *function* $h : [0,1] \to [0,\infty]$ *such that* $h(1) = 0$, $m_1, m_2 \in \mathcal{M}$, *and* $k \in (0,1)$ *such that the following implication holds for every* $p, q \in S$ *and for every* $t > 0$:

$$h \circ F_{p,q}(m_2(t)) < m_1(t) \quad \Rightarrow \quad h \circ F_{fp,fq}(m_2(kt)) < m_1(kt). \tag{5.2}$$

If $m_1(s) = m_2(s) = s$, and $h(s) = 1 - s$ for every $s \in [0,1]$, we obtain the Hicks definition.

Theorem 5.3 *Let* $(S, \mathcal{F}, T)$ *be a complete Menger space with t-norm* T *such that* $\sup_{a<1} T(a,a) = 1$ *and let* $f : S \to S$ *be a generalized* C*-contraction such that* $h(0) \in \mathbb{R}$. *Then* $x = \lim_{n\to\infty} f^n p$ *is the unique fixed point of the mapping* f *for an arbitrary* $p \in S$.

Proof. First, we shall prove that f is uniformly continuous. Let $\varepsilon > 0$ and $\lambda \in (0,1)$. We have to prove that there exists $N(\bar{\varepsilon}, \bar{\lambda}) = \{(p,q) \mid (p,q) \in S \times S,\ F_{p,q}(\bar{\varepsilon}) > 1 - \bar{\lambda}\}$ such that

$$(p,q) \in N(\bar{\varepsilon}, \bar{\lambda}) \quad \Rightarrow \quad F_{fp,fq}(\varepsilon) > 1 - \lambda.$$

Let $s \in \mathbb{R}$ be such that

$$m_2(ks) < \varepsilon, \qquad m_1(ks) < h(1 - \lambda). \tag{5.3}$$

Since m_1 and m_2 are continuous at 0, and $m_1(0) = m_2(0) = 0$ such a number s exists. We prove that $\bar{\varepsilon} = m_2(s)$, $\bar{\lambda} = 1 - h^{-1}(m_1(s))$. If $(p,q) \in N(\bar{\varepsilon}, \bar{\lambda})$ we have

$$\begin{aligned} F_{p,q}(m_2(s)) &> 1 - (1 - h^{-1}(m_1(s))) \\ &= h^{-1}(m_1(s)). \end{aligned}$$

Since h is decreasing it follows that $h \circ F_{p,q}(m_2(s)) < m_1(s)$. Hence,

$$h \circ F_{fp,fq}(m_2(ks)) < m_1(ks).$$

Using (5.3) we conclude that

$$h \circ F_{fp,fq}(m_2(ks)) < h(1 - \lambda)$$

and since h is decreasing we have

$$F_{fp,fq}(\varepsilon) \geq F_{fp,fq}(m_2(ks)) > 1 - \lambda.$$

Therefore $(fp, fq) \in N(\varepsilon, \lambda)$ if $(p,q) \in N(\bar{\varepsilon}, \bar{\lambda})$.

We prove that for every $\varepsilon > 0$ and $\lambda \in (0,1)$ there exists $n_0(\varepsilon, \lambda) \in \mathbb{N}$ such that for every $p, q \in S$

$$n > n_0(\varepsilon, \lambda) \quad \Rightarrow \quad F_{f^n p, f^n q}(\varepsilon) > 1 - \lambda.$$

Since $h(0) \in \mathbb{R}$ and $\lim_{s\to\infty} m_1(s) = \infty$, it follows that there exists $s \in \mathbb{R}$ such that $h(0) < m_1(s)$. From $F_{p,q}(m_2(s)) \geq 0$ it follows that

$$h \circ F_{p,q}(m_2(s)) \leq h(0) < m_1(s),$$

which implies that $h \circ F_{fp,fq}(m_2(ks)) < m_1(ks)$ and continuing in this way we obtain that for every $n \in \mathbb{N}$

$$h \circ F_{f^n p, f^n q}(m_2(k^n s)) < m_1(k^n s).$$

Let $n_0(\varepsilon, \lambda)$ be such a natural number that $m_2(k^n s) < \varepsilon$, $m_1(k^n s) < h(1-\lambda)$, for every $n \geq n_0(\varepsilon, \lambda)$. Then $n > n_0(\varepsilon, \lambda)$ implies that

$$F_{f^n p, f^n q}(\varepsilon) \geq F_{f^n p, f^n q}(m_2(k^n s)) > 1 - \lambda. \tag{5.4}$$

If $q = f^m p$, from (5.4) we obtain that

$$F_{f^n p, f^{n+m} p}(\varepsilon) > 1 - \lambda \quad \text{for every} \quad n > n_0(\varepsilon, \lambda) \quad \text{and every} \quad m \in \mathbb{N}. \tag{5.5}$$

Relation (5.5) means that $(f^n p)_{n\in\mathbb{N}}$ is a Cauchy sequence, and since S is complete there exists $x = \lim_{n\to\infty} f^n p$, which is obviously a fixed point of f since f is continuous.

For every $p \in S$ and $q \in S$ such that $fp = p$ and $fq = q$ we have for every $n \in \mathbb{N}$ that $f^n p = p$, $f^n q = q$, and therefore from (5.4) we have $F_{p,q}(\varepsilon) > 1 - \lambda$ for every $\lambda \in (0, 1)$, and $\varepsilon > 0$. This implies that $F_{p,q}(\varepsilon) = 1$ for every $\varepsilon > 0$ and therefore $p = q$. □

Remark 5.4 If $f : S \to S$ is a generalized C-contraction and $(S, \mathcal{F}, T)$ is a Menger space such that $T \geq T_h$ (h is the additive generator of T_h), then f is a usual metric contraction in the metric

$$d_{m_1,m_2}(p, q) = \{t \mid t \geq 0,\ m_1(t) \leq h \circ F_{p,q}(m_2(t))\}.$$

In this case the following equivalence holds:

$$d_{m_1,m_2}(p, q) < s \quad \Longleftrightarrow \quad h \circ F_{p,q}(m_2(s)) < m_1(s).$$

Hence if $T \geq T_h$ f is a generalized C-contraction if and only if f is a contraction with respect to the metric d_{m_1,m_2}.

Remark 5.5 In the paper [110] a fixed point result is obtained for a more general class of mappings f, where (5.2) is replaced by the implication

$$h \circ F_{p,q}(m_2(t)) < m_1(t) \quad \Rightarrow \quad h \circ F_{fp,fq}(m_2(\psi(t))) < m_1(\psi(t)),$$

and $\psi : [0, \infty) \to [0, \infty)$ is such that $\lim_{n\to\infty} \psi^n(t) = 0$ for every $t > 0$.

If $f : S \to S$ is a C-contraction and $T \not\geq T_{\mathbf{L}}$ then f is not necessarily a metric contraction with respect to the metric

$$\beta(p,q) = \inf\{h \mid h \in \mathbb{R},\ F(h^+) > 1 - h\},$$

as was shown in [262] by the following example.

Example 5.6 Let $T \not\geq T_{\mathbf{L}}$. Then there exist $a, b \in (0,1)$ such that $0 < a \leq b < 1$ and

$$0 \leq T(a,b) < T_{\mathbf{L}}(a,b) = a + b - 1.$$

Let $S = \{p,q,r\}$ and define $\mathcal{F} : S \times S \to \mathcal{D}^+$ by

$$F_{p,q}(x) = \begin{cases} 0 & \text{if } x \leq 0, \\ a & \text{if } 0 < x \leq 2, \\ 1 & \text{if } 2 < x, \end{cases}$$

$$F_{q,r}(x) = \begin{cases} 0 & \text{if } x \leq 0, \\ b & \text{if } 0 < x \leq 2, \\ 1 & \text{if } 2 < x, \end{cases}$$

and

$$\begin{aligned} F_{p,r}(x) &= \tau_T(F_{p,q}, F_{q,r})(x) \\ &= \begin{cases} 0 & \text{if } x \leq 0, \\ T(a,b) & \text{if } 0 < x \leq 2, \\ b & \text{if } 2 < x \leq 4, \\ 1 & \text{if } 4 < x. \end{cases} \end{aligned}$$

It can be verified that $(S, \mathcal{F}, T)$ is a Menger space, but

$$\begin{aligned} \beta(p,q) + \beta(q,r) &= 1 - a + 1 - b \\ &= 1 - T_{\mathbf{L}}(a,b) \\ &< 1 - T(a,b) \\ &= \beta(p,r). \end{aligned}$$

In [262] a comparison of the C-contraction and the Sehgal and Bharucha-Reid contraction is made.

If $f : S \to S$ is a C-contraction on a complete Menger space $(S, \mathcal{F}, T)$, where $T \geq T_{\mathbf{L}}$, then (S, β) is a complete metric space and f is a metric contraction with respect to the metric β. Hence by the Banach contraction principle f has a unique fixed point. But since this is not true for Sehgal's and Bharucha-Reid's contraction (probabilistic q-contraction) we conclude that a probabilistic q-contraction need not be a C-contraction.

The following example shows that a C-contraction need not be a probabilistic q-contraction.

Example 5.7 Let $S = \{0, 1, 2, \ldots\}$. Let $\mathcal{F} : S \times S \to \mathcal{D}^+$ be defined by

$$F_{p,q}(x) = F_{q,p}(x) = \begin{cases} 0 & \text{if } x \leq 2^{-\min\{p,q\}}, \\ 1 - 2^{-\min\{p,q\}} & \text{if } 2^{-\min\{p,q\}} < x \leq 1, \\ 1 & \text{if } 1 < x \end{cases}$$

for $p \neq q$, and for $p = q$, $F_{p,p} = H_0$, $(p, q) \in S \times S$. Then $(S, \mathcal{F}, T_{\mathbf{M}})$ is a Menger space. Let $f(r) = r + 1$, $r \in \mathbb{N} \cup \{0\}$. Then it is obvious that f is not a probabilistic γ-contraction but f is a C-contraction. Indeed, let $\gamma \in (0, 1)$ and $x \in (1, 1/\gamma)$. Then $\gamma x < 1$ and

$$F_{f(0),f(1)}(\gamma x) = F_{1,2}(\gamma x) \leq \frac{1}{2} < 1 = F_{0,1}(x),$$

which implies that f is not a probabilistic γ-contraction.

On the other hand, since $T_{\mathbf{M}} > T_{\mathbf{L}}$ and $\beta(fp, fq) = \beta(p, q)/2$ for every $(p, q) \in S \times S$, f is a C-contraction.

Under some additional conditions on $\mathcal{F}$ every probabilistic γ-contraction on $(S, \mathcal{F})$ is a C-contraction [262].

Lemma 5.8 *Let $(S, \mathcal{F})$ be a probabilistic metric space and $f : S \to S$ a probabilistic γ-contraction. If for every $p, q \in S$, $F_{fp,fq}$ is strictly increasing on $[0, 1]$, then $\beta(fp, fq) < \beta(p, q)$ for every $p, q \in S$.*

Proof. Let $\eta \in \left(0, \dfrac{1-\gamma}{\gamma}\beta(p, q)\right)$, where γ is the contraction constant. Then $\beta(p, q) > \gamma(\beta(p, q) + \eta)$ and therefore

$$\begin{aligned} F_{fp,fq}(\beta(p,q)) &> F_{fp,fq}(\gamma(\beta(p,q)+\eta)) \\ &\geq F_{p,q}(\beta(p,q)+\eta) \\ &\geq F_{p,q}(\beta(p,q)^+) \\ &> 1 - \beta(p,q). \end{aligned}$$

Using the definition of β, we conclude that $\beta(fp, fq) < \beta(p,q)$ for every $p, q \in S$. $\square$

Theorem 5.9 *Let $(S, \mathcal{F})$ be a probabilistic metric space such that* Range$(\mathcal{F})$ *is finite and that each element of* Range$(\mathcal{F}) \setminus \{H_0\}$ *is strictly increasing on* $[0,1]$. *Then every probabilistic γ-contraction on $(S, \mathcal{F})$ is a C-contraction.*

Proof. Let f be a probabilistic γ-contraction on $(S, \mathcal{F})$. Since, by Lemma 5.8 $\beta(fp, fq) < \beta(p,q)$ for every $p, q \in S$ there exists $\gamma_{p,q} \in (0,1)$ such that

$$\beta(fp, fq) < \gamma_{p,q}\beta(p,q).$$

Since Range$(\mathcal{F})$ is finite there exists a $\gamma \in (0,1)$ such that for every $(p,q) \in S \times S$, $\gamma_{p,q} < \gamma$. Thus $\beta(fp, fq) \leq \gamma\beta(p,q)$ for all $p, q \in S$ and f is a C-contraction. $\square$

In general it is not true that every probabilistic γ-contraction is a C-contraction. Next example shows that in Theorem 5.9 we cannot get rid of the condition that each $F_{p,q} \neq H_0$ is strictly increasing on $[0,1]$.

Example 5.10 [262]. Let $S = \{p, q, r\}$ and let $\mathcal{F} : S \times S \to \mathcal{D}^+$ be defined by

$$F_{p,r}(x) = F_{r,p}(x) = F_{r,q}(x) = F_{q,r}(x) = \begin{cases} 0 & \text{if } x \leq 0, \\ 1/2 & \text{if } 0 < x \leq 2, \\ 1 & \text{if } x > 2, \end{cases}$$

and

$$F_{p,q}(x) = F_{q,p}(x) = \begin{cases} 0 & \text{if } x \leq 0, \\ 1/2 & \text{if } 0 < x \leq 3/2, \\ 1 & \text{if } x > 3/2. \end{cases}$$

In this example, $F_{p,q}$ is not strictly increasing. Then $(S, \mathcal{F}, T_{\mathbf{M}})$ is a Menger space. Let $f(p) = f(q) = p$, $f(r) = q$. Since $F_{p,q}(3x/4) = F_{p,r}(x)$ for all x, it follows that f is a probabilistic 3/4-contraction. Since $\beta(fp, fr) = \beta(p,q) = 1/2 = \beta(p,r), f$ is not a C-contraction.

In the next example f is a probabilistic 1/2-contraction, Range$(\mathcal{F})$ is not finite and f is not a C-contraction.

Example 5.11 [262]. For every $n \in \mathbb{N}$, let $p_n : (0,1) \to \mathbb{R}^+$ be given by $p_n(t) = 2^{-n}(1-t)\cdot t^{-1}$, $t \in (0,1)$. Let $S = \{p_n \mid n \in \mathbb{N}\}$. If $\mathbf{m}_0$ is the Lebesgue measure on $(0,1)$ and $x \geq 0$ let

$$\begin{aligned} F_{p_n,p_m}(x) &= \mathbf{m}_0(\{t \mid t \in (0,1),\ |p_n(t) - p_m(t)| < x\}) \\ &= \frac{x}{x + |2^{-n} - 2^{-m}|} \quad (n, m \in \mathbb{N}). \end{aligned}$$

Then $(S, \mathcal{F}, T_{\mathbf{L}})$ is a Menger space and this is also the case if $T_{\mathbf{L}}$ is replaced by $T_{\mathbf{M}}$. Let $f : S \to S$ be defined by $f(p_n) = p_{n+1}$ $(n \in \mathbb{N})$. Since

$$F_{fp_n,fp_m}(x/2) = F_{p_{n+1},p_{m+1}}(x/2) = F_{p_n,p_m}(x),$$

f is a probabilistic $\frac{1}{2}$-contraction.

We shall prove that f is not a C-contraction. Suppose, on the contrary, that there exists a $\gamma \in (0,1)$ such that $\beta(fp_n, fp_m) < \gamma\beta(p_n, p_m)$ for all $n, m \in \mathbb{N}$. We have that

$$\beta(p_n, p_m) = \frac{1}{2}\left(\sqrt{|2^{-n} - 2^{-m}|^2 + 4|2^{-n} - 2^{-m}|} - |2^{-n} - 2^{-m}|\right)$$

and

$$\beta(fp_n, fp_m) = \frac{1}{2}\left(\sqrt{|2^{-n-1} - 2^{-m-1}|^2 + 4|2^{-n-1} - 2^{-m-1}|} - |2^{-n-1} - 2^{-m-1}|\right).$$

Hence from $\beta(fp_n, fp_m) < \gamma\beta(p_n, p_m)$ it follows that

$$\gamma > \frac{\sqrt{|2^{-n-1} - 2^{-m-1}|^2 + 4|2^{-n-1} - 2^{-m-1}|} - |2^{-n-1} - 2^{-m-1}|}{\sqrt{|2^{-n} - 2^{-m}|^2 + 4|2^{-n} - 2^{-m}|} - |2^{-n} - 2^{-m}|}. \tag{5.6}$$

If $m \to \infty$ in (5.6) we obtain

$$\gamma > \frac{\sqrt{|2^{-2n-2} + 2^{-n+1}} - 2^{-n-1}}{\sqrt{2^{-2n} + 2^{-n+2}} - 2^{-n}} = \frac{\sqrt{1 + 2^{n+2}} + 1}{\sqrt{1 + 2^{n+3}} + 1}, \tag{5.7}$$

and if $n \to \infty$ in (5.7) it follows that $\gamma \geq 1$, which is a contradiction.

In this case $(S, \mathcal{F}, T_{\mathbf{L}})$ is a pseudo-metrically generated space, where the generating pseudo-metrics are given by the family $(\delta_t)_{t\in(0,1)}$

$$\delta_t(p_n, p_m) = |p_n(t) - p_m(t)| = |2^{-n} - 2^{-m}|\frac{(1-t)}{t}, \quad t \in (0,1).$$

Then

$$\begin{aligned} \delta_t(f(p_n), f(p_m)) &= |p_{n+1}(t) - p_{m+1}(t)| \\ &= \frac{1}{2}|p_n(t) - p_m(t)| \\ &= \frac{1}{2}\delta_t(p_n, p_m). \end{aligned}$$

Thus f is a $1/2$-contraction on each of the pseudo-metric spaces (S, δ_t).

Let (S, d) be a metric space, $G \in \mathcal{D}^+$ and $\alpha > 0$. If for every $p, q \in S$

$$F_{p,q}(x) = G\Big(\frac{x}{d(p,q)^\alpha}\Big) \quad (x \in \mathbb{R})$$

then $(S, \mathcal{F})$ is, as we know, an α-simple space, which is denoted by (S, G, d, α) and $(S, \mathcal{F}, T)$ is a Menger space for an arbitrary t-norm T.

Theorem 5.12 *Let (S, G, d, α) be an α-simple space with a strictly increasing function G. Then $f : S \to S$ is a probabilistic γ-contraction on (S, G, d, α) if and only if f is a $\gamma^{1/\alpha}$-contraction on (S, d).*

Proof. Let $\gamma \in (0, 1)$ and $x > 0$. Then

$$G\Big(\frac{\gamma x}{d(fp, fq)^\alpha}\Big) = F_{fp,fq}(\gamma x) \geq F_{p,q}(x) = G\Big(\frac{x}{d(p,q)^\alpha}\Big)$$

if and only if

$$d(fp, fq) \leq \gamma^{1/\alpha} d(p, q).$$

□

Theorem 5.13 *Let (S, G, d, α) be an α-simple space with a strictly increasing function G. Then $f : S \to S$, which is a C-contraction on (S, G, d, α), is a contraction on (S, d).*

Proof. If $F_{p,q}$ is strictly increasing then the quasi-inverse $F_{p,q}^{(-1)}$ is continuous and $\beta(p, q)$ is the unique solution of the equation $x = F_{p,q}^{(-1)}(1 - x)$, i.e.,

$$\beta(p,q) = F_{p,q}^{(-1)}(1 - \beta(p,q)). \tag{5.8}$$

If $F_{p,q}(x) = G(x/d(p,q)^\alpha)$ then $F_{p,q}^{(-1)}(x) = d(p,q)^\alpha G^{(-1)}(x)$, and from (5.8) it follows that

$$\beta(p,q) = d(p,q)^\alpha G^{(-1)}(1 - \beta(p,q)). \tag{5.9}$$

Suppose that f is a C-contraction, i.e., that

$$\beta(fp, fq) \leq \gamma\beta(p,q) \quad \text{for every } p, q \in S, \tag{5.10}$$

where $\gamma \in (0, 1)$. Then (5.9) and (5.10) imply

$$\begin{aligned} d(fp, fq)^\alpha &= \frac{\beta(fp, fq)}{G^{(-1)}(1 - \beta(fp, fq))} \\ &\leq \frac{\gamma\beta(p,q)}{G^{(-1)}(1 - \gamma\beta(p,q))} \\ &\leq \frac{\gamma\beta(p,q)}{G^{(-1)}(1 - \beta(p,q))} \\ &= \gamma d(p,q)^\alpha, \end{aligned}$$

which means that f is a contraction on (S, d). □

The converse of Theorem 5.13 is not true.

Example 5.14 [262] Let $G(x) = 0$ for $x \leq 0$ and $G(x) = \dfrac{x}{x+1}$ if $x > 0$ and let (S, d) be the metric space $(\mathbb{R}, |\cdot|)$.

Then $(S, G, d, 1)$ is a 1-simple space and define $f : S \to S$ by $f(p) = p/2$ $(p \in \mathbb{R})$. Obviously f is a contraction on (S, d) but f is not a C-contraction.

Suppose that f is a C-contraction, i.e., that there exists $\gamma \in (0,1)$, such that for every $x > 0$

$$F_{p,q}(x) > 1 - x \quad \Rightarrow \quad F_{fp,fq}(\gamma x) > 1 - \gamma x.$$

Since

$$F_{p,q}(x) > 1 - x \quad \Longleftrightarrow \quad |p - q| < \frac{x^2}{1-x}$$

and

$$F_{fp,fq}(\gamma x) > 1 - \gamma x \quad \Longleftrightarrow \quad |p - q| < \frac{2\gamma^2 x^2}{1 - \gamma x},$$

f is a C-contraction if and only if for every $p, q \in \mathbb{R}$ and $x \in (0,1)$

$$|p - q| < \frac{x^2}{1-x} \quad \Rightarrow \quad |p - q| < \frac{2\gamma^2 x^2}{1 - \gamma x}. \tag{5.11}$$

From (5.11) we conclude that

$$\frac{x^2}{1-x} \leq \frac{2\gamma^2 x^2}{1 - \gamma x} \tag{5.12}$$

and (5.12) is equivalent to $1 - \gamma x \leq 2\gamma^2(1 - x)$. If $x \to 1$ we obtain $\gamma \geq 1$. Hence f is not a C-contraction.

A generalization of the notion of C-contraction is given in [116].

Definition 5.15 *A mapping $f : S \to S$ is a generalized C-contraction of Kras-noselski's type if for each pair of real numbers (a, b), with $0 < a < b$, there exists $L(a, b) \in (0,1)$ such that for every $p, q \in S$ if*

$$a \leq 1 - F_{p,q}(a) \quad \textit{and} \quad 1 - F_{p,q}(b^+) \leq b,$$

then the following implication holds for every $x > 0$:

$$F_{p,q}(x) > 1 - x \quad \Rightarrow \quad F_{fp,fq}(L(a,b)x) > 1 - L(a,b)x. \tag{5.13}$$

Lemma 5.16 *In every probabilistic semi-metric space* $(S,\mathcal{F})$

$$\mathbf{K}(p,q)=\sup\{t \mid t \le 1-F_{p,q}(t)\}$$

is the only non-negative real number k *with the property*

$$1-F_{p,q}(k^+) \le k \le 1-F_{p,q}(k).$$

Lemma 5.17 *Let* $(S,\mathcal{F},T)$ *be a Menger space, where* $T \ge T_{\mathbf{L}}$. *If* $f: S \to S$ *is a generalized* C*-contraction of Krasnoselski's type then*

$$(\forall a,b>0)(\forall x,y \in S)(a \le \mathbf{K}(x,y) \le b \quad \Rightarrow \quad \mathbf{K}(fx,fy) \le L(a,b)\mathbf{K}(x,y)).$$

Proof. Let $\mathbf{K}(p,q) \in [a,b]$, $0<a<b$. Then we have $\mathbf{K}(p,q)=k=(1-t)a+tb$ for some $t \in [0,1]$. The inequality $a \le k$ implies $a \le 1-F_{p,q}(k) \le 1-F_{p,q}(a)$, and therefore $a \le 1-F_{p,q}(a)$.

On the other hand, the inequality $k \le b$ implies $1-F_{p,q}(b^+) \le 1-F_{p,q}(k^+) \le k \le b$, and therefore $1-F_{p,q}(b^+) \le b$.

Since f is a generalized C-contraction of Krasnoselski's type we have that (5.13) holds. For every $d>0$ we have

$$1-F_{p,q}(k+d) \le 1-F_{p,q}(k^+) \le k < k+d,$$

and therefore

$$F_{fp,fq}(L(a,b)(k+d)) > 1-L(a,b)(k+d),$$

which means that

$$\mathbf{K}(fp,fq) \le L(a,b)(k+d) \quad \text{for every } d>0.$$

This implies that

$$\mathbf{K}(fp,fq) \le L(a,b)\mathbf{K}(p,q),$$

i.e., f is a generalized contraction of Krasnoselski's type in the metric space $(S,\mathbf{K})$. □

Theorem 5.18 *Let* $(S,\mathcal{F},T)$ *be a complete Menger space, where* $T \ge T_{\mathbf{L}}$. *If* $f: S \to S$ *is a generalized* C*-contraction of Krasnoselski's type, then there exists a unique fixed point of the mapping* f *which is globally attractive.*

Proof. Theorem 5.18 follows immediately by Theorem 3.62 and Lemma 5.17. □

Open problem: Does Theorem 5.18 holds if we assume on T only the condition $\sup\limits_{a<1} T(a,a)=1$.

Remark 5.19 From Theorems 3.70 and 5.18 it follows that a generalized B-contracti need not be a generalized C-contraction of Krasnoselski's type.

5.2 Multi-valued generalizations of Hicks' contraction principle

As a multi-valued generalization of the notion of a C-contraction we introduced in [224] the notion of a (Ψ, C)-contraction, where $\Psi : [0, \infty) \to [0, \infty)$.

Definition 5.20 *Let $(S, \mathcal{F})$ be a probabilistic metric space and $f : S \to 2^S$. The mapping f is called a (Ψ, C)-contraction, where $\Psi : \mathbb{R}_+ \to \mathbb{R}_+$, if for every $p, q \in S$ and every $x > 0$ the following implication holds:*

$$F_{p,q}(x) > 1 - x \quad \Rightarrow \quad (\forall u \in fp)(\exists v \in fq) F_{u,v}(\Psi(x)) > 1 - \Psi(x).$$

If $\Psi(x) = kx$, $x > 0$, $k \in (0,1)$, then a (Ψ, C)-contraction $f : S \to S$ is a C-contraction. We shall give an example of a (Ψ, C)-contraction.

Example 5.21 Let (M, d) be a separable metric space, $(\Omega, \mathcal{A}, P)$ a probability space, and S the space of all the equivalence classes of measurable mappings from Ω into M. Let

$$d(X, Y) = \sup\{u \mid u \geq 0,\ u < P\{\omega \mid \omega \in \Omega,\ d(X(\omega), Y(\omega)) > u\}\}$$

for every $X, Y \in S$ (the Ky Fan metric in S). The topology induced by the metric d coincides with the (ε, λ)-topology in the Menger space $(S, \mathcal{F}, T_L)$, where

$$F_{X,Y}(u) = P\{\omega \mid \omega \in \Omega,\ d(X(\omega), Y(\omega)) < u\},\ X, Y \in S, u \geq 0.$$

It is obvious that for every $X, Y \in S$

$$d(X, Y) = \sup\{u \mid u > 0, F_{X,Y}(u) < 1 - u\}.$$

Let $f : S \to CB(S)$ be such that

$$D(fX, fY) \leq \Psi(d(X, Y)) \text{ for every } X, Y \in S$$

where D is the Hausdorff metric induced by d.

If $\Psi : [0, \infty) \to [0, \infty)$ is strictly increasing we prove that f is a (Ψ, C)-contraction.

Suppose that for some $u > 0$, $F_{X,Y}(u) > 1 - u$. Then from the definition of metric d it follows that $d(X, Y) < u$. Since Ψ is strictly increasing we have that $\Psi(d(X, Y)) < \Psi(u)$, which implies that $D(fX, fY) < \Psi(u)$, i.e., that

$$\sup_{U \in fX} \inf_{V \in fY} d(U, V) < \Psi(u) \text{ and } \sup_{U \in fY} \inf_{V \in fX} d(U, V) < \Psi(u).$$

Therefore for every $U \in fX$ there exists $V \in fY$ such that $d(U, V) < \Psi(u)$, which means that $F_{U,V}(\Psi(u)) > 1 - \Psi(u)$.

Two fixed point theorems for a multi-valued (Ψ, C)-contraction are proved in this section. The first fixed point theorem is based on a result from the metric fixed point theory and the second one on results from section 1.8 on infinitary operations.

Theorem 5.22 *Let $(S, \mathcal{F}, T)$ be a complete Menger space such that $T \geq T_{\mathbf{L}}$, M a nonempty, closed subset of S, $f : M \to CB(M)$ a (Ψ, C)-contraction such that the mapping Ψ satisfies the following conditions:*

(a) Ψ is right continuous;
(b) for every $q > 0$

$$\sup\{u \mid u \leq \Psi(u) + q\} = m(q) < \infty;$$

(c) for every $q > 0$

$$\lim_{n\to\infty} \Psi^n(m(q)) = 0;$$

(d) for every $u \geq 0$ and $v \geq 0$

$$\Psi(u) + \Psi(v) \leq \Psi(u + v).$$

Then there exists $x \in M$ such that $x \in fx$.

Proof. We know that in the case $T \geq T_{\mathbf{L}}$ the function $d : S \times S \to [0, \infty)$ defined by

$$d(x, y) = \sup\{u \mid F_{x,y}(u) \leq 1 - u\}$$

is a metric on S which generates the (ε, λ)-topology. We prove that the mapping f satisfies the inequality

$$D(fx, fy) \leq \Psi(d(x, y)) \text{ for every } x, y \in M,$$

where D is the Hausdorff metric

$$D(A, B) = \max\{\sup_{a\in A} \inf_{b\in B} d(a, b), \sup_{b\in B} \inf_{a\in A} d(a, b)\}.$$

Suppose that $u > d(x, y)$, and let us prove that $D(fx, fy) < \Psi(u)$. From $u > d(x, y)$, it follows that $F_{x,y}(u) > 1 - u$, and since the mapping f is a (Ψ, C)-contraction, for every $p \in fx$ there exists $q(p) \in fy$ such that

$$F_{p,q(p)}(\Psi(u)) > 1 - \Psi(u).$$

Therefore $d(p, q(p)) < \Psi(u)$, which implies that

$$\sup_{p\in fx} \inf_{q\in fy} d(p, q) \leq \Psi(u),$$

and similarly

$$\sup_{p\in fy}\inf_{q\in fx} d(p,q)\le \Psi(u).$$

Hence $D(fx,fy)\le \Psi(u)$, and since the mapping Ψ is right continuous we obtain $D(fx,fy)\le \Psi(d(x,y))$ for every $x,y\in M$. The mapping f satisfies conditions a), b), c) and d) from [331], and therefore it follows that there exists $x\in M$ such that $x\in fx$. □

If $(x_n)_{n\in\mathbb{N}}$ is a sequence from the interval $[0,\infty)$ such that $\lim\limits_{n\to\infty} x_n=0$ then for some $n_0\in\mathbb{N}$ we have $x_n\in[0,1)$ for every $n\ge n_0$ and so $1-x_n\in(0,1]$ for every $n\ge n_0$. Hence, in this case, $\lim\limits_{n\to\infty}\mathop{\large\mathrm{T}}\limits_{i=1}^{\infty}(1-x_{n+i})$ means $\lim\limits_{n\ge n_0}\mathop{\large\mathrm{T}}\limits_{i=1}^{\infty}(1-x_{n+i})$.

Theorem 5.23 *Let $(S,\mathcal{F},T)$ be a complete Menger space, T a t-norm such that $\sup\limits_{a<1} T(a,a)=1$, $M\in C(S)$ and $f: M\to C(M)$ a (Ψ,C)-contraction, where the series $\sum\limits_{n=1}^{\infty}\Psi^n(s)$ is convergent for some $s>1$. If f is weakly demicompact or*

$$\lim_{n\to\infty}\mathop{\large\mathrm{T}}_{i=1}^{\infty}(1-\Psi^{n+i-1}(s))=1 \tag{5.14}$$

then there exists at least one element $x\in M$ such that $x\in fx$.

Proof. Let $x_0\in M$ and $x_1\in fx_0$. Since $s>1$ and $F_{x_1,x_0}(u)\ge 0$ for every $u\ge 0$ it follows that $F_{x_1,x_0}(s)>1-s$. The mapping f is a (Ψ,C)-contraction and therefore there exists $x_2\in fx_1$ such that

$$F_{x_2,x_1}(\Psi(s))>1-\Psi(s).$$

Continuing in this way we obtain a sequence $(x_n)_{n\in\mathbb{N}}$ from M such that for every $n\ge 2$ $x_n\in fx_{n-1}$ and

$$F_{x_n,x_{n-1}}(\Psi^{n-1}(s))>1-\Psi^{n-1}(s). \tag{5.15}$$

Since the series $\sum\limits_{n=1}^{\infty}\Psi^n(s)$ is convergent we have

$$\lim_{n\to\infty}\Psi^n(s)=0. \tag{5.16}$$

From (5.15) and (5.16) we infer for every $\varepsilon>0$ that

$$\lim_{n\to\infty}F_{x_n,x_{n-1}}(\varepsilon)=1. \tag{5.17}$$

Indeed, if $\varepsilon > 0$ and $\lambda \in (0,1)$ are given, and $n_0 = n_0(\varepsilon, \lambda) \in \mathbb{N}$ is such that for every $n \geq n_0$, $\psi^n(s) \leq \min(\varepsilon, \lambda)$ then

$$F_{x_{n+1},x_n}(\varepsilon) \geq F_{x_{n+1},x_n}(\psi^n(s)) > 1 - \psi^n(s) > 1 - \lambda \text{ for every } n \geq n_0.$$

If f is weakly demicompact (5.17) implies that there exists a convergent subsequence $(x_{n_k})_{k\in\mathbb{N}}$. Suppose now that (5.14) holds and prove that $(x_n)_{n\in\mathbb{N}}$ is a Cauchy sequence. This means that for every $\varepsilon > 0$ and every $\lambda \in (0,1)$ there exists $n_1(\varepsilon, \lambda) \in \mathbb{N}$ such that

$$F_{x_{n+p},x_n}(\varepsilon) > 1 - \lambda \tag{5.18}$$

for every $n \geq n_1(\varepsilon, \lambda)$ and every $p \in \mathbb{N}$.

Let $\varepsilon > 0$ and $\lambda \in (0,1)$ be given. From (5.16) it follows that there exists $n_2 \in \mathbb{N}$ such that $\Psi^n(s) < 1$ for every $n \geq n_2$. Let $n_3(\varepsilon) \in \mathbb{N}$ be such that

$$\sum_{n \geq n_3(\varepsilon)} \Psi^n(s) < \varepsilon.$$

Since $\sum\limits_{n=1}^{\infty} \Psi^n(s)$ is a convergent series such a natural number $n_3(\varepsilon)$ exists. Hence for every $p \in \mathbb{N}$ and every $n \geq n_3(\varepsilon)$ we have that

$$F_{x_{n+p+1},x_n}(\varepsilon) \geq \mathop{\mathrm{T}}_{i=1}^{p+1} F_{x_{n+i},x_{n+i-1}}(\Psi^{n+i-1}(s)),$$

and (5.15) implies that

$$F_{x_{n+p+1},x_n}(\varepsilon) \geq \mathop{\mathrm{T}}_{i=1}^{p+1} (1 - \Psi^{n+i-1}(s))$$

for every $n \geq \max(n_2, n_3(\varepsilon))$ and every $p \in \mathbb{N}$.

For every $p \in \mathbb{N}$ and $n \geq \max(n_2, n_3(\varepsilon))$

$$\mathop{\mathrm{T}}_{i=1}^{p+1} (1 - \Psi^{n+i-1}(s)) \geq \mathop{\mathrm{T}}_{i=1}^{\infty} (1 - \Psi^{n+i-1}(s)),$$

and therefore for every $p \in \mathbb{N}$ and every $n \geq \max(n_2, n_3(\varepsilon))$,

$$F_{x_{n+p+1},x_n}(\varepsilon) \geq \mathop{\mathrm{T}}_{i=1}^{\infty} (1 - \Psi^{n+i-1}(s)). \tag{5.19}$$

From (5.14) it follows that there exists $n_4(\lambda) \in \mathbb{N}$ such that

$$\mathop{\mathrm{T}}_{i=1}^{\infty} (1 - \Psi^{n+i-1}(s)) > 1 - \lambda, \tag{5.20}$$

for every $n \geq n_4(\lambda)$. The conditions (5.19) and (5.20) imply that (5.18) holds for $n_1(\varepsilon, \lambda) = \max(n_2, n_3(\varepsilon), n_4(\lambda))$ and every $p \in \mathbb{N}$. This means that $(x_n)_{n\in\mathbb{N}}$ is a Cauchy sequence, and since S is complete there exists $\lim\limits_{n\to\infty} x_n$. Hence in both cases there exists $(x_{n_k})_{k\in\mathbb{N}}$ such that $\lim\limits_{k\to\infty} x_{n_k} = x$.

It remains to prove that $x \in fx$. Since $fx = \overline{fx}$ it is enough to prove that $x \in \overline{fx}$, i.e., that for every $\varepsilon > 0$ and $\lambda \in (0,1)$ there exists $b_{\varepsilon,\lambda} \in fx$ such that

$$F_{x,b_{\varepsilon,\lambda}}(\varepsilon) > 1 - \lambda. \tag{5.21}$$

Since $\sup\limits_{x<1} T(x,x) = 1$, for $\lambda \in (0,1)$ there exists $\delta(\lambda) \in (0,1)$ such that

$$T(1-\delta(\lambda), 1-\delta(\lambda)) > 1-\lambda.$$

If $\delta'(\lambda)$ is such that

$$T(1-\delta'(\lambda), 1-\delta'(\lambda)) > 1-\delta(\lambda)$$

and $\delta''(\lambda) = \min(\delta(\lambda), \delta'(\lambda))$ we have that

$$\begin{aligned} T(1-\delta''(\lambda), T(1-\delta''(\lambda), 1-\delta''(\lambda))) &\geq T(1-\delta(\lambda), T(1-\delta'(\lambda), 1-\delta'(\lambda))) \\ &\geq T(1-\delta(\lambda), 1-\delta(\lambda)) \\ &> 1-\lambda. \end{aligned}$$

Since $\lim\limits_{k\to\infty} x_{n_k} = x$ there exists $k_1 \in \mathbb{N}$ such that $F_{x,x_{n_k}}(\varepsilon/3) > 1-\delta''(\lambda)$ for every $k \geq k_1$. Let $k_2 \in \mathbb{N}$ be such that

$$F_{x_{n_k},x_{n_k+1}}(\varepsilon/3) > 1-\delta''(\lambda) \text{ for every } k \geq k_2.$$

The existence of such a k_2 follows by (5.17).

Let $t_0 \in \mathbb{R}_+$ be such that $\Psi(t_0) < \min\{\varepsilon/3, \delta''(\lambda)\}$ and $k_3 \in \mathbb{N}$ such that $F_{x_{n_k},x}(t_0) > 1 - t_0$, for every $k \geq k_3$. Since f is a (Ψ, C)-contraction there exists $b_{\varepsilon,\lambda,k} \in fx$ such that

$$F_{x_{n_k+1},b_{\varepsilon,\lambda,k}}(\Psi(t_0)) > 1-\Psi(t_0) \text{ for every } k \geq k_3.$$

Therefore for every $k \geq k_3$

$$\begin{aligned} F_{x_{n_k+1},b_{\varepsilon,\lambda,k}}(\varepsilon/3) &\geq F_{x_{n_k+1},b_{\varepsilon,\lambda,k}}(\Psi(t_0)) \\ &> 1-\Psi(t_0) \\ &> 1-\delta''(\lambda). \end{aligned}$$

If $k \geq \max(k_1, k_2, k_3)$ we have

$$\begin{aligned} F_{x,b_{\varepsilon,\lambda,k}}(\varepsilon) &\geq T(F_{x,x_{n_k}}(\varepsilon/3), T(F_{x_{n_k},x_{n_k+1}}(\varepsilon/3), F_{x_{n_k+1},b_{\varepsilon,\lambda,k}}(\varepsilon/3))) \\ &\geq T(1-\delta''(\lambda), T(1-\delta''(\lambda), 1-\delta''(\lambda))) \\ &> 1-\lambda \end{aligned}$$

and (5.21) is proved for $b_{\varepsilon,\lambda} = b_{\varepsilon,\lambda,k}$, $k \geq \max(k_1, k_2, k_3)$. Hence $x \in \overline{fx} = fx$, which means that x is a fixed point of the mapping f. □

Remark 5.24 Theorem 5.23 generalizes Theorem 1 from [224], where $T \in \mathcal{T}^H$, since in this case (5.14) is satisfied.

Corollary 5.25 *Let $(S, \mathcal{F}, T)$ be a complete Menger space such that T is a strict t-norm with a multiplicative generator θ, $M \in C(S)$ and $f : M \to C(M)$ a (Ψ, C)-contraction such that the series $\sum_{n=1}^{\infty} \Psi^n(s)$ is convergent for some $s > 1$. If f is weakly demicompact or*

$$\lim_{n\to\infty} \prod_{i=1}^{\infty} \theta(1 - \Psi^{n+i-1}(s)) = 1 \tag{5.22}$$

then there exists at least one element $x \in M$ such that $x \in fx$.

Proof. Since θ^{-1} is continuous, the conclusion follows by Theorem 5.23 and the equality

$$\lim_{n\to\infty} \mathbf{T}_{i=1}^{\infty} (1 - \Psi^{n+i-1}(s)) = \theta^{-1}\left(\lim_{n\to\infty} \prod_{i=1}^{\infty} \theta(1 - \Psi^{n+i-1}(s))\right) = 1. \qquad \square$$

Corollary 5.26 *Let $(S, \mathcal{F}, T_\lambda^{\mathbf{D}})$ be a complete Menger space such that $\lambda > 0, M \in C(S)$, and $f : M \to C(M)$ be a (Ψ, C)-contraction. Assume that one of the following two conditions are satisfied:*

(i) $\lambda \geq 1$ and there exists $s > 1$ such that $\sum_{n=1}^{\infty} \psi^n(s)$ is convergent;

(ii) $\lambda \in (0, 1)$ and there exists $s > 1$ such that $\sum_{n=1}^{\infty} (\psi^n(s))^\lambda$ is convergent.

Then there exists $x \in M$ such that $x \in fx$.

Corollary 5.27 *Let $(S, \mathcal{F}, T_\lambda^{\mathbf{AA}})$ be a complete Menger space such that $\lambda > 0, M \in C(S)$, and $f : M \to C(M)$ be a (Ψ, C)-contraction. Assume that one of the following two conditions are satisfied:*

(i) $\lambda \geq 1$ and there exists $s > 1$ such that $\sum_{n=1}^{\infty} \psi^n(s)$ is convergent;

(ii) $\lambda \in (0, 1)$ and there exists $s > 1$ such that $\sum_{n=1}^{\infty} (\psi^n(s))^\lambda$ is convergent.

Then there exists $x \in M$ such that $x \in fx$.

Corollary 5.28 *Let $(S, \mathcal{F}, T_\lambda^{\mathbf{SW}})$ be a complete Menger space, $T_\lambda^{\mathbf{SW}}, \lambda \in (-1, \infty]$, $M \in C(S)$ and $f : M \to C(M)$ a (Ψ, C)-contraction, where the series $\sum_{n=1}^{\infty} \Psi^n(s)$ is convergent for some $s > 1$. Then there exists at least one element $x \in M$ such that $x \in fx$.*

Let us recall that

$$\mathcal{T}_0 = \bigcup_{\lambda\in(0,\infty)} \{T_\lambda^{\mathbf{D}}\} \bigcup \bigcup_{\lambda\in(0,\infty)} \{T_\lambda^{\mathbf{AA}}\} \bigcup T^H \bigcup_{\lambda\in(-1,\infty]} \{T_\lambda^{\mathbf{SW}}\}$$

If $f : M \to C(M)$ is a (Ψ, C)-contraction, where $\Psi(s) = ks$ ($k \in [0, 1)$) for every $s > 0$, we say, as in the case of a single-valued mapping, that f is a C-contraction.

Corollary 5.29 *Let $(S, \mathcal{F}, T)$ be a complete Menger space such that $T \in \mathcal{T}_0, M \in C(S)$, and $f : M \to C(M)$ be a C-contraction. Then there exists $x \in M$ such that $x \in fx$.*

Proof. In this case $\Psi^n(s) = k^n s$ for every $n \in \mathbb{N}$ and $s \geq 0$. Corollary 5.29 follows from Corollaries 5.26, 5.27,5.28. □

Theorem 5.23 can be applied to the operator equations of the form

$$x(\omega) = f(x(\omega), \omega) \text{ a.e.},$$

where $(\Omega, \mathcal{A}, m)$ is a measure space with a σ-algebra $\mathcal{A}$, and a $\mathbf{S}$-decomposable measure m, (M, d) a separable metric space, and $f : \Omega \times M \to M$ a random operator, which means that for every measurable mapping $X : \Omega \to M$ the mapping $\omega \mapsto f(\omega, X(\omega))$ is measurable.

Let $(\widehat{f}X)(\omega) = f(\omega, X(\omega))$ for every $\omega \in \Omega$, $\widehat{X} \in S$, where S is the set of all equivalence classes of measurable mappings $X : \Omega \to M$. Then $\widehat{f} : S \to S$.

Theorem 5.30 *Let $(\Omega, \mathcal{A}, m)$ be as in Proposition 2.82, where $\mathbf{S} = \mathbf{S}_\lambda^{\mathbf{SW}}$ for some $\lambda \in (-1, \infty]$, (M, d) a complete separable metric space, $f : \Omega \times M \to M$ a random operator, and $\Psi : [0, \infty) \to [0, \infty)$ such that the series $\sum_{n=1}^{\infty} \Psi^n(s)$ is convergent for some $s > 1$. If for every $\widehat{X}, \widehat{Y} \in S$ and $u > 0$ we have*

$$m\{d(X, Y) < u\} > 1 - u \quad \Rightarrow \quad m\{d(\widehat{f}X, \widehat{f}Y) < \Psi(u)\} > 1 - \Psi(u), \qquad (5.23)$$

then there exists a measurable mapping $X : \Omega \to M$ such that

$$X(\omega) = f(\omega, X(\omega)) \text{ a.s. .}$$

Proof. By Propositions 2.82 and 2.84 it follows that $(S, \mathcal{F}, T)$ is a complete Menger space, where t-norm $T = T_{\lambda}^{\mathbf{SW}}$, for some $\lambda \in (-1, \infty]$. Relation (5.23) means that $\widehat{f} : S \to S$ is a (Ψ, C)-contraction with respect to $\mathcal{F}$ defined by

$$F_{\widehat{X},\widehat{Y}}(u) = m\{d(X,Y) < u\}, u \geq 0 \quad (X \in \widehat{X}, Y \in \widehat{Y}).$$

Hence by Theorem 5.23 it follows that there exists $\widehat{X} \in S$ such that $\widehat{f}\widehat{X} = \widehat{X}$. This means that $f(\omega, X(\omega)) = X(\omega)$ a.s. $(X \in \widehat{X})$ with respect to m. □

As a consequence of Theorem 5.23 we prove a result on the existence of a fixed point for a class of multi-valued mappings in fuzzy metric spaces.

Theorem 5.31 *Let (X, d, L, R) be a complete fuzzy metric space and*

$$\lim_{u\to\infty} d(x,y)(u) = 0 \textit{ for all } x, y \in X, \quad \lim_{a\to 0^+} R(a,a) = 0.$$

Let $f : X \to C(X)$ and $\Psi : [0, \infty) \to [0, \infty)$ be such that the series $\sum\limits_{n=1}^{\infty} \Psi^n(s_1)$ is convergent for some $s_1 > 1$ and the following implication holds:

For every $x, y \in X$ and every $u \in fx$ there exists $v \in fy$ such that for every $s > 0$

$$d(x,y)(s) < s \quad \Rightarrow \quad 1 - F_{u,v}(\Psi(s)) < \Psi(s). \tag{5.24}$$

If f is weakly demicompact or

$$\lim_{n\to\infty} \mathop{\mathsf{R}}_{i=n}^{\infty} \Psi^i(s_1) = 0, \tag{5.25}$$

then there exists $x \in X$ such that $x \in fx$.

Proof. We shall prove that (5.24) implies that f is a (Ψ, C)-contraction. Suppose that $s > 1 - F_{x,y}(s)$. If $s \geq \lambda_1(x,y)$ then from $1 - F_{x,y}(s) = d(x,y)(s) < s$ and (5.24) we have that $\Psi(s) > 1 - F_{u,v}(\Psi(s))$. If $s < \lambda_1(x,y)$ then $F_{x,y}(s) = 0$ and from $s > 1 - F_{x,y}(s)$ it follows that $s > 1$. Hence $d(x,y)(s) < s$ which implies that $\Psi(s) > 1 - F_{u,v}(\Psi(s))$. Therefore f is a (Ψ, C)-contraction. Since (5.25) implies

$$\begin{aligned} \lim_{n\to\infty} \mathop{\mathsf{T}}_{i=n}^{\infty} (1 - \Psi^i(s_1)) &= 1 - \lim_{n\to\infty} \mathop{\mathsf{R}}_{i=n}^{\infty} \Psi^i(s_1) \\ &= 1, \end{aligned}$$

from Theorem 5.23 it follows that there exists $x \in X$ such that $x \in fx$. □

Corollary 5.32 *Let (X, d, L, R) be a complete fuzzy metric space such that*

$$\lim_{u\to\infty} d(x,y)(u) = 0 \textit{ for all } x, y \in X, \qquad \lim_{a\to 0^+} R(a,a) = 0$$

Let $f : X \to C(X)$ and $\Psi : [0,\infty) \to [0,\infty)$ be such that the series $\sum_{n=1}^{\infty} \Psi^n(s_1)$ is convergent for some $s_1 > 1$ and the following implication holds:

For every $x, y \in X$ and every $u \in fx$ there exists $v \in fy$ such that for every $s > 0$

$$d(x,y)(s) < s \quad \Rightarrow \quad 1 - F_{u,v}(\Psi(s)) < \Psi(s).$$

If f is weakly demicompact or R is a strict t-conorm with a multiplicative generator ξ such that

$$\lim_{n\to\infty} \prod_{i=n}^{\infty} \xi(\Psi^i(s_1)) = 1$$

then there exists $x \in X$ such that $x \in fx$.

Chapter 6

Fixed point theorems in topological vector spaces and applications to random normed spaces

This chapter contains some results from the fixed point theory in topological vector spaces which are of special interest for the fixed point theory in random normed spaces. Namely, a random normed space $(S, \mathcal{F}, T)$ with a continuous t-norm T is a topological vector space which is not necessarily a locally convex space. It is known that a random normed space $(S, \mathcal{F}, T)$ is a locally convex space when T is a continuous t-norm of H-type. In the fixed point theory in a not necessarily locally convex topological vector spaces a very useful notion is that of an admissible subset which was introduced by Klee. Many important function spaces are admissible. If $(\Omega, \mathcal{A}, P)$ is a probability measure space and $S(\Omega, \mathcal{A}, P)$ the space of all classes of equivalence of measurable mappings $X : \Omega \to M$, where (M, d) is the separable metric space $(\mathbb{R}^n, \|\cdot\|_{\mathbb{R}^n})$, then $S(\Omega, \mathcal{A}, P)$ is an admissible topological vector space in the topology of convergence in probability, see [174], and $(S(\Omega, \mathcal{A}, P), \mathcal{F}, T_{\mathbf{L}})$ is a random normed space, where

$$\mathcal{F}_{X,Y}(u) = P(\|X - Y\|_{\mathbb{R}^n} < u)\ (X, Y \in S(\Omega, \mathcal{A}, P), u \geq 0).$$

By the KKM principle of Ky Fan many results from Nonlinear Analysis can be obtained [286]. In section 6.1 Tychonoff's and Browder's fixed point theorems are proved by the KKM principle of Ky Fan and applied to mappings on random normed spaces. The notion of an admissible subset of a topological vector space ([160]) is defined in section 6.2. Some fixed point theorems for mappings defined on admissible subsets are proved and applications to random normed spaces are given. Several fixed point theorems of Krasnoselski's type are proved in section 6.3.

Section 6.4 deals with continuous dependence of the fixed point on parameters and in section 6.5 a degree theory for mappings in topological vector spaces is given.

Throughout the chapter we assume for the probabilistic metric spaces $(S, \mathcal{F}, \tau)$ that Range$(\mathcal{F}) \subset \mathcal{D}^+$.

6.1 Tychonoff's and Browder's fixed point theorems

We start this section with the well known Brouwer fixed point theorem, [24].

Theorem 6.1 *Every non-empty convex and compact subset of* $\mathbb{R}^n$ *has the fixed point property.*

This means that every continuous mapping $f : K \to K$, where K is a non-empty convex and compact subset of $\mathbb{R}^n$, has at least one fixed point.

The Brouwer fixed point theorem is generalized in many directions. J. Schauder generalized the Brouwer fixed point theorem to normed spaces and A. Tychonoff to locally convex spaces. Further generalizations in normed and locally convex spaces can be found in [53].

Ky Fan proved in [59] the KKM-maps principle, which is one of the fundamental results of Nonlinear Analysis, see [286]. We shall give the proof of the KKM-maps principle together with some applications to the fixed point theory in topological vector spaces.

Definition 6.2 *Let E be a vector space and X a nonempty subset of E. A function $G : X \to 2^E$ is called a Knaster–Kuratowski–Mazurkiewicz map (shortly, KKM-map) provided*

$$\mathrm{co}\{x_1, x_2, \ldots, x_n\} \subset \bigcup_{i=1}^{n} G(x_i),$$

for each finite subset $\{x_1, x_2, \ldots, x_n\} \subset X$.

Theorem 6.3 (The KKM-maps principle) *Let E be a topological vector space, X a nonempty subset of E and $G : X \to 2^E$ a KKM-map such that $G(x)$ is closed for every $x \in X$. If there exists an $x_0 \in X$ such that $G(x_0)$ is compact then*

$$\bigcap_{x \in X} G(x) \neq \varnothing.$$

Proof. We prove that the family $(G(x))_{x \in X}$ has the finite intersection property, i.e., that for every finite subset $\{x_1, x_2, \ldots, x_s\} \subset X$ we have

$$\bigcap_{i=1}^{s} G(x_i) \neq \varnothing.$$

Suppose, on the contrary, that there exists a finite subset $\{x_1, x_2, \ldots, x_s\} \subset X$ such that

$$\bigcap_{i=1}^{s} G(x_i) = \varnothing. \tag{6.1}$$

Put $L = \text{Lin}\,\{x_1, x_2, \ldots, x_s\}$, i.e., let L be the finite dimensional flat spanned by $\{x_1, x_2, \ldots, x_s\}$. It is well known that the induced topology on L is the Euclidean topology, and let d be the Euclidean metric on L. Since $G(x_i)$ is closed in E for every $i = 1, 2, \ldots, s$, the set $L \cap G(x_i)$ is closed in L for every $i = 1, 2, \ldots, s$, and

$$d(x, L \cap G(x_i)) = 0 \quad \Longleftrightarrow \quad x \in L \cap G(x_i).$$

From (6.1) it follows that

$$\bigcap_{i=1}^{s} (L \cap G(x_i)) = \varnothing.$$

which implies that for every $u \in L$

$$\sum_{i=1}^{s} d(u, L \cap G(x_i)) > 0. \tag{6.2}$$

Let $M = \text{co}\{x_1, x_2, \ldots, x_s\}$ and put $\lambda(u) = \sum\limits_{i=1}^{s} d(u, L \cap G(x_i))$, for every $u \in M$. From (6.2) it follows that $\lambda : M \to (0, \infty)$ is a continuous mapping and that $f : M \to M$, defined by

$$f(u) = \frac{1}{\lambda(u)} \sum_{i=1}^{s} d(u, L \cap G(x_i)) x_i, \quad u \in M,$$

is a continuous mapping.

From the Brouwer fixed point theorem it follows that there exists $z \in M$ such that $z = f(z)$. Let $J = \{j \mid j \in \{1, 2, \ldots, s\},\ d(z, L \cap G(x_j)) > 0\}$. From (6.2) it follows that $J \neq \varnothing$ and from $z = f(z)$ we have

$$z = \frac{1}{\lambda(z)} \sum_{j \in J} d(z, L \cap G(x_j)) x_j \in \text{co}\{x_j \mid j \in J\}.$$

On the other hand, since G is a KKM-map it follows that

$$z \in \text{co}\{x_j \mid j \in J\} \subseteq \bigcup_{j \in J} G(x_j).$$

This means that there exists at least one $j_0 \in J$ such that $z \in G(x_{j_0})$, i.e., that $d(z, L \cap G(x_{j_0})) = 0$, which is a contradiction. Hence the family $\{G(x)\}_{x \in X}$ has the finite intersection property. Then $\bigcap\limits_{x \in X} G(x) \neq \varnothing$. Namely, from $\bigcap\limits_{x \in X} G(x) = \varnothing$ it follows that $\bigcup_{x\in X}((G(x))^c\cap G(x_0)) = G(x_0)$, which implies that $((G(x))^c\cap G(x_0))_{x\in X}$ is an open cover of the compact set $G(x_0)$. Hence there exists a finite cover $((G(u_i))^c\cap G(x_0))_{i=1}^m$ of the set $G(x_0)$ implying $\bigcap_{i=1}^m G(u_i)\cap G(x_0) = \varnothing$, which contradicts the finite intersection property of the family $(G(x)\cap G(x_0))_{x\in X}$. □

The KKM-maps principle is generalized in many directions, see, e.g., [286]. Since many results from KKM theory are formulated for topological vector spaces they can be applied to random normed spaces.

For the purpose of this paragraph we shall give only two applications of the KKM-maps principle. The first one is the well known Tychonoff fixed point theorem and the second one is the fixed point theorem of F. Browder [22].

Theorem 6.4 *Let E be a Hausdorff locally convex topological vector space and K a nonempty, convex and compact subset of E. Then every continuous mapping $f : K\to K$ has at least one fixed point.*

Proof. Let $(p_j)_{j\in J}$ be the family of semi-norms which defines the topology in E. A point $x_0\in K$ is a fixed point of the mapping f if and only if

$$p_j(x_0 - fx_0) = 0 \quad\text{for every}\quad j\in J. \tag{6.3}$$

Let $A_j = \{y \mid y\in K,\ p_j(y-fy) = 0\}$, $j\in J$. From (6.3) it follows that $x_0\in K$ is a fixed point of the mapping f if and only if $x_0\in\bigcap_{j\in J} A_j$. Hence in order to prove that the set of fixed points of the mapping f is non-empty we have to prove that

$$\bigcap_{j\in J} A_j \neq \varnothing. \tag{6.4}$$

From the compactness of the set K it follows that we need to prove only the finite intersection property of the family $(A_j)_{j\in J}$. Let $\{i_1, i_2, \ldots, i_s\}$ be an arbitrary finite subset of J and define the mapping $G : K\to 2^E$ by

$$G(x) = \left\{y \mid y\in K,\ \sum_{k=1}^{s} p_{i_k}(y-fy) \le \sum_{k=1}^{s} p_{i_k}(x-fy)\right\}, \quad x\in K.$$

If $y_0\in K$ is such that $y_0\in\bigcap_{x\in K} G(x)$ then

$$\sum_{k=1}^{s} p_{i_k}(y_0 - fy_0) \le \sum_{k=1}^{s} p_{i_k}(x - fy_0) \quad \text{for every } x\in K,$$

which implies that $\sum_{k=1}^{s} p_{i_k}(y_0 - fy_0) = 0$, since $fy_0\in K$.

Hence

$$y_0 \in \bigcap_{x\in K} G(x) \quad \Rightarrow \quad y_0 \in \bigcap_{k=1}^{s} A_{i_k}.$$

We shall prove that $\bigcap_{x\in K} G(x) \neq \emptyset$ using the KKM-maps principle.

For every $x \in K$, $x \in G(x)$ and therefore $G(x) \neq \emptyset$ for every $x \in K$. Since f is a continuous mapping, $G(x)$ is closed for every $x \in K$.

It remains to prove that G is a KKM-map. Suppose, on the contrary, that G is not a KKM-map. Then there exists a finite set $\{x_1, x_2, \ldots, x_m\} \subset K$ such that $\mathrm{co}\{x_1, x_2, \ldots, x_m\}$ is not a subset of $\bigcup_{i=1}^{m} G(x_i)$, which implies the existence of an element $y \in \mathrm{co}\{x_1, x_2, \ldots, x_m\}$ such that

$$y \notin \bigcup_{j=1}^{m} G(x_j). \tag{6.5}$$

From $y \in \mathrm{co}\{x_1, x_2, \ldots, x_m\}$ it follows that $y = \sum_{j=1}^{m} \lambda_j x_j$ for some non-negative λ_j, $j = 1, 2, \ldots, m$, such that $\sum_{j=1}^{m} \lambda_j = 1$, and (6.5) implies the inequalities

$$\sum_{k=1}^{s} p_{i_k}(y - fy) > \sum_{k=1}^{s} p_{i_k}(x_j - fy), \quad j = 1, 2, \ldots, m. \tag{6.6}$$

Hence

$$\begin{aligned} \sum_{j=1}^{m} \lambda_j \sum_{k=1}^{s} p_{i_k}(y - fy) &= \sum_{k=1}^{s} p_{i_k}(y - fy) \\ &> \sum_{k=1}^{s} p_{i_k}\Big(\sum_{j=1}^{m} \lambda_j (x_j - fy)\Big), \end{aligned}$$

and therefore

$$\sum_{k=1}^{s} p_{i_k}(y - fy) > \sum_{k=1}^{s} p_{i_k}(y - fy),$$

which is a contradiction.

By the KKM-maps principle we conclude that $\bigcap_{x\in K} G(x) \neq \emptyset$. □

Since a random normed space $(S, \mathcal{F}. T)$, where T is a continuous t-norm of H-type, is a Hausdorff locally convex topological vector space, from Tychonoff's fixed point theorem we obtain the following corollary.

Corollary 6.5 *Let $(S, \mathcal{F}, T)$ be a random normed space, where t-norm T is continuous and of H-type. Then every non-empty convex and compact subset of S has the fixed point property.*

Using the KKM-maps principle the following fixed point theorem of Browder for multi-valued mappings can be proved.

Theorem 6.6 *Let K be a non-empty convex and compact subset of a topological vector space E and $F : K \to 2^K$ a multi-valued mapping such that:*

(i) $F^{-1}(y) = \{x \mid x \in K,\ y \in F(x)\}$ is open for every $y \in K$;
(ii) $F(x)$ is nonempty and convex for every $x \in K$.

Then there exists a point $x_0 \in K$ such that $x_0 \in F(x_0)$.

Proof. Let $G : K \to 2^K$ be defined by $G(x) = K \setminus F^{-1}(x)$ for every $x \in K$. Then $\bigcap_{x \in K} G(x) = \varnothing$. Indeed, $F(x) \neq \varnothing$ for every $x \in K$, implies that $K = \bigcup_{y \in K} F^{-1}(y)$ and thus $\bigcap_{x \in K} G(x) = \varnothing$. By the KKM-maps principle we conclude that G is not a KKM-map. This means that there exists a finite subset $\{x_1, x_2, \ldots, x_m\} \subset K$ such that $\operatorname{co}\{x_1, x_2, \ldots, x_m\}$ is not a subset of $\bigcup_{i=1}^{m} G(x_i)$. Let $y \in \operatorname{co}\{x_1, x_2, \ldots, x_m\}$ and $y \notin G(x_i)$ for every $i = 1, 2, \ldots, m$. Then $y \in F^{-1}(x_i)$ for every $i = 1, 2, \ldots, m$, which means that $x_i \in F(y)$ for every $i = 1, 2, \ldots, m$. Since $F(y)$ is convex

$$\operatorname{co}\{x_1, x_2, \ldots, x_m\} \subset F(y),$$

and thus $y \in F(y)$. □

Corollary 6.7 *Let $(S, \mathcal{F}, T)$ be a random normed space where t-norm T is continuous, and K a non-empty convex and compact subset of S. If $F : K \to 2^K$ is such that (i) and (ii) in Theorem 6.6 hold, then there exists a point $x_0 \in K$ such that $x_0 \in F(x_0)$.*

By 2^S_{co} we shall denote the family of all non-empty and convex subsets of S.

Theorem 6.8 *Let $(S, \mathcal{F}, T)$ be a complete random normed space with a continuous t-norm T, M a non-empty convex and compact subset of S, H a linear mapping from S into S, $\Phi : M \to 2^S_{\text{co}}$ such that $HM + \Phi M \subseteq M$, and assume that the following conditions are satisfied:*

(i) one of the following two conditions is satisfied:

(a) For every $x \in M$ there exists $n(x) \in \mathbb{N}$ such that for every $y \in M$ and every $\varepsilon > 0$

$$F_{H^{n(x)}(x) - H^{n(x)}(y)}(q\varepsilon) \geq F_{x-y}(\varepsilon),$$

where $q \in (0, 1)$.

(b) Φ is closed and there exists $n_0 \in \mathbb{N}$ such that for every $x, y \in M$ and $\varepsilon > 0$

$$F_{H^{n_0}x - H^{n_0}y}(\varepsilon) \geq F_{x-y}(\varepsilon).$$

(ii) for every $y \in S$, the set $\Phi^{-1}(y)$ is open.

Then there exists at least one fixed point of the mapping $H + \Phi$.

Proof. For every $y \in \overline{\Phi(M)}$ define the mapping $G_y : M \to M$ by

$$G_y(x) = Hx + y, \quad x \in M.$$

Then for every $n \in \mathbb{N}$, $u \in M$, and $y \in \overline{\Phi(M)}$ there holds

$$G_y^n u = H^n u + \sum_{k=0}^{n-1} H^k y,$$

and (i) (a) implies that

$$F_{G_y^{n(x)}(x) - G_y^{n(x)}(u)}(q\varepsilon) \geq F_{x-u}(\varepsilon)$$

for every $x \in M$, $y \in \overline{\Phi(M)}$, $\varepsilon > 0$, and $u \in M$. Since t-norm T is continuous it follows that the compact set M is probabilistic bounded in the (ε, λ)-topology. By Theorem 3.54 it follows that for every $y \in \overline{\Phi(M)}$ there exists one and only one element $Ry \in M$ such that

$$Ry = HRy + y.$$

Since the set M is compact we can prove that the mapping $R : \overline{\Phi(M)} \to M$ is continuous. Let $(y_\alpha)_{\alpha \in \mathcal{A}}$ be a net from $\overline{\Phi(M)}$ such that $\lim_{\alpha \in \mathcal{A}} y_\alpha = y$ and $Ry_\alpha = HRy_\alpha + y_\alpha$ for every $\alpha \in \mathcal{A}$. Since M is a compact set there exists a convergent subnet $(Ry_\beta)_{\beta \in \mathcal{B}}$ of the net $(Ry_\alpha)_{\alpha \in \mathcal{A}}$ ($\mathcal{B}$ is a cofinal subset of $\mathcal{A}$) and therefore

$$\lim_{\beta \in \mathcal{B}} Ry_\beta = H(\lim_{\beta \in \mathcal{B}} Ry_\beta) + \lim_{\beta \in \mathcal{B}} y_\beta = H(\lim_{\beta \in \mathcal{B}} Ry_\beta) + y.$$

This implies that $\lim_{\beta \in \mathcal{B}} Ry_\beta = Ry$ for every cofinal subnet $(Ry_\beta)_{\beta \in \mathcal{B}}$ of the net $(Ry_\alpha)_{\alpha \in \mathcal{A}}$ and thus $\lim_{\alpha \in \mathcal{A}} Ry_\alpha = Ry$

Next we define the mapping $R^* : M \to 2^M$ by $R^*x = \bigcup_{y \in \Phi x} Ry, \quad x \in M$ and prove that R^*x is a convex set for every $x \in M$. Since the mapping H is affine it follows that for every $x \in M$ and every $\alpha, \beta \geq 0$ such that $\alpha + \beta = 1$, we have

$$R(\alpha y_1 + \beta y_2) = \alpha R y_1 + \beta R y_2, \quad y_1, y_2 \in \Phi(x). \tag{6.7}$$

Since Φ is a convex valued mapping, using the relation (6.7) it is easy to show that R^*x is a convex set. The mapping R is a one to one mapping and so there exists $R^{-1} : R(\Phi(M)) \to \Phi(M)$. Furthermore, $(R^*)^{-1}y = \{x \mid y \in R^*x\}$, and we shall prove that

$$(R^*)^{-1}y = \Phi^{-1}(R^{-1}y).$$

First, we show that $(R^*)^{-1}y \subseteq \Phi^{-1}(R^{-1}y)$. To that end suppose that $x \in (R^*)^{-1}y$. Then $y \in R^*x$, which means that $y \in \bigcup_{z\in\Phi x} Rz$. Thus there exists $z \in \Phi x$ such that $y = Rz$. Then we have

$$R^{-1}y = z \in \Phi(x)$$

which implies $x \in \Phi^{-1}(R^{-1}y)$.

Suppose now that $x \in \Phi^{-1}(R^{-1}y)$. Then $R^{-1}y \in \Phi(x)$, i.e., $R^{-1}y = z$ and $z \in \Phi(x)$. This implies that $y = Rz$, $z \in \Phi x$, and so $y \in R^*x$, which means that $x \in (R^*)^{-1}y$, and since $\Phi^{-1}y$ is open for every $y \in S$ we conclude by Corollary 6.7 that $\text{Fix}(R^*) \neq \varnothing$. Since $\text{Fix}(R^*) \subseteq \text{Fix}(H + \Phi)$ it follows that $\text{Fix}(H + \Phi) \neq \varnothing$.

Now suppose that (i) (b) is true. For every $n \in \mathbb{Z}^+$ we shall define the mappings H_n and Φ_n by

$$H_n x = \lambda_n Hx, \qquad \Phi_n x = \lambda_n \Phi(x) + (1 - \lambda_n)x_0 \quad \text{for every } x \in M,$$

where $(\lambda_n)_{n\in n\mathbb{N}}$ is a sequence from the interval $(0, 1)$ such that $\lim\limits_{n\to\infty} \lambda_n = 1$. Then for every $m \in \mathbb{N}$ and $x \in M$, $H_m^{n_0} x = \lambda_m^{n_0} H^{n_0} x$ and therefore

$$\begin{aligned} F_{H_m^{n_0}x - H_m^{n_0}y}(\lambda_m^{n_0}\varepsilon) &= F_{\lambda_m^{n_0}(H^{n_0}x - H^{n_0}y)}(\lambda_m^{n_0}\varepsilon) \\ &= F_{H^{n_0}x - H^{n_0}y}(\varepsilon) \\ &\geq F_{x-y}(\varepsilon). \end{aligned}$$

Since M is convex we have

$$H_n M + \Phi_n M = \lambda_n(HM + \Phi M) + (1 - \lambda_n)x_0 \subset M.$$

For every $y \in S$ and every $n \in \mathbb{N}$, $\Phi_n^{-1}y$ is open since

$$\begin{aligned} \Phi_n^{-1}y &= \{x \mid y \in \Phi_n x\} \\ &= \{x \mid y \in \lambda_n \Phi x + (1 - \lambda_n)x_0\} \\ &= \Big\{x \mid \frac{y - (1 - \lambda_n)x_0}{\lambda_n} \in \Phi x\Big\} \\ &= \Phi^{-1}\Big(\frac{y - (1 - \lambda_n)x_0}{\lambda_n}\Big). \end{aligned}$$

Applying (i)(a) we conclude that for every $n \in \mathbb{N}$ there exists $x_n \in M$ such that $x_n \in H_n x_n + \Phi_n x_n$. This means that there exists an element $y_n \in \Phi x_n$ such that

$$x_n = \lambda_n H x_n + \lambda_n y_n + (1 - \lambda_n)x_0.$$

Then we have

$$\begin{aligned}\lim_{n\to\infty} x_n - Hx_n - y_n &= \lim_{n\to\infty} (\lambda_n - 1)Hx_n + (\lambda_n - 1)y_n + (1-\lambda_n)x_0 \\ &= \lim_{n\to\infty} (\lambda_n - 1)(Hx_n + y_n) + (1-\lambda_n)x_0 \\ &= 0\end{aligned}$$

since $Hx_n + y_n \subseteq M$ and M is bounded. By the compactness of M there exists a subsequence $(n_k)_{k\in\mathbb{N}}$ such that

$$\lim_{k\to\infty} Hx_{n_k} + y_{n_k} = y^*$$

and therefore $\lim_{k\to\infty} x_{n_k} = y^*$. From $y_{n_k} \in \Phi x_{n_k}$ $(k \in \mathbb{N})$ and $\lim_{k\to\infty} y_{n_k} = y^* - Hy^*$ we have $y^* - Hy^* \in \Phi(y^*)$ since Φ is a closed mapping. This means that $y^* \in \mathrm{Fix}(\Phi + H)$ and the proof is complete. □

6.2 Admissible subsets of topological vector spaces and their application on the fixed point theory

Schauder's conjecture which states *every compact convex subset of a linear metric space has the fixed point property,* is one of the most resistant open problems in the fixed point theory of non-locally convex linear metric space. In fact Schauder posed the problem in The Scottish Book [184] in 1935. but it remains open, despite efforts by many analysts and topologists for many decades. For instance, it is not known whether a compact and convex subset of L^p $(0 \le p < 1)$ has the fixed point property. Schauder's problem is one of the most important problems in the fixed point theory. Up to now only some partial answers to Schauder's problem have been obtained. In [209] the authors have shown that all Roberts spaces have the fixed point property, where a Roberts space is any compact convex set with no extreme point.

Whether Tychonoff's fixed point theorem holds in a general topological vector space is still an open problem. However. there are many papers which deal with the fixed point problem in not necessarily locally convex topological vector spaces [96].

The notion of an admissible subset of a topological vector space is introduced in [160], and this notion turned out to be very important in the fixed point theory.

Hahn and Pötter proved [123] a generalization of the Tychonoff fixed point theorem in a general topological vector space using the notion of an admissible subset $K \subset E$, and we shall apply their result in this chapter.

First, we give the definition of an admissible subset of a topological vector space. All topological vector spaces will be assumed to be Hausdorff.

Definition 6.9 *Let X be a topological vector space and $\mathcal{V}$ be the fundamental system of neighbourhoods of zero in X. A subset $K \subseteq X$ is said to be admissible if for every compact subset A of K and every $U \in \mathcal{V}$ there exists a continuous mapping $h : A \to K$ such that:*

1. *$\dim Lin\, h(A) < \infty$ (where Lin stands for linear hull);*
2. *for every $x \in A$, $x - h(x) \in U$.*

If $K = X$ we say that X is an admissible topological vector space.

M. Nagumo proved in [204] that every convex subset of a locally convex topological vector space is admissible.

There are some non-locally convex topological vector spaces which are admissible. J. Mach [180] proved the admissibility of function spaces $L_{\phi,k}$ and L_ϕ. T. Riedrich proved the admissibility of the space $L^p(0,1)(0 < p < 1)$. The admissibility of some function spaces are proved by J. Ishii in [142] and by C. Krauthausen in [174]. Some sufficient conditions for the admissibility are given by Klee in [160]. An admissible non-locally convex topological vector space is the space $S(0,1)$ of all the equivalence classes of real finite measurable mappings on $(0,1)$. More generally (see [174]), if $(\Omega, \mathcal{A}, P)$ is a probability measure space and $S(\Omega, \mathcal{A}, P)$ is the space of all classes of measurable functions $X : \Omega \to \mathbb{R}^n$, then $(S(\Omega, \mathcal{A}, P), \mathcal{F}, T_{\mathbf{L}})$ is an admissible topological vector space and a random normed space, where for every $\varepsilon > 0$ and every $X, Y \in S(\Omega, \mathcal{A}, P)$

$$F_{X,Y}(\varepsilon) = P(\{\omega \mid \omega \in \Omega, \|X(\omega) - Y(\omega)\|_{\mathbb{R}^n} < \varepsilon\}).$$

In this section the proof of the admissibility of the random normed space $S(0,1)$ is given, see [256].

It is still an open question whether there exists a convex non-admissible subset of a topological vector space.

An example of a non-convex non-admissible subset of ℓ_2 is $\{(t^n)_{n\in\mathbb{N}} \mid 0 \le t \le 1/2\}$ [174].

Let X be a topological space, Y a topological space, and $f : X \to Y$. The mapping f is said to be *compact* if f is continuous and $\overline{f(X)}$ is compact. A mapping $h : X \to E$, where E is a topological vector space, is said to be *finite* if h is compact and $h(X)$ is a subset of a finite-dimensional subspace of E.

It is easy to see that the following proposition is valid.

Proposition 6.10 *A topological vector space E is admissible if and only if for every compact mapping $f : A \to E$, where A is a topological space, and for every neighbourhood V of zero in E there exists a finite mapping $h_V : A \to E$ such that*

$$f(x) - h_V(x) \in V \text{ for every } x \in A.$$

In [123] S. Hahn and K.F. Pötter proved the following fixed point theorem.

Theorem 6.11 *Let X be a topological vector space, W a closed neighbourhood of zero in X, K a convex closed and admissible subset of X, $0 \in K$, and $f : W \cap K \to K$ a continuous mapping such that $\overline{f(W \cap K)}$ is compact. If the following implication holds*

$$(\forall x \in \partial W \cap K)\ (\forall a \in \mathbb{R})\ f(x) = ax \quad \Rightarrow \quad a \leq 1$$

then there exists $x \in W \cap K$ so that $x = f(x)$.

Remark 6.12 Theorem 6.11 is a generalization of Tychonoff's fixed point theorem ($W = E$) since every convex subset of a locally convex space is admissible and the fixed point property is invariant with respect to the translation. If $K = E$ we obtain the fixed point theorem from [122]. From Theorem 6.11 the fixed point theorems of J.Schauder [261], E.Rothe [258], M.Altman [7] and M.Landsberg [176] can be obtained as corollaries. Later on in this section we shall give the proof of a more general fixed point theorem for multivalued mappings.

Remark 6.13 Nguyen To Nhu [210] introduced the notion of weakly admissible compact convex subsets of a linear metric space and proved the following theorem: *Any weakly admissible compact convex subset has the fixed point property.* He posed two open problems:

Question 1. Is every compact, convex subset weakly admissible;

Question 2. Is every weakly admissible compact convex set

(i) an absolute retract?

(ii) admissible?

An application of Theorem 6.11 to the integral equations in $S(0,1)$ is given by Hahn and Pötter in [123] and we shall give this result next.

The elements of $S(0,1)$ are denoted by $\widehat{x}, \widehat{y}$, and the metric d in $S(0,1)$ is defined by

$$d(\widehat{x}, \widehat{y}) = \int_0^1 \frac{|x(t) - y(t)|}{1 + |x(t) - y(t)|}\, \mathrm{m}_0(dt)$$

where $\{x(t)\}$ is an element of the class $\widehat{x}$ and m_0 is the Lebesgue measure.

The convergence in the metric space $(S(0,1), d)$ is the convergence in the measure m_0, i.e., in the (ε, λ)-topology. It is known that $S(0,1)$ is a complete, non-locally convex topological vector space, which is also a random normed space.

In [256] the admissibility of $S(0,1)$ was proved, and we shall give here only the sketch of the proof.

We write $\int_E f(t)\,dt$ instead of $\int_E f(t)\,dm_0(t)$. For every $n \in \mathbb{N}$ and $k \in \{1,2,\dots,n\}$ let

$$I_{k,n} = \left[\frac{k-1}{n}, \frac{k}{n}\right], \quad \chi_{k,n}(t) = \begin{cases} 1 & \text{if } t \in I_{k,n}, \\ & \qquad\qquad (0 \le t \le 1) \\ 0 & \text{if } t \notin I_{k,n}, \end{cases}$$

and $F_n = \operatorname{Lin}\{\chi_{1,n}, \chi_{2,n}, \dots, \chi_{n,n}\}$. Further on, for every nonnegative $f \in S(0,1)$, every $n \in \mathbb{N}$ and every $k \in \{1,2,\dots,n\}$ the number $s_{k,n}(f)$ is defined by

$$s_{k,n}(f) = \left(\int_{I_{k,n}} \frac{f(t)}{1+f(t)}\,dt\right)\left(\int_{I_{k,n}} \frac{1}{1+f(t)}\,dt\right)^{-1}.$$

Let $s_n(f) = \sum_{k=1}^{n} s_{k,n}(f)\chi_{k,n}$ and for an arbitrary $g \in S(0,1)$

$$s_n(g) = s_n(g^+) - s_n(g^-).$$

Then $s_n : S(0,1) \to F_n$, for every $n \in \mathbb{N}$ and in [256] the following result is obtained:

1) The mapping s_n is a retract of $S(0,1)$ on F_n, for every $n \in \mathbb{N}$;
2) The sequence $(s_n)_{n\in\mathbb{N}}$ is equicontinuous;
3) For every $f \in S(0,1)$
$$\lim_{n\to\infty} s_n(f) = f.$$

Using 2) and 3) it is easy to prove that $S(0,1)$ is admissible.

Let K be an arbitrary compact subset of $S(0,1)$ and $\varepsilon > 0$. Since the sequence $(s_n)_{n\in\mathbb{N}}$ is equicontinuous, there exists a $\delta = \delta(\varepsilon, f) > 0$ (($f \in S(0,1)$) such that for every $n \in \mathbb{N}$

$$\overline{f} \in S(0,1),\ d(\overline{f}, f) \le \delta \quad \Rightarrow \quad d(s_n(f),\ s_n(\overline{f})) \le \frac{\varepsilon}{3}\,.$$

Choose that $\delta < \frac{\varepsilon}{3}$ and let $N = N(f,\varepsilon) \in \mathbb{N}$ be such that

$$d(f, s_n(f)) \le \frac{\varepsilon}{3} \quad \text{for every } n \ge N.$$

Using the inequality

$$d(\overline{f}, s_n(\overline{f})) \le d(f, \overline{f}) + d(s_n(f), s_n(\overline{f})) + d(f, s_n(f))$$

we conclude that for every $\bar{f}$ which belongs to the set

$$U(f) = \{\bar{f} \mid \bar{f} \in S(0,1), d(\bar{f}, f) \leq \delta\}$$

we have

$$d(\bar{f}, s_n(\bar{f})) \leq \frac{\varepsilon}{3} + \frac{\varepsilon}{3} + \frac{\varepsilon}{3} = \varepsilon.$$

Since K is compact there exists a finite set $\{f_1, f_2, \ldots, f_r\}$ from $S(0,1)$ such that

$$K \subseteq \bigcup_{i=1}^{r} U(f_i).$$

Let $N_0 = \max\{N(f_i, \varepsilon) \mid i \in \{1, 2, \ldots, r\}\}$ and $g \in K$. Then there exists $i_0 \in \{1, 2, \ldots, r\}$ such that $g \in U(f_{i_0})$. Then

$$d(g, s_n(g)) \leq \varepsilon \text{ for every } n \geq N_0.$$

This implies that $S(0,1)$ is admissible. The admissibility of $S(\Omega, \mathcal{A}, P)$ is proved in [174].

Proposition 6.14 *Let $G : [0,1] \times [0,1] \to \mathbb{R}$ be a continuous mapping such that $|G(s,t)| \leq 1$, for every $s, t \in [0,1]$. Then for every $\lambda \in [-\frac{1}{2}, \frac{1}{2}]$ and every $y \in S(0,1)$ so that $|y(s)| \leq k/2 < 1$, for every $s \in [0,1]$, there exists a solution $x_0 \in S(0,1)$ of the integral equation*

$$x(s) = y(s) + \lambda \cdot \int_0^1 G(s,t) \frac{x(t)}{1 + |x(t)|} \, dt, s \in [0,1]$$

such that $d(x_0, 0) \leq k$.

Proof. Let $W = \{x \mid x \in S(0,1), d(x,0) \leq k\}$ and $F : W \to S(0,1)$ is defined by

$$(F(\hat{x}))(s) = y(s) + \lambda \cdot \int_0^1 G(s,t) \frac{x(t)}{1 + |x(t)|} \, dt, \quad s \in [0,1], \ \hat{x} \in W.$$

Then F is a continuous mapping from W into $S(0,1)$, which can be easily proved by the Lebesgue theorem. Let us prove that $\overline{F(W)}$ is compact. First, we prove, by the Arzela–Ascoli theorem, that $F'(W)$ is relatively compact in the space of continuous functions $C(0,1)$, where $F' : W \to C(0,1)$ is defined by

$$(F'(\hat{x}))(s) = \int_0^1 G(s,t) \frac{x(t)}{1 + |x(t)|} \, dt, \quad s \in [0,1], \hat{x} \in W.$$

In $C(0,1)$ the topology is defined by the norm

$$\|\tilde{x}\| = \sup_{t \in [0,1]} |x(t)|, \qquad \tilde{x} = \{x(t)\}, \tilde{x} \in C(0,1).$$

It follows that $|(F'(\hat{x}))(s)| \leq 1$, $s \in [0,1]$, and

$$|(F'(\hat{x}))(s_1) - (F'(\hat{x}))(s_2)| \leq k \max_{t\in[0,1]} |G(s_1,t) - G(s_2,t)| \qquad (s_1, s_2 \in [0,1])$$

and since the function G is uniformly continuous it follows that $F'(W)$ is relatively compact in $C(0,1)$ and so in $S(0,1)$. Since $\overline{F'(W)}$ is compact in $C(0,1)$ it follows that F is a compact mapping from W into $S(0,1)$. We prove that $x \in \partial W$ and $F(x) = \alpha x$ implies that $\alpha \leq 1$. This can be easily obtained from the following inequalities:

$$\begin{aligned} |\alpha||x(s)| &\leq |y(s)| + |\lambda| \cdot \int_0^1 |G(s,t)| \frac{|x(t)|}{1+|x(t)|}\, dt \\ &\leq \frac{k}{2} + 2 \cdot |\lambda| \cdot \frac{k}{2} \\ &\leq k \leq k(1 + |x(s)|). \end{aligned}$$

This implies that

$$|\alpha| k = |\alpha| \cdot \int_0^1 \frac{|x(s)|}{1+|x(s)|}\, ds \leq k$$

and so $|\alpha| \leq 1$. Thus all the conditions of Theorem 6.11 are satisfied and there exists an $x_0 \in S(0,1)$ such that $d(x_0, 0) \leq k$ and

$$y(s) = x_0(s) - \lambda \cdot \int_0^1 G(s,t) \frac{x_0(t)}{1+|x_0(t)|}\, dt, \quad s \in [0,1].$$

□

In the last 30 years there has been an increasing interest in the fixed point theory for multi-valued mappings in topological vector spaces. The fundamental results in this theory are the fixed point theorem of S. Kakutani [154] and the results of the paper of J.V. Neumann [208] in which some applications of the fixed point theory for multi-valued mappings in the theory of games are given.

The generalizations of Kakutani's fixed point theorem in Banach spaces are given by H.F. Bohnenblust and S. Karlin [20] and I.L. Glicksberg [73]. K. Fan proved in [59] a generalization of Tychonoff's fixed point theorem. We mention also results from the papers of F.E. Browder [22], K. Fan [59], C.J. Himmelberg [135], A. Idzik [138], E. Tarafdar and T. Husain [306], E. Tarafdar [305], A.W. Kaplan [157], S. Hahn [119, 120], M. Lassonde [177] and O. Hadžić [87, 88, 89, 90].

A bibliography on the degree theory of multi-valued mappings can be found in the book [96]. Some fixed point theorems for densifying multi-valued mappings are proved in [44, 137, 212, 237].

Here we shall give a result of S. Hahn proved in [120]. The proof of this result is based of Kakutani's fixed point theorem. From Hahn's fixed point theorem we can obtain, as corollaries, many well known fixed point theorems for multi-valued mappings. In [151] further generalizations of Hahn's results are obtained.

First, we shall give some definitions and notations. Let X and Y be two Hausdorff topological spaces and $\mathcal{S}(Y)$ a system of non-empty subsets of Y. We recall that a mapping $F : X \to \mathcal{S}(Y)$ is said to be *upper semi-continuous on* X if F is upper semi-continuous at each point of X. If $F : X \to \mathcal{S}(Y)$ is upper semi-continuous on X and $\overline{F(X)}$ is compact in Y, F is said to be *compact*.

Let A be a subset of a topological vector space E. By $\mathcal{R}(A)$ we denote the system of all nonempty, convex, closed (in the induced topology) subsets of E.

The following generalization of Tychonoff's fixed point theorem for multi-valued mappings in locally convex topological vector spaces is well known, see [53].

Theorem 6.15 *Let K be a non-empty, convex and compact subset of a locally convex space E and $F : K \to \mathcal{R}(K)$ an upper semi-continuous mapping. Then there exists a fixed point of the mapping F.*

If $E = \mathbb{R}^n$ from Theorem 6.15 the Kakutani's fixed point theorem follows.

In [120] S. Hahn introduced the notion of an $(\mathcal{S})$-admissible subset of a topological vector space.

Definition 6.16 *Let E be a topological vector space, $Z = \overline{Z} \subseteq E$, and $\mathcal{S}(Z)$ a non-empty system of subsets of Z. The set Z is said to be $(\mathcal{S})$-admissible if for each compact mapping $F : A \to \mathcal{S}(Z)$, where A is a topological space, and for each neighbourhood V of zero in E there exists a finite-dimensional vector subspace E_V of E and a compact mapping $F_V : A \to \mathcal{S}(Z)$ such that $F_V(A) \subseteq E_V$ and $F_V(x) \subseteq F(x) + V$, for every $x \in A$. If $Z = E$ then E is called an $(\mathcal{S})$-admissible topological vector space.*

If $\mathcal{S}(Z) = \bigcup_{x \in Z} \{x\}$ then Z is $(\mathcal{S})$-admissible if it is admissible.

It is known that each closed convex subset of a locally convex space is $(\mathcal{R})$-admissible [119]. An example of $(\mathcal{R})$-admissible subsets of a topological vector space is given in the next proposition.

First, we shall introduce the notion of a subset of Zima's type.

Definition 6.17 *Let E be a topological vector space, $K \subseteq E$, and $\mathcal{V}$ the fundamental system of neighbourhoods of zero in E. The set K is said to be of Zima's type if for every $V \in \mathcal{V}$ there exists $U \in \mathcal{V}$ such that*

$$\mathrm{co}(U \cap (K - K)) \subseteq V.$$

Every convex subset of a locally convex space is of Zima's type.

Definition 6.18 *Let (E, p) be a para-normed space and $\varnothing \neq K \subset E$. The set K satisfies the Zima condition if there exists $C(K) > 0$ such that for every $\lambda \in [0, 1]$*

$$p(\lambda(x - y)) \leq C(K)p(x - y) \quad \textit{for every} \ \ x, y \in K.$$

It is easy to see that every subset K of a para-normed space which satisfies the Zima's condition is of Zima's type.

We shall give a non-trivial example of a subset of Zima's type in the para-normed space $S(\Omega, \mathcal{A}, P)$.

Example 6.19 Let $E = S(\Omega, \mathcal{A}, P)$ and

$$p(\widehat{x}) = \int\limits_{\Omega} \frac{\|x(\omega)\|_{\mathbb{R}^n}}{1 + \|x(\omega)\|_{\mathbb{R}^n}} \, dP, \tag{6.8}$$

where $\{x(t)\} \in \widehat{x} \in E$. If $s > 0$ let K_s be defined by

$$K_s = \{\widehat{x} \mid \widehat{x} \in E, \ \|x(\omega)\|_{\mathbb{R}^n} \leq s \ \text{ for every } \omega \in \Omega\}.$$

We prove that

$$p(\lambda(\widehat{x} - \widehat{y})) \leq (1 + 2s)\lambda p(\widehat{x} - \widehat{y})$$

for every $\widehat{x}, \widehat{y} \in K_s$ and every $\lambda \in [0, 1]$. If $\hat{x}, \hat{y} \in K_s$ then $\|x(\omega)\|_{\mathbb{R}^n} \leq s$, $\|y(\omega)\|_{\mathbb{R}^n} \leq s$ for every $\omega \in \Omega$ and therefore

$$\begin{aligned} 1 + \|x(\omega) - y(\omega)\|_{\mathbb{R}^n} &\leq (1 + 2s) + (1 + 2s)\lambda\|x(\omega) - y(\omega)\|_{\mathbb{R}^n} \\ &= (1 + 2s)(1 + \lambda\|x(\omega) - y(\omega)\|_{\mathbb{R}^n}). \end{aligned}$$

This implies that

$$\begin{aligned} p(\lambda(\hat{x} - \hat{y})) &= \int_{\Omega} \frac{\lambda\|x(\omega) - y(\omega)\|_{\mathbb{R}^n}}{1 + \lambda\|x(\omega) - y(\omega)\|_{\mathbb{R}^n}} dt \\ &\leq (1 + 2s)\lambda \int_{\Omega} \frac{\|x(\omega) - y(\omega)\|_{\mathbb{R}^n}}{1 + \|x(\omega) - y(\omega)\|_{\mathbb{R}^n}} dt \\ &= (1 + 2s)\lambda p(\hat{x} - \hat{y}). \end{aligned}$$

Hence $C(K_s) = 1 + 2s > 1$.

A convex subset $K \neq \varnothing$ of a para-normed space (E, p), which satisfies the Zima condition, is admissible [96] since it is of Zima's type. Namely, the following proposition holds.

Proposition 6.20 *Let E be a topological vector space and K a closed and convex subset of Zima's type of E. Then K is $(\mathcal{R})$-admissible.*

Proof. Let $\mathcal{V}$ be the fundamental system of neighbourhoods of zero in E and A be a topological space. If $F : A \to \mathcal{R}(K)$ is a compact mapping we have to prove that for every $V \in \mathcal{V}$ there exists a finite-dimensional compact mapping $F_V : A \to \mathcal{R}(K)$ (dim Lin $(F_V(A)) < \infty$) such that

$$F_V(x) \subseteq F(x) + V \ \text{ for every } x \in A. \tag{6.9}$$

Since $\overline{F(A)}$ is a compact set there exists a finite set $\{x_1, x_2, \ldots, x_n\} \subseteq F(A)$ such that $F(A) \subseteq \bigcup_{i=1}^{n}\{x_i + U\}$, where $U \in \mathcal{V}$, such that $\overline{\text{co}}(U \cap (K - K)) \subseteq V$. For every $x \in A$ define $F_V(x)$ by

$$F_V(x) = [F(x) + \overline{\text{co}}(U \cap (K - K))] \cap \overline{\text{co}}\{x_1, x_2, \ldots, x_n\}.$$

We prove that F_V is a finite-dimensional compact mapping from A into $\mathcal{R}(K)$. Since $F(x) \in \mathcal{R}(K)$ it follows that $F_V(x) \in \mathcal{R}(K)$ for every $x \in A$, and since $F_V(K) \subseteq \overline{\text{co}}\{x_1, x_2, \ldots, x_n\}$ we have that F_V is a finite-dimensional mapping. In order to prove that F_V is a closed mapping choose a convergent net $(x_\alpha)_{\alpha \in \mathcal{A}}$ in A such that $\lim_{\alpha \in \mathcal{A}} x_\alpha = x$, and if $y_\alpha \in F_V(x_\alpha)(\alpha \in \mathcal{A})$ with $\lim_{\alpha \in \mathcal{A}} y_\alpha = y$ we have to prove that $y \in F_V(x)$. Since $y_\alpha \in F_V(x_\alpha)$, $\alpha \in \mathcal{A}$, it follows that for each $\alpha \in \mathcal{A}$ there exists $z_\alpha \in F(x_\alpha)$ and $u_\alpha \in \overline{\text{co}}(U \cap (K - K))$ such that $y_\alpha = z_\alpha + u_\alpha \in \overline{\text{co}}\{x_1, x_2, \ldots, x_n\}$. From the compactness of the set $\overline{F(A)}$ it follows that there exists a convergent subnet (z_{α_δ}), say $\lim_\delta z_{\alpha_\delta} = z$. Since F is closed and $\lim_\delta x_{\alpha_\delta} = x$ we conclude that $z \in F(x)$. Furthermore, $\lim_\delta y_{\alpha_\delta} = y$ and so $\lim_\delta u_{\alpha_\delta} = u = y - z$ is in $\overline{\text{co}}(U \cap (K - K))$. Therefore we have that $y = u + z \in F_V(x)$, and since $F_V(x) \subseteq F(x) + \overline{\text{co}}(U \cap (K - K))$ for every $x \in A$, (6.9) is proved. □

Remark 6.21 If K is a compact convex subset of a Hausdorff topological vector space then K is of Z-type if and only if K is affinely embeddable in a locally convex topological vector space. Further results on subsets of Z-type, convex, totally bounded, and strongly convex totally bounded subsets are obtained in [100, 313, 314].

In [120] S. Hahn introduced the notion of a quasi-compact mapping.

Definition 6.22 *Let E be a topological vector space, K a closed and convex subset of E and M a closed subset of E so that $M \subseteq K$. An upper semi-continuous mapping F from M into a system $\mathcal{S}(K)$ of non-empty subsets of K is called quasi-compact if for each $b \in M$ there exists a closed, convex subset $K_0 \subseteq E$ such that $b \in K_0, F(M \cap K_0) \subseteq K_0 \cap K, F(M \cap K_0)$ is relatively compact and $K \cap K_0$ is an $(\mathcal{S})$-admissible subset of E.*

Proposition 6.23 *Let E be an admissible topological vector space and F a compact mapping from M into E, where M is a closed subset of E. Then F is quasi-compact.*

Proof. It is enough to take $K_0 = K = E$ in Definition 6.22. □

Proposition 6.24 *Let E be a topological vector space and K a closed convex and admissible subset of E. Then each compact mapping F from M into K, where M is a closed subset of E such that $M \subseteq K$, is quasi-compact.*

Proof. It is enough to take in Definition 6.22 that $K_0 = K$. □

Proposition 6.25 *Let E be a locally convex space, K a closed and convex subset of E, M a closed subset of E such that $M \subseteq K$ and $F : M \to \mathcal{R}(K)$ a compact mapping. Then F is quasi-compact.*

Proof. Since K is $(\mathcal{R})$-admissible, if we take $K_0 = K$ in Definition 6.22 we obtain that F is quasi-compact. □

The following fixed point theorem is proved by S. Hahn [120].

Theorem 6.26 *Let E be a topological vector space, W a closed neighbourhood of a point $b \in E, K$ a closed and convex subset of E such that $b \in K$ and $\mathcal{R}_0(K) \in \{\mathcal{R}(K), \cup_{x\in K}\{x\}\}$. If $F : W \cap K \to \mathcal{R}_0(K)$ is a quasi-compact mapping such that $\beta x + (1-\beta)b \notin F(x)$ for every $x \in \partial W \cap K$ and $\beta > 1$, then there exists a point $x_0 \in W \cap K$ such that $x_0 \in F(x_0)$.*

Proof. We shall define a compact mapping $G : K \cap K_0 \to \mathcal{R}_0(K \cap K_0)$, so that

$$x \in W \cap K,\ x \notin F(x) \quad \Rightarrow \quad x \notin G(x),\ x \in K \cap K_0, \tag{6.10}$$

where K_0 is such that $K_0 = \overline{\text{co}}K_0, b \in K_0, F(W \cap K \cap K_0) \subseteq K \cap K_0, K \cap K_0$ is $(\mathcal{R}_0)$-admissible and $F(W \cap K \cap K_0)$ is relatively compact.

Such a subset K_0 exists since F is a quasi-compact mapping. Let

$$X_0 = \{x \mid x \in W \cap K \cap K_0, \text{ there exists } t \in [0,1] \text{ such that } x \in tF(x) + (1-t)b\}.$$

Then X_0 is a compact subset such that $\partial W \cap K \cap K_0 \cap X_0 = \varnothing$. The set $\partial W \cap K \cap K_0$ is closed and E is a complete regular topological space and there exists a continuous mapping $\lambda : E \to [0,1]$ such that

$$\lambda(x) = \begin{cases} 0 & \text{if } x \in X_0, \\ 1 & \text{if } x \in \partial W \cap K \cap K_0. \end{cases}$$

The mapping $G : K \cap K_0 \to \mathcal{R}_0(K \cap K_0)$ is defined in the following way:

$$G(x) = \begin{cases} (1-\lambda(x))F(x) + \lambda(x)b & \text{if } x \in W \cap K \cap K_0, \\ b & \text{if } x \in (K \cap K_0) \setminus W. \end{cases}$$

Then G is a compact mapping and it is obvious that the implication (6.10) holds. Therefore it remains to be proved that the mapping G has at least one fixed point.

Suppose that $x \notin G(x)$, for every $x \in K \cap K_0$. It is obvious that there exists a neighbourhood V of zero in E such that

$$(x - G(x)) \cap V = \emptyset \quad \text{for every } x \in K \cap K_0 \tag{6.11}$$

since

$$\bigcap_{x \in K \cap K_0} \{x - G(x)\}$$

is a closed subset of E. Since $K \cap K_0$ is an $(\mathcal{R}_0)$-admissible set there exists a finite-dimensional subspace E_V of E and a compact mapping $G_V : K \cap K_0 \to \mathcal{R}_0(K \cap K_0)$ such that $G_V(K \cap K_0) \subseteq E_V$ and $G_V(x) \subseteq G(x) + V$ for every $x \in K \cap K_0$. From Kakutani's fixed point theorem it follows that there exists $x_0 \in K \cap K_0 \cap E_V$ such that $x_0 \in G_V(x_0)$. Since $G_V(x) \subseteq G(x) + V$, for every $x \in K \cap K_0$ it follows that there exists $z \in G(x_0)$ such that $x_0 \in z + V$, but this is a contradiction since (6.11) holds. □

If E is a locally convex space from Theorem 6.26 a result of G. Kayser [158] follows.

Corollary 6.27 *Let E be a topological vector space, K a closed, convex and $(\mathcal{R}_0)$-admissible subset of E and $F : K \to \mathcal{R}_0(K)$ a compact mapping. Then there exists at least one element $x_0 \in K$ such that $x_0 \in F(x_0)$.*

Theorem 6.11 is a corollary of Theorem 6.26. Using Proposition 6.20 and Corollary 6.27 we obtain the following corollary.

Corollary 6.28 *Let E be a topological vector space, K a closed and convex subset of Zima's type of E, and $F : K \to \mathcal{R}_0(K)$ a compact mapping. Then there exists at least one element $x_0 \in K$ such that $x_0 \in F(x_0)$.*

Remark 6.29 In [87] it is proved that in Corollary 6.28 we can suppose that $F(K)$ is of Zima's type (see [96], Proposition 2, II.5).

We end this section with a fixed point theorem in a complete random normed space.

Theorem 6.30 *Let $(S, \mathcal{F}, T)$ be a complete random normed space with continuous t-norm T, A a probabilistic bounded closed and convex subset of S, and $f : A \to A$ a continuous mapping such that the following condition is satisfied:*

$$(\forall C \subseteq A)(\alpha_C \geq \alpha_{\overline{co}f(C)} \quad \Rightarrow \quad \alpha_C = H_0). \tag{6.12}$$

If every closed and convex subset of A is admissible then f has a fixed point.

Proof. Let $p \in A$. We shall prove that the set K, which is the set of limit points of $\{f^n(p) \mid n \in \mathbb{N}\}$, is a nonempty and compact subset of A such that $K \subseteq f(K)$. Let $O(p,f) = \{f^n(p) \mid n \in \mathbb{N}\}$. Then $\overline{O(p,f)}$ is compact, since from the relation $\alpha_{O(p,f)} \neq H_0$ it follows that

$$\alpha_{O(p,f)} < \alpha_{\overline{\mathrm{co}}f(O(p,f))},$$

which is a contradiction because

$$\begin{aligned} \alpha_{O(p,f)} &= \alpha_{\{f(p)\}\cup f(O(p,f))} \\ &= \min\{\alpha_{\{f(p)\}}, \alpha_{f(O(p,f))}\} \\ &= \alpha_{f(O(p,f))} \\ &\geq \alpha_{\overline{\mathrm{co}}f(O(p,f))}. \end{aligned}$$

It is easy to see that $K \neq \emptyset$ and that $K \subseteq f(K)$. Let

$$X = \{C \mid C \subseteq A,\ C = \overline{\mathrm{co}}C,\ K \subseteq C,\ f(C) \subseteq C\}.$$

It is obvious that $X \neq \emptyset$, since $A \in X$. If ξ is a chain in the ordered set $(X, \subseteq)$ then $\bigcap\limits_{C\in\xi} C$ is a lower bound of ξ, which can be easily verified. Hence by Zorn's lemma ξ has a minimal element C_0. From $f(C_0) \subseteq C_0$, since C_0 is closed and convex, it follows that the set $\overline{\mathrm{co}}f(C_0)$ is a subset of C_0. So we have that

$$f(\overline{\mathrm{co}}f(C_0)) \subseteq f(C_0) \subseteq \overline{\mathrm{co}}f(C_0).$$

From $K \subseteq \overline{\mathrm{co}}f(C_0)$ it follows that the set $\overline{\mathrm{co}}f(C_0)$ is in ξ. Since C_0 is a minimal element of ξ it follows that $\overline{\mathrm{co}}f(C_0) = C_0$ and so $\alpha_{\overline{\mathrm{co}}f(C_0)} = \alpha_{C_0}$. This implies that C_0 is precompact. Since S is complete and C_0 is closed it follows that C_0 is compact, and so $f|_{C_0}$ is a compact mapping of C_0 into C_0. Hence all the conditions of the Hahn and Pötter fixed point theorem are satisfied, and thus there exists $x_0 \in C_0$ such that $x_0 = f(x_0)$. □

Corollary 6.31 *Let A be a closed and convex subset of $K_s \subset S(\Omega, \mathcal{A}, P)$ for some $s > 0$, $f : A \to A$ be a continuous mapping, and (6.12) hold. Then f has a fixed point.*

Proof. The set A is probabilistic bounded and every closed convex subset of A is admissible. Hence, there exists $\psi \in A$ such that $\psi(\omega) = (f\psi)(\omega)$ a.e. □

Corollary 6.32 *Let $(S, \mathcal{F}, T_{\mathbf{M}})$ be a complete random normed space, A a probabilistic bounded, closed and convex subset of S and $f : A \to A$ a continuous mapping so that the following condition is satisfied:*

$$C \subseteq A,\ \alpha_C \geq \alpha_{f(C)} \Rightarrow \alpha_C = H_0.$$

Then f has a fixed point.

Proof. If in Theorem 6.30 $T = T_{\mathbf{M}}$ then S is a locally convex space, which implies that every closed and convex subset of A is admissible and $\alpha_C = \alpha_{\overline{\text{co}}C}$ for every subset C of A. Hence all the conditions of Theorem 6.30 are satisfied and f has a fixed point. □

6.3 Fixed point theorems of Krasnoselski's type

In this section we shall give some probabilistic versions of the well known Krasnoselski's fixed point theorem.

Theorem 6.33 *Let $(X, \|\cdot\|)$ be a Banach space, K a non-empty closed and convex subset of X, $f : K \to X$ a q-contraction, $0 < q < 1$, and $g : K \to X$ a compact mapping such that $f(K)+g(K) \subseteq K$. Then there exists $x \in K$ such that $x = fx+gx$.*

Remark 6.34 Using the theory of densifying operators Theorem 6.33 can be proved under the condition $(f+g)(K) \subseteq K$.

Let $(S, \mathcal{F}, T)$ be a Menger space, Λ a compact topological space, and $\mathcal{C}(\Lambda, S)$ the set of all continuous mappings from Λ into S. We shall prove the following Lemma.

Lemma 6.35 *If $(S, \mathcal{F}, T)$ is a Menger space with a continuous t-norm T and the mapping*

$$\widetilde{\mathcal{F}} : \mathcal{C}(\Lambda, S) \times \mathcal{C}(\Lambda, S) \to \mathcal{D}^+$$

is defined by

$$\widetilde{F}_{f,g}(x) = \sup_{\delta<x} \inf_{\lambda\in\Lambda} F_{f(\lambda),g(\lambda)}(\delta), \quad f, g \in \mathcal{C}(\Lambda, S), \ x \in \mathbb{R},$$

then $(\mathcal{C}(\Lambda, S), \ \widetilde{\mathcal{F}}, T)$ is a Menger space.

Proof. It is easy to see that $\widetilde{F}_{f,g} \in \mathcal{D}^+$ for every $f, g \in \mathcal{C}(\Lambda, S)$. We shall prove only that

$$\lim_{x\to\infty} \widetilde{F}_{f,g}(x) = 1 \text{ for every } f, g \in \mathcal{C}(\Lambda, S).$$

Since Λ is compact and S is a Hausdorff topological space, in the (ε, λ)-topology, the sets

$$A = \{f(\lambda) \mid \lambda \in \Lambda\}, \quad B = \{g(\lambda) \mid \lambda \in \Lambda\}$$

are compact and so probabilistic bounded. Hence $D_{A\cup B} \in \mathcal{D}^+$ and since

$$\widetilde{F}_{f,g}(x) \geq \sup_{\delta<x} \inf_{p,q\in A\cup B} F_{p,q}(\delta) = D_{A\cup B}(x)$$

it follows that $\widetilde{F}_{f,g} \in \mathcal{D}^+$. We shall prove that for every $f, g, h \in \mathcal{C}(\Lambda, S)$ and every $u, v > 0$

$$\widetilde{F}_{f,h}(u+v) \geq T(\widetilde{F}_{f,g}(u), \ \widetilde{F}_{g,h}(v)) \tag{6.13}$$

Indeed, we have that for every $\lambda \in \Lambda$ and every $\delta_1, \delta_2 > 0$

$$F_{f(\lambda),h(\lambda)}(\delta_1 + \delta_2) \geq T(F_{f(\lambda),g(\lambda)}(\delta_1), F_{g(\lambda),h(\lambda)}(\delta_2))$$

and therefore

$$\inf_{\lambda\in\Lambda} F_{f(\lambda),h(\lambda)}(\delta_1 + \delta_2) \geq T\left(\inf_{\lambda\in\Lambda} F_{f(\lambda),g(\lambda)}(\delta_1), \inf_{\lambda\in\Lambda} F_{g(\lambda),h(\lambda)}(\delta_2)\right)$$

which implies, since T is continuous, that

$$\begin{aligned}\sup_{\delta<u+v} \inf_{\lambda\in\Lambda} F_{f(\lambda),h(\lambda)}(\delta) &\geq \sup_{\substack{\delta_1<u\\ \delta_2<v}} T\left(\inf_{\lambda\in\Lambda} F_{f(\lambda),g(\lambda)}(\delta_1), \inf_{\lambda\in\Lambda} F_{g(\lambda),h(\lambda)}(\delta_2)\right)\\ &= T\left(\sup_{\delta_1<u} \inf_{\lambda\in\Lambda} F_{f(\lambda),g(\lambda)}(\delta_1), \sup_{\delta_2<v} \inf_{\lambda\in\Lambda} F_{g(\lambda),h(\lambda)}(\delta_2)\right).\end{aligned}$$

Hence the triangle inequality (6.13) is proved. □

It is easy to prove that $(\mathcal{C}(\Lambda, S), \widetilde{\mathcal{F}}, T)$ is complete if $(S, \mathcal{F}, T)$ is complete.

Similarly, if $(S, \mathcal{F}, T)$ is a random normed space, where the t-norm T is continuous, then $(\mathcal{C}(\Lambda, S), \widetilde{\mathcal{F}}, T)$ is a random normed space where

$$\widetilde{F}_f(x) = \sup_{\delta<x} \inf_{\lambda\in\Lambda} F_{f(\lambda)}(\delta), \quad f \in \mathcal{C}(\Lambda, S),\ x \in \mathbb{R}.$$

Theorem 6.36 *Let $(S, \mathcal{F}, T)$ be a complete random normed space with a continuous t-norm T, M a closed convex probabilistic bounded, and admissible subset of S, $f : M \to S$ a probabilistic q-contraction and $g : M \to S$ a compact mapping such that $fx + gy \in M$ for every $x, y \in M$. Then there exists $x \in M$ such that*

$$fx + gx = x. \tag{6.14}$$

Proof. Since, for every $y \in \overline{g(M)}$, the mapping $x \mapsto fx+y$ $(x \in M)$ is a probabilistic q-contraction and M is probabilistic bounded there exists one and only one element $Ry \in M$ such that

$$Ry = f(Ry) + y, \quad y \in \overline{g(M)}. \tag{6.15}$$

We prove that the mapping $y \mapsto Ry$ $(y \in \overline{g(M)})$ is continuous. Let $\mathcal{C}(\overline{g(M)}, M)$ be the set of all continuous mappings from $\overline{g(M)}$ into M and $\mathcal{C}(\overline{g(M)}, S)$ the set of all continuous mappings from $\overline{g(M)}$ into S. Let $\widetilde{x} \in \mathcal{C}(\overline{g(M)}, S)$, $\varepsilon > 0$, and by definition

$$\widetilde{F}_{\widetilde{x}}(\varepsilon) = \sup_{\delta<\varepsilon} \inf_{y\in\overline{g(M)}} F_{\widetilde{x}(y)}(\delta). \tag{6.16}$$

If $\widetilde{\mathcal{F}} : \mathcal{C}(\overline{g(M)}, S) \times \mathcal{C}(\overline{g(M)}, S) \to \mathcal{D}^+$ is defined by $\widetilde{\mathcal{F}}(\widetilde{x}, \widetilde{y}) = \widetilde{F}_{\widetilde{x}-\widetilde{y}}$ and $\widetilde{F}$ is given in (6.16) then $(\mathcal{C}(\overline{g(M)}, S), \widetilde{\mathcal{F}}, T)$ is a complete Menger space.

First, we shall prove that the set $\mathcal{C}(\overline{g(M)}, M)$ is a probabilistic bounded subset of $\mathcal{C}(\overline{g(M)}, S)$, i.e., that $\widetilde{D}_{\mathcal{C}(\overline{g(M)},M)} \in \mathcal{D}^+$, where

$$\widetilde{D}_{\mathcal{C}(\overline{g(M)},M)}(x) = \sup_{\varepsilon<x} \inf_{\widetilde{x}_1,\widetilde{x}_2\in\mathcal{C}(\overline{g(M)},M)} \widetilde{F}_{\widetilde{x}_1-\widetilde{x}_2}(\varepsilon).$$

Since for every $\widetilde{x}_1, \widetilde{x}_2 \in \mathcal{C}(\overline{g(M)}, M)$

$$\widetilde{F}_{\widetilde{x}_1-\widetilde{x}_2}(\varepsilon) = \sup_{\delta<\varepsilon} \inf_{y\in\overline{g(M)}} F_{x_1(y)-x_2(y)}(\delta)$$

we have

$$\begin{aligned}\widetilde{F}_{\widetilde{x}_1-\widetilde{x}_2}(\varepsilon) &= \sup_{\delta<\varepsilon} \inf_{y\in\overline{g(M)}} F_{x_1(y)-x_2(y)}(\delta)\\ &\geq \sup_{\delta<\varepsilon} \inf_{u,v\in M} F_{u,v}(\delta)\\ &= D_M(\varepsilon).\end{aligned}$$

Therefore

$$\inf_{\widetilde{x}_1,\widetilde{x}_2\in\mathcal{C}(\overline{g(M)},M)} \widetilde{F}_{\widetilde{x}_1-\widetilde{x}_2}(\varepsilon) \geq D_M(\varepsilon)$$

and

$$\sup_{\varepsilon<x} \inf_{\widetilde{x}_1,\widetilde{x}_2\in\mathcal{C}(\overline{g(M)},M)} \widetilde{F}_{\widetilde{x}_1-\widetilde{x}_2}(\varepsilon) \geq \sup_{\varepsilon<x} D_M(\varepsilon) = D_M(x).$$

This means that $\widetilde{D}_{\mathcal{C}(\overline{g(M)},M)}(x) \geq D_{\overline{g(M)}}(x)$, for every $x \in \mathbb{R}$, which implies that $\widetilde{D}_{\mathcal{C}(\overline{g(M)},M)} \in \mathcal{D}^+$.

Let $L : \mathcal{C}(\overline{g(M)}, M) \to \mathcal{C}(\overline{g(M)}, M)$ be defined by

$$(L\widetilde{x})(y) = f(\widetilde{x}(y)) + y,\ y \in \overline{g(M)},\quad \widetilde{x} \in \mathcal{C}(\overline{g(M)}, M).$$

Then, for every $\varepsilon > 0$ and $\widetilde{x}_1, \widetilde{x}_2 \in \mathcal{C}(\overline{g(M)}, M)$

$$\begin{aligned}\widetilde{F}_{L\widetilde{x}_1-L\widetilde{x}_2}(\varepsilon) &= \sup_{\delta<\varepsilon} \inf_{y\in\overline{g(M)}} F_{(L\widetilde{x}_1)(y)-(L\widetilde{x}_2)(y)}(\delta)\\ &\geq \sup_{\delta<\varepsilon} \inf_{y\in\overline{g(M)}} F_{f(\widetilde{x}_1(y))-f(\widetilde{x}_2(y))}(\delta)\\ &\geq \sup_{\delta<\varepsilon} \inf_{y\in\overline{g(M)}} F_{\widetilde{x}_1(y)-\widetilde{x}_2(y)}(\delta/q)\\ &= \widetilde{F}_{\widetilde{x}_1-\widetilde{x}_2}(\varepsilon/q),\end{aligned}$$

which means that L is a probabilistic q-contraction. Hence there exists one and only one element $\widetilde{x} \in \mathcal{C}(\overline{g(M)}, M)$ such that $L\widetilde{x} = \widetilde{x}$ and so

$$(L\widetilde{x})(y) = f(\widetilde{x}(y)) + y = \widetilde{x}(y) \quad \text{for every} \quad y \in \overline{g(M)}.$$

This means that $Ry = \widetilde{x}(y)$, for every $y \in \overline{g(M)}$ and since $\widetilde{x}$ is continuous it follows that R is continuous. Then $R \circ g : M \to M$ satisfies all the conditions of Hahn's and Pötter's fixed point theorem which implies the existence of an element $x \in M$ such that $(R \circ g)(x) = x$. Using (6.15) we conclude that (6.14) holds. □

The following theorem is a generalization of Theorem 2 from [35].

Theorem 6.37 *Let $(S, \mathcal{F}, T)$ be a complete random normed space such that T is a continuous t-norm of H-type, M a probabilistic bounded closed and convex subset of S, $f_1 : M \to S$ be a generalized probabilistic B-contraction, and $f_2 : M \to S$ a compact mapping such that $f_1x + f_2y \in M$ for every $x, y \in M$. Then there exists $x \in M$ such that*

$$x = f_1x + f_2x.$$

Proof. From Theorem 3.69 it follows that for every $y \in M$ there exists a unique $x(y) \in M$ such that

$$x(y) = f_1(x(y)) + f_2y.$$

We prove that the mapping $K : M \to M$ which is defined by $Ky = x(y)$, $y \in M$, is continuous. Suppose that $(y_n)_{n\in\mathbb{N}}$ is a convergent sequence from M such that $\lim\limits_{n\to\infty} y_n = y$ and prove that

$$\lim_{n\to\infty} x(y_n) = x(y). \tag{6.17}$$

If (6.17) does not hold then there exist $\varepsilon_0 > 0$ and $\mu_0 \in (0,1)$ such that for every $k \in \mathbb{N}$ there exists $n_k \in \mathbb{N}$ such that $n_1 < n_2 < \ldots$, and

$$F_{x(y_{n_k})-x(y)}(\varepsilon_0) \leq 1 - \mu_0 < 1. \tag{6.18}$$

Since M is probabilistic bounded, for every $\lambda \in (0,1)$ there exists $\delta_\lambda > 0$ such that

$$F_{x(y_{n_k})-x(y)}(\delta_\lambda) > 1 - \lambda > 0, \quad \text{for every } k \in \mathbb{N}. \tag{6.19}$$

Let $\lambda_0 \in (0,1)$ be fixed. Then (6.18) and (6.19) imply that for every $u > 0$

$$F_{f_1(x(y_{n_k}))-f_1(x(y))}(u) \geq F_{x(y_{n_k})-x(y)}\left(\frac{u}{L(\delta_{\lambda_0}, \varepsilon_0)}\right).$$

Let $\overline{L} = L(\delta_{\lambda_0}, \varepsilon_0)$ and $\overline{L} < r < 1$. Then, for every $u > 0$,

$$\begin{aligned} F_{x(y_{n_k})-x(y)}(u) &= F_{f_1(x(y_{n_k}))+f_2y_{n_k}-f_1(x(y))-f_2y}(u) \\ &\geq T\left(F_{f_1(x(y_{n_k}))-f_1(x(y))}(ru), F_{f_2y_{n_k}-f_2y}((1-r)u)\right) \\ &\geq T\left(F_{x(y_{n_k})-x(y)}\left(\frac{ru}{\overline{L}}\right), F_{f_2y_{n_k}-f_2y}((1-r)u)\right) \end{aligned}$$

Let

$$G(u) = \liminf F_{x(y_{n_k})-x(y)}(u), u \geq 0.$$

Clearly G is non-decreasing. Since T is continuous and non-decreasing in each place and $\lim_{k\to\infty} F_{f_2y_{n_k}-f_2y}((1-r)u) = 1$ (by the continuity of f_2), we obtain that

$$G(u) = G(\frac{ru}{L}) \quad \text{for every } u > 0,$$

which implies (recall that G is non-decreasing) that G is a constant function on $(0,\infty)$. From the inequality (6.19) we see that $G(\delta_\lambda) \geq 1-\lambda$, thus $G(u) \geq 1-\lambda$ for every $u > 0$. As λ is an arbitrary number from $(0,1)$ one has $G(u) = 1$ for every $u > 0$, in contradiction with (6.18). □

Applying Tychonoff's fixed point theorem for $K : M \to M$ it follows that Fix $(K) \neq \emptyset$. Since Fix $(K) \subseteq$ Fix $(f_1 + f_2)$ the proof is complete. □

Theorem 6.38 *Let $(S,\mathcal{F},T)$ be a complete random normed space with a continuous t-norm $T \geq T_{\mathbf{L}}$, K a non-empty closed convex and admissible subset of S, $f_1, f_2 : K \to S$ and the following conditions are satisfied:*

1. *f_1 is a generalized C-contraction of Krasnoselski's type;*
2. *f_2 is a compact mapping;*
3. *$f_1(K) + f_2(K) \subseteq K$.*

Then there exists $x \in K$ such that $x = f_1x + f_2x$.

Proof. We know that $d : S \times S \to [0,\infty)$, defined by

$$d(x,y) = \sup\{t \mid F_{x,y}(t) \leq 1-t\},$$

is a bounded translation invariant metric which defines the (ε,λ)-topology. From Lemma 5.17 it follows that for every $a, b > 0$ and every $x, y \in K$

$$a \leq d(x,y) \leq b \quad \Rightarrow \quad d(f_1x, f_1y) \leq L(a,b)d(x,y).$$

Then for every $y \in K$ there exists one and only one fixed point $x(y) \in K$ of the mapping $x \mapsto f_1x + f_2y$ $(x, y \in K)$ since for every $a, b > 0$ and every $u, v, y \in K$

$$(a \leq d(u,v) \leq b \quad \Rightarrow \quad d(f_1u + f_2y, f_1v + f_2y) = d(f_1u, f_1v) \leq L(a,b)d(u,v)).$$

We shall prove that the mapping $y \mapsto x(y)$ $(y \in K)$ is continuous.

Suppose that $(y_n)_{n\in\mathbb{N}}$ is a sequence from K such that $\lim_{n\to\infty} y_n = y$ and prove that $\lim_{n\to\infty} x(y_n) = x(y)$, where $x(y_n) = f_1(x(y_n)) + f_2(y_n)$ $(n \in \mathbb{N})$, $x(y) = f_1(x(y)) + f_2y$. If, on the contrary, this is not the case there exist $\varepsilon > 0$ and an increasing sequence $(n(k))_{k\in\mathbb{N}}$ from $\mathbb{N}$ such that

$$\|x(y_{n(k)}) - x(y)\|^* \geq \varepsilon \quad \text{for every } k \in \mathbb{N}, \tag{6.20}$$

where $\|z\|^* = d(0,z)$, for every $z \in S$. Then

$$\|x(y_{n(k)}) - x(y)\|^* \leq L(\varepsilon,1)\|x(y_{n(k)}) - x(y)\|^* + \|f_2(y_{n(k)}) - f_2 y\|^*$$

and therefore

$$\varepsilon \leq \|x(y_{n(k)}) - x(y)\|^* \leq \frac{\|f_2(y_{n(k)}) - f_2 y\|^*}{1 - L(\varepsilon,1)}.$$

Since $\lim_{k\to\infty} \|y_{n(k)} - y\|^* = 0$ and f_2 is continuous we obtain a contradiction. Using Hahn and Pötter's fixed point theorem similarly as in Theorem 6.37, it follows that there exists $x \in K$ such that $x = f_1(x) + f_2(x)$. □

Corollary 6.39 *Let $S = S(\Omega, \mathcal{A}, P)$ and $f_1, f_2 : S \to S$. If f_1 is a generalized C-contraction of Krasnoselski's type and f_2 is a compact mapping, then there exists an element $\psi \in S$ such that $\psi(\omega) = (f_1\psi)(\omega) + (f_2\psi)(\omega)$ a.e..*

Proof. $(S, \mathcal{F}, T_{\mathbf{L}})$ is an admissible random normed space of E-type and hence all the conditions of Theorem 6.38 are verified for $K = S$. □

Corollary 6.40 *Let $S = S(\Omega, \mathcal{A}, P)$, M be a non-empty closed and convex subset of K_s for some $s > 0$, $f_1, f_2 : M \to S$, and the conditions 1, 2, 3 from Theorem 6.38 hold for $K = M$. Then there exists an element $\psi \in M$ such that $\psi(\omega) = (f_1\psi)(\omega) + (f_2\psi)(\omega)$ a.e.*

Proof. The set M is of Zima's type and then admissible. All the conditions of Theorem 6.38 hold. □

At the end of this section we shall prove a fixed point theorem in random para-normed spaces.

In general, a random para-normed space $(E, \mathcal{F}, T)$ $(T \geq T_{\mathbf{L}})$ is not a locally convex topological vector space, even in the case when T is of the H-type.

First, we shall introduce the probabilistic Zima condition.

Definition 6.41 *Let $(E, \mathcal{F}, T)$ be a random para-normed space and $\emptyset \neq K \subset E$. The set K satisfies the probabilistic Zima condition if there exists $C(K) > 0$ so that for every $\lambda \in (0,1)$, every $\varepsilon > 0$ and every $x, y \in K$*

$$F_{\lambda(x-y)}(\lambda\varepsilon) \geq F_{x-y}(\varepsilon/C(K)).$$

Example 6.42 Let $(\Omega, \mathcal{A}, P)$ be a probability measure space and E the vector space of all the equivalence classes of measurable mappings $\widehat{X} : \Omega \to S(0,1)$. Then $(E, \mathcal{F}, T_{\mathbf{L}})$ is a random para-normed space, where

$$F_{\widehat{X}}(\varepsilon) = P(\{\omega \mid \omega \in \Omega,\, p(X(\omega)) < \varepsilon\}) \qquad (\{X(\omega)\} \in \widehat{X},\ \varepsilon > 0),$$

where p is defined by (6.8). Let $s > 0$ and $\widetilde{K}_s \subset E$ be defined by

$$\widetilde{K}_s = \{\widehat{X} \mid \widehat{X} \in E,\ X(\omega) \in K_s \quad \text{for every} \quad \omega \in \Omega\}.$$

Then for every $\widehat{X}, \widehat{Y} \in \widetilde{K}_s, \lambda \in (0,1)$ and $\varepsilon > 0$

$$F_{\lambda(\widehat{X}-\widehat{Y})}(\varepsilon\lambda) \geq F_{\widehat{X}-\widehat{Y}}\left(\frac{\varepsilon}{1+2s}\right). \tag{6.21}$$

Indeed, we have for every $\omega \in \Omega$

$$p(\lambda(X(\omega) - Y(\omega))) \leq (1+2s)\lambda p(X(\omega) - Y(\omega)),$$

and therefore

$$\begin{aligned} &P(\{\omega \mid \omega \in \Omega, p(X(\omega) - Y(\omega)) < \varepsilon/(1+2s)\}) \\ &\qquad \leq P(\{\omega \mid \omega \in \Omega,\ p(\lambda(X(\omega) - Y(\omega))) < \varepsilon\lambda\}), \end{aligned}$$

which means (6.21).

In [98] the following result is obtained.

Proposition 6.43 *Let $(E, \mathcal{F}, T)$ be a random para-normed space with continuous t-norm T and K a non-empty and convex subset of E which satisfies the probabilistic Zima condition. If the t-norm T is of H-type then K is admissible.*

Proof. Let A be a compact subset of K, $\varepsilon > 0$ and $\lambda \in (0,1)$. We have to prove that there exists a continuous mapping $h_{\varepsilon,\lambda} : A \to K$ such that for every $x \in A$

$$F_{x-h_{\varepsilon,\lambda}(x)}(\varepsilon) > 1 - \lambda, \qquad \dim \operatorname{Lin}\,(h_{\varepsilon,\lambda}(A)) < \infty.$$

Let $\delta(\lambda) \in (0,1)$ be such that for every $n \in \mathbb{N}$

$$u > 1 - \delta(\lambda) \quad \Rightarrow \quad u_T^{(n)} > 1 - \lambda.$$

Since the set A is compact, there exists a finite set $\{u_1, u_2, \ldots, u_m\} \subseteq A$ such that

$$A \subseteq \bigcup_{r=1}^{m} N_{u_r}\left(\frac{\varepsilon}{C(K)},\ \delta(\lambda)\right).$$

Let $\eta_r : A \to \mathbb{R}^+$ $(r \in \{1, 2, \ldots, m\})$ be a family of functions with the property

$$\eta_r(x) \neq 0 \quad \Rightarrow \quad F_{x-u_r}(\varepsilon/C(K)) > 1 - \delta(\lambda)$$

and $\sum_{r=1}^{m} \eta_r(x) = 1$, for every $x \in A$. Since E is metrizable, such a family exists. Let $h_{\varepsilon,\lambda} : A \to K$ be defined in the following way:

$$h_{\varepsilon,\lambda}(x) = \sum_{i=1}^{m} \eta_i(x) u_i, \quad x \in A.$$

Since K is convex and $h_{\varepsilon,\lambda}(A) \subseteq \text{co}\,\{u_1, u_2, \ldots, u_m\}$, it follows that

$$\dim \text{Lin}\,(h_{\varepsilon,\lambda}(A)) < \infty, \qquad h_{\varepsilon,\lambda}(A) \subset K.$$

Suppose that $x \in A$ and

$$\eta_i(x) = \begin{cases} 0 & \text{if } i \in \{1, 2, \ldots, m\} \setminus \{i_1, i_2, \ldots, i_s\}, \\ \neq 0 & \text{if } i \in \{i_1, i_2, \ldots, i_s\}. \end{cases}$$

Then we have

$$\begin{aligned} F_{x-h_{\varepsilon,\lambda}(x)}(\varepsilon) &= F_{\sum_{k=1}^{s} \eta_{i_k}(x)\cdot x - \sum_{k=1}^{s} \eta_{i_k}(x) u_{i_k}} \left(\sum_{k=1}^{s} \eta_{i_k}(x)\varepsilon \right) \\ &\geq \underbrace{T(T \ldots T}_{(\text{s}-1)-\text{times}}(F_{\eta_{i_1}(x)\cdot x - \eta_{i_1}(x)\cdot u_{i_1}}(\eta_{i_1}(x)\varepsilon), \\ &\qquad F_{\eta_{i_2}(x)x - \eta_{i_2}(x)u_{i_2}}(\eta_{i_2}(x)\varepsilon), \ldots, F_{\eta_{i_s}(x)x - \eta_{i_s}(x)u_{i_s}}(\eta_{i_s}(x)\varepsilon)) \\ &\geq (\min_{1 \leq k \leq s} \{F_{x-u_{i_k}}(\varepsilon/C(K))\})_T^{(s-1)} \\ &> 1 - \lambda \end{aligned}$$

since

$$F_{x-u_{i_k}}(\varepsilon/C(K)) > 1 - \delta(\lambda) \quad \text{for every} \quad k \in \{1, 2, \ldots, s\}.$$

This means that K is an admissible subset of E. □

Theorem 6.44 *Let $(S, \mathcal{F}, T)$ be a complete random para-normed space with a continuous t-norm T of H-type, M a closed and convex subset of S, $f : M \to S$ a probabilistic q-contraction and $g : M \to S$ a compact mapping such that $fx + gy \in M$, for every $x, y \in M$. If M satisfies the probabilistic Zima condition then there exists $x \in M$ such that $fx + gx = x$.*

Proof. The proof is similar to the proof of Theorem 6.36. Since t-norm T is of H-type it follows that M is an admissible subset of S. On the other hand the mapping $L : \mathcal{C}(\overline{g(M)}, M) \to \mathcal{C}(\overline{g(M)}, M)$ defined by

$$(L\tilde{z})(y) = f(\tilde{z}(y)) + y, \qquad y \in \overline{g(M)},\ \tilde{z} \in \mathcal{C}(\overline{g(M)}, M)$$

has a unique fixed point $\widetilde{x} \in \mathcal{C}(\overline{g(M)}, M)$, since the t-norm T is of H-type. Hence $L\widetilde{x} = \widetilde{x}$ and so

$$(L\widetilde{x})(y) = f(\widetilde{x}(y)) + y = \widetilde{x}(y) \quad \text{for every} \quad y \in \overline{g(M)}.$$

The rest of the proof is analogous to the proof of Theorem 6.36. □

6.4 Continuous dependence of the fixed points on parameters of $(\alpha, \mathbf{g})$-condensing mappings

Using the function of Kuratowski we prove in this section a theorem on continuous dependence of the fixed points on parameters of (α, g)-condensing mappings in probabilistic metric spaces [109]. An application of the theorem is also given.

In Lemma 6.45 Φ is the set of all monotone increasing functions $g : [0, \infty) \to [0, \infty)$.

Lemma 6.45 *Let $(S, \mathcal{F}, T)$ be a Menger space with a continuous t-norm T, M a non-empty closed and probabilistic bounded subset of S, Λ a metric space and $G : M \times \Lambda \to M$ so that the following conditions are satisfied:*

(i) there exists $g \in \Phi$ such that for every $B \subseteq M$, every $u > 0$ and every $\lambda \in \Lambda$

$$\alpha_{G(B,\lambda)}(g(u)) \geq \alpha_B(u);$$

(ii) the mapping $\lambda \mapsto G(x, \lambda)$ $(\lambda \in \Lambda)$ is continuous on Λ uniformly with respect to $x \in M$.

Then for every compact subset Λ_0 of Λ and every $B \subseteq M$

$$\alpha_{G(B,\Lambda_0)}(g(u)) \geq \alpha_B(u), \quad \text{for every} \quad u > 0. \tag{6.22}$$

Proof. In order to prove (6.22) we prove that for every $u > 0$, every $s \in (0, u)$ and every $B \subseteq M$

$$\alpha_{G(B,\Lambda_0)}(g(u)) \geq \alpha_B(u - s). \tag{6.23}$$

Since the function $\alpha_B(\cdot)(B \subseteq M)$ is left continuous, (6.23) implies that

$$\lim_{s \to 0} \alpha_B(u - s) = \alpha_B(u) \leq \alpha_{G(B,\Lambda_0)}(g(u)), \quad u > 0.$$

We prove the following implication:

$$0 < r < \alpha_B(u - s) \quad \Rightarrow \quad r \leq \alpha_{G(B,\Lambda_0)}(g(u)).$$

Suppose that $0 < r < \alpha_B(u - s)$ and prove that for every $h \in (0, r)$

$$r - h \leq \alpha_{G(B,\Lambda_0)}(g(u)).$$

The mapping $(u,v) \mapsto (u, T(r,v))$ is continuous and since $T(1, T(r,1)) = r$, it follows that there exists $\bar{h} \in (0,1)$ so that

$$u, v \in (1-\bar{h}, 1] \quad \Rightarrow \quad T(u, T(r,v)) > r - h.$$

The mapping $\lambda \mapsto G(x,\lambda)$ is continuous, on Λ uniformly with respect to x, and therefore there exists for every $\bar{\lambda} \in \Lambda_0$ a $\rho(\bar{\lambda}) > 0$ with the property that

$$d(\lambda, \bar{\lambda}) < \rho(\bar{\lambda}) \quad \Rightarrow \quad F_{G(x,\lambda),G(x,\bar{\lambda})}\left(\frac{g(u)-g(u-s)}{8}\right) > \eta \quad \text{for every } x \in M,$$

where $T(\eta,\eta) > 1-\bar{h}$. Let $L(a,v)$ be the ball with the center $a \in \Lambda$ and the radius v. Since $\Lambda_0 \subseteq \bigcup\limits_{\lambda \in \Lambda_0} L\left(\lambda, \frac{\rho(\lambda)}{2}\right)$ and Λ_0 is a compact set, there exists $\{\lambda_1, \lambda_2, \ldots, \lambda_k\} \subseteq \Lambda_0$ such that

$$\Lambda_0 \subseteq \bigcup_{i=1}^{k} L\Big(\lambda_i, \frac{\rho(\lambda_i)}{2}\Big). \tag{6.24}$$

We prove the following implication:

$$x \in M,\ \lambda', \lambda'' \in \Lambda_0, d(\lambda', \lambda'') < \rho \quad \Rightarrow \quad F_{G(x,\lambda'),G(x,\lambda'')}\left(\frac{g(u)-g(u-s)}{4}\right) > 1-\bar{h},$$

where $\rho = \min_{1 \le i \le k}\{\rho(\lambda_i)2^{-1}\}$. Suppose that $\lambda', \lambda'' \in \Lambda_0$ and $d(\lambda', \lambda'') < \rho$.

From relation (6.24) it follows that there exists λ_i such that $d(\lambda', \lambda_i) < \rho(\lambda_i)2^{-1}$. Then

$$d(\lambda'', \lambda_i) \le d(\lambda'', \lambda') + d(\lambda', \lambda_i) \le \rho(\lambda_i),$$

which implies that

$$\begin{aligned} F_{G(x,\lambda'),G(x,\lambda'')}\left(\frac{g(u)-g(u-s)}{4}\right) &\ge T\Big(F_{G(x,\lambda'),G(x,\lambda_i)}\Big(\frac{g(u)-g(u-s)}{8}\Big), \\ &\qquad F_{G(x,\lambda_i),G(x,\lambda'')}\Big(\frac{g(u)-g(u-s)}{8}\Big)\Big) \\ &\ge T(\eta,\eta) \\ &> 1-\bar{h}. \end{aligned}$$

Let $\Lambda_0 = \bigcup\limits_{i=1}^{n} S_i$ be such that $\operatorname{diam} S_i < \rho$ and $\bar{\lambda}_i \in S_i$ for every $i \in \{1,2,\ldots,n\}$. We prove that for every $i \in \{1,2,\ldots,n\}$ and $h \in (0,r)$

$$\alpha_{G(B,S_i)}(g(u)) \ge r - h. \tag{6.25}$$

If there exists a finite family $(A_j)_{j=1}^{l(i)}$ such that

$$G(B,S_i) \subseteq \bigcup_{j=1}^{l(i)} A_j \quad \text{and} \quad \alpha_{A_j}(g(u)) \ge r-h, \quad j \in \{1,2,\ldots,l(i)\}, \tag{6.26}$$

then (6.25) holds. We prove that (6.26) implies (6.25).

From (6.26) we obtain that

$$\begin{aligned}\alpha_{G(B,S_i)}(g(u)) &\geq \alpha_{\bigcup_{j=1}^{l(i)} A_j}(g(u)) \\ &= \min_{1\leq j\leq l(i)} \alpha_{A_j}(g(u)) \\ &\geq r-h.\end{aligned}$$

We prove that (6.26) holds. Since $0 < r < \alpha_B(u-s) \leq \alpha_{G(B,\bar{\lambda}_i)}(g(u-s))$ from the definition of the function α, it follows that there exists a finite family $\{B_1, B_2, \ldots, B_{l(i)}\}$ in S such that

$$G(B,\bar{\lambda}_i) = \bigcup_{j=1}^{l(i)} B_j, \qquad D_{B_j}(g(u-s)) > r, \quad j \in \{1,2,\ldots,l(i)\}. \tag{6.27}$$

Therefore

$$\sup_{c<g(u-s)} \inf_{v,w\in B_j} F_{v,w}(c) > r,$$

which implies that $F_{v,w}(g(u-s)) > r$, for every $v,w \in B_j$ and for every $j \in \{1,2,\ldots,l(i)\}$. Let $A_j = B_j \cup C_j$, $j \in \{1,2,\ldots,l(i)\}$, where

$$C_j = \left\{x \mid x \in G(B,S_i), \exists z \in B_j \text{ such that } F_{z,x}\left(\frac{g(u)-g(u-s)}{4}\right) > 1-\bar{h}\right\}.$$

If $x \in G(B,S_i)$ then there exist $y \in B$ and $\lambda \in S_i$ such that $x = G(y,\lambda)$. Since $(\lambda,\bar{\lambda}_i) \in S_i \times S_i$ we have $d(\lambda,\bar{\lambda}_i) < \rho$ and

$$F_{G(y,\lambda),G(y,\bar{\lambda}_i)}\left(\frac{g(u)-g(u-s)}{4}\right) > 1-\bar{h}. \tag{6.28}$$

From (6.27) it follows that $z = G(y,\bar{\lambda}_i) \in B_j$ for some $j \in \{1,2,\ldots,l(i)\}$, and therefore (6.28) implies that $G(y,\lambda) \in C_j$. This means that $x = G(y,\lambda) \in A_j$. Therefore $G(B,S_i) \subseteq \bigcup_{j=1}^{l(i)} A_j$ for every $j \in \{1,2,\ldots,l(i)\}$, and (6.26) is satisfied if for every $j \in \{1,2,\ldots,l(i)\}$

$$D_{A_j}(g(u)) = \sup_{c<g(u)} \inf_{v,w\in A_j} F_{v,w}(c) \geq r-h. \tag{6.29}$$

We prove that for every $v,w \in A_j$

$$F_{v,w}(2^{-1}(g(u)+g(u-s))) \geq r-h,$$

which implies (6.29), since the mapping g is monotone increasing and so $g(u) > 2^{-1}(g(u)+g(u-s))$.

There are the following three cases:

1) Let $v, w \in B_j$. Since $2^{-1}(g(u)+g(u-s)) > g(u-s)$, we have that (6.27) implies

$$F_{v,w}(2^{-1}(g(u)+g(u-s))) \geq F_{v,w}(g(u-s)) > r > r-h.$$

2) Let $v \in B_j$ and $w \in C_j$. Then there exists $z \in B_j$ such that

$$F_{w,z}\Big(\frac{g(u)-g(u-s)}{4}\Big) > 1-\bar{h},$$

which implies that

$$\begin{aligned} F_{v,w}(2^{-1}(g(u)+g(u-s))) &\geq T(F_{v,z}(g(u-s)), F_{z,w}(2^{-1}(g(u)-g(u-s)) \\ &\geq T(F_{v,z}(g(u-s)), T(1, F_{z,w}(4^{-1}(g(u)-g(u-s)))) \\ &\geq T(1, T(r, F_{z,w}(4^{-1}(g(u)-g(u-s)))) \\ &> r-h. \end{aligned}$$

3) Let $v, w \in C_j$. Then there exist $\widetilde{v} \in B_j$ and $\widetilde{w} \in B_j$ such that

$$F_{\widetilde{v},v}(4^{-1}(g(u)-g(u-s))) > 1-\bar{h},$$

$$F_{\widetilde{w},w}(4^{-1}(g(u)-g(u-s))) > 1-\bar{h}.$$

Then we have that

$$\begin{aligned} F_{v,w}(2^{-1}(g(u)+g(u-s))) &\geq T(F_{\widetilde{v},v}(4^{-1}(g(u)-g(u-s))), \\ &\qquad T(F_{\widetilde{v},\widetilde{w}}(g(u-s)), F_{\widetilde{w},w}(4^{-1}(g(u)-g(u-s)))) \\ &> r-h, \end{aligned}$$

and (6.29) is satisfied. Since $\alpha_{A_j}(g(u)) \geq D_{A_j}(g(u))$ we obtain that for every $j \in \{1, 2, \ldots, l(i)\}$ $\alpha_{A_j}(g(u)) \geq r-h$, and so $\alpha_{G(B,S_i)}(g(u)) \geq r-h$ for every $i \in \{1, 2, \ldots, n\}$. It is obvious that this implies

$$\begin{aligned} \alpha_{G(B,\Lambda_0)}(g(u)) &= \alpha_{G(B,\bigcup_{i=1}^n S_i)}(g(u)) \\ &= \min_{1\leq i\leq n} \alpha_{G(B,S_i)}(g(u)) \\ &\geq r-h. \end{aligned}$$

Since h is an arbitrary number from $(0, r)$, we obtain that $\alpha_{G(B,\Lambda_0)}(g(u)) \geq r$. □

Using Lemma 6.45 we can prove the following theorem on the continuous dependence of the fixed points on parameters.

Theorem 6.46 *Let $(S, \mathcal{F}, T)$ be a complete Menger space with a continuous t-norm T, M a non-empty closed and probabilistic bounded subset of S, Λ a complete metric space and $G : M \times \Lambda \to M$ such that the following conditions are satisfied:*

(i) for every $\lambda \in \Lambda$ the mapping $x \mapsto G(x, \lambda)$ is continuous $(x \in M)$, and for every $x \in M$ the mapping $\lambda \mapsto G(x, \lambda)$ is continuous $(\lambda \in \Lambda)$ uniformly with respect to $x \in M$;

(ii) for each $\lambda \in \Lambda$ the equation $x = G(x, \lambda)$ has a solution in M;

(iii) for every $B \subseteq M$, every $u > 0$ and every $\lambda \in \Lambda$

$$\alpha_{G(B,\lambda)}(g(u)) \geq \alpha_B(u),$$

where $g \in \Phi$ and $\lim\limits_{n\to\infty} (g^{-1})^n u = \infty$ for every $u > 0$.

Then $\lambda \mapsto F(\lambda)$ is upper semi-continuous at each $\lambda \in \Lambda$, where

$$F(\lambda) = \{x \mid x \in M, \ x = G(x, \lambda)\}.$$

Proof. First, we prove that for every compact set $\Lambda_0 \subseteq \Lambda$ and $B \subseteq M$ the following implication holds:

$$\alpha_B \neq H_0 \quad \Rightarrow \quad \text{there exists } u_0 > 0 \text{ such that } \alpha_{G(B,\Lambda_0)}(u_0) > \alpha_B(u_0).$$

If we suppose that $\alpha_B \neq H_0$ and $\alpha_{G(B,\Lambda_0)}(u) \leq \alpha_B(u)$ for every $u > 0$, then from Lemma 6.45 we have that

$$\alpha_B(u) \leq \alpha_{G(B,\Lambda_0)}(g(u)) \leq \alpha_B(g(u)) \quad \text{for every } u > 0$$

and therefore

$$\alpha_B(u) \geq \alpha_B(g^{-1}(u)) \quad \text{for every } u > 0.$$

Since the function $\alpha_B(\cdot)$ is such that $\lim\limits_{u\to\infty} \alpha_B(u) = 1$, from the condition $\lim\limits_{n\to\infty} (g^{-1})^n(u) = \infty$ it follows that for every $u > 0$, $\alpha_B(u) = 1$, which implies that $\alpha_B = H_0$.

This is a contradiction and the above implication is proved.

Now we can prove that the mapping F is upper semi-continuous. Suppose, in the opposite, that F is not an upper semi-continuous mapping at some $\lambda_0 \in \Lambda$. Then there exists an open set $O \supset F(\lambda_0)$ such that for every $\delta > 0$ there exists $\lambda(\delta) \in \Lambda$ such that $d(\lambda(\delta), \lambda_0) < \delta$ and $F(\lambda(\delta)) \not\subset O$. Let $\delta_1 > \delta_2 > \ldots$, $\lim\limits_{n\to\infty} \delta_n = 0$. Then $\lambda(\delta_n) = \lambda_n \to \lambda_0$ and let $x_n \in F(\lambda_n) \setminus O$, for every $n \in \mathbb{N}$. We shall prove that there is a convergent subsequence of the sequence $(x_n)_{n\in\mathbb{N}}$. Let $\Lambda_0 = \{\lambda_n \mid n \in \mathbb{N}\}$ and $B = \{x_n \mid n \in \mathbb{N}\}$. Suppose that $\alpha_B \neq H_0$. For every $n \in \mathbb{N}$ we have that $x_n = G(x_n, \lambda_n)$, and therefore for every $u > 0$

$$\begin{aligned} \alpha_B(u) &= \alpha_{\{G(x_n,\lambda_n) \mid n\in\mathbb{N}\}}(u) \\ &\geq \alpha_{G(B,\Lambda_0)}(u) \\ &\geq \alpha_{G(B,\bar{\Lambda}_0)}(u). \end{aligned}$$

Since there exists $u_0 > 0$ such that

$$\alpha_{G(B,\tilde{\Lambda}_0)}(u_0) > \alpha_B(u_0)$$

we obtain that $\alpha_B(u_0) > \alpha_B(u_0)$, which is a contradiction. Therefore $\alpha_B = H_0$, which implies that B is a relatively compact set and there exists a convergent subsequence $(x_{n_k})_{k\in\mathbb{N}}$. Suppose that $\lim_{k\to\infty} x_{n_k} = x_0 \in M$. Then from $\lim_{k\to\infty} \lambda_{n_k} = \lambda_0$ we obtain that the subsequence $(x_{n_k})_{k\in\mathbb{N}}$ given by $x_{n_k} = G(x_{n_k}, \lambda_{n_k})$ tends to $G(x_0, \lambda_0)$. This implies that $x_0 = G(x_0, \lambda_0)$ and so $x_0 \in F(\lambda_0)$. On the other hand, x_n belongs, for every $n \in \mathbb{N}$, to the complement of O which implies that $x_0 \notin O \supset F(\lambda_0)$.

We obtain a contradiction, which implies that F is an upper semi-continuous mapping. □

Proposition 6.47 *Let $(S, \mathcal{F}, T)$ be a complete Menger space with a continuous t-norm T, M a non-empty closed and probabilistic bounded subset of S, $\widetilde{\Lambda}$ a complete metric space, $Q : M \to \widetilde{\Lambda}$ a compact mapping, and $G : M \times \overline{Q(M)} \to M$ such that all the conditions of Lemma 6.45 are satisfied for $\Lambda = \overline{Q(M)}$. Then the mapping $x \mapsto G(x, Qx)$ $(x \in M)$ is an (α, g)-condensing mapping.*

Proof. We have to prove that for every subset $B \subseteq M$ and every $u > 0$

$$\alpha_{\{G(x,Qx)|x\in B\}}(g(u)) \geq \alpha_B(u).$$

From Lemma 6.45 it follows that for every $u > 0$

$$\alpha_{G(B,\overline{Q(M)})}(g(u)) \geq \alpha_B(u),$$

and since

$$\alpha_{\{G(x,Qx)|x\in B\}}(u) \geq \alpha_{G(B,\overline{Q(M)})}(u),$$

we obtain that

$$\alpha_{\{G(x,Qx)|x\in B\}}(g(u)) \geq \alpha_{G(B,\overline{Q(M)})}(g(u)) \geq \alpha_B(u),$$

for every $u > 0$. □

Applying Proposition 6.47 we obtain the following theorem.

Theorem 6.48 *Let $(S, \mathcal{F}, T_M)$ be a complete random normed space, M a non-empty closed convex and probabilistic bounded subset of S, $\widetilde{\Lambda}$ a complete metric space, $Q : M \to \widetilde{\Lambda}$ a compact mapping and $G : M \times \overline{Q(M)} \to M$ such that all the conditions of Lemma 6.45 are satisfied for $\Lambda = \overline{Q(M)}$. If $\lim_{n\to\infty} (g^{-1})^n(u) = \infty$ for every $u > 0$, then the mapping $f : M \to M$ defined by $f(x) = G(x, Qx)$ $(x \in M)$ has a fixed point.*

Proof. From Proposition 6.47 it follows that the mapping f is an (α, g)-condensing mapping. It is easy to prove that the following implication holds

$$C \subset M, \alpha_C \geq \alpha_{f(C)} \Rightarrow \alpha_C = H_0.$$

Indeed, suppose that $C \subset M$ is such that $\alpha_C \geq \alpha_{f(C)}$. Since f is an (α, g)-condensing mapping we have that $\alpha_{f(C)}(u) \geq \alpha_C(g^{-1}(u))$ and so

$$\alpha_C(u) \geq \alpha_{f(C)}(u) \geq \alpha_C(g^{-1}(u)), \quad u \in [0, \infty).$$

This implies that for every $n \in \mathbb{N}$ and every $u \in [0, \infty)$

$$\alpha_C(u) \geq \alpha_C((g^{-1})^n(u)).$$

Since $\lim_{n\to\infty} (g^{-1})^n(u) = \infty$ for every $u > 0$, we conclude that $\alpha_C = H_0$. By Corollary 6.32 it follows that f has a fixed point. □

6.5 A degree theory in topological vector spaces

In [29] a degree theory in a random normed space $(S, \mathcal{F}, T)$, where $T(x, x) \geq x$, for every $x \in [0, 1]$ is developed. Since the condition $T(a, a) \geq a$ for every $a \in [0, 1]$, implies that $T = T_{\mathbf{M}}$ and in this case S is, in the (ε, λ)-topology, a locally convex space the degree theory in locally convex spaces can also be used. Hence it is of interest for applications on more general random normed space to develop a degree theory in topological vector spaces.

In this section we shall give some results from Kaballo's paper [153] in which a generalization of the degree theory for an arbitrary topological vector space is given. A degree theory of a compact field in a not necessarily locally convex topological vector space is given in [124]. Some fundamental results of degree theory in $\mathbb{R}^n$ can be found in [257].

First, we give some notations and definitions. Let X be a topological space and E be a topological vector space. By $C(X, E)$ we denote the set of all compact mappings from X into E.

Definition 6.49 *Let E be a topological vector space, G a set and Σ a family of mappings from G into E. A mapping $f : G \to E$ is said to be uniformly Σ-approachable if and only if for every neighbourhood V of zero in E there exists $h_V \in \Sigma$ so that for every $x \in G$*

$$f(x) - h_V(x) \in V.$$

If $B \subseteq E$ then $f : G \to E$ is said to be uniformly Σ-approachable in B if for every neighbourhood of zero V in E there exists $h_V \in \Sigma$ so that $h_V(G) \subseteq B$ and $f(x) - h_V(x) \in V$ for every $x \in G$.

A mapping $f : X \to E$ is said to be approachable if it is uniformly Σ-approachable, where Σ is the set of all finite mappings of X into E.

Definition 6.50 *Let X be a topological space and E a topological vector space. By $CA(X,E)$ we denote the set of all approachable mappings $f : X \to E$, $f \in C(X,E)$.*

It is easy to see that a topological vector space E is admissible if for every topological space X is $C(X,E) = CA(X,E)$.

Let $G \subseteq \mathbb{R}^n$ be open, $f : \overline{G} \to \mathbb{R}^n$ continuous, $a \in \mathbb{R}^n$, $a \notin f(\partial G)$, and $(f - I)(\overline{G})$ bounded, where ∂G is the boundary of G. For such a triple (a, G, f) Nagumo defined and investigated the topological degree $\deg^n(a, G, f) \in Z$ (the set $\{0, \pm 1, \pm 2, \dots\}$) [203]. Further results on the theory of topological degree can be found in [8, 65, 69, 70, 179, 211, 234, 288].

The topological degree $\deg^n(a, G, f)$ has the following properties:

1. If $a \in G$ then $\deg^n(a, G, I) = 1$ and if $a \notin \overline{G}$ then $\deg^n(a, G, I) = 0$, where I is the identity mapping.

2. If $\deg^n(a, G, f) \neq 0$ then there exists $x \in G$ so that $f(x) = a$.

3. Let $G_1, G_2, \dots, G_k$ be open subsets of $\mathbb{R}^n$ so that

$$\bigcup_{i=1}^{k} G_i \subseteq G, \ \bigcup_{i=1}^{k} \overline{G_i} = \overline{G}, \ G_i \cap G_j = \varnothing \ \text{for } i \neq j$$

and $a \notin f(\partial G_i)$, for $i \in \{1, 2, \dots, k\}$. Then

$$\deg^n(a, G, f) = \sum_{i=1}^{k} \deg^n(a, G_i, f).$$

4. Let $f : [0,1] \times \overline{G} \to \mathbb{R}^n$ be continuous, $(f - \overline{I})([0,1] \times \overline{G})$ bounded, where $\overline{I}(t,x) = x$, $(t,x) \in [0,1] \times \overline{G}$, $a : [0,1] \to \mathbb{R}^n$ continuous and for every $t \in [0,1]$, $a(t) \notin f_t(\partial G)$ where $f_t : \overline{G} \to \mathbb{R}^n$ is defined by $f_t(x) = f(t,x)$ $(t \in [0,1],\ x \in \overline{G})$. Then $\deg^n(a(t), G, f_t)$ is constant on $[0,1]$.

5. Let $a \notin f(\partial G)$, $X = \{x \mid x \in G,\ f(x) = a\}$ and $G_0 \subseteq \mathbb{R}^n$ an open set such that $X \subseteq G_0 \subseteq G$. Then

$$\deg^n(a, G, f) = \deg^n(a, G_0, f).$$

Let E be a real topological vector space, G be an open subset of E, $f \in CA(\overline{G}, E)$ and $a \notin (I + f)(\partial G)$. Since $a \notin (I + f)(\partial G)$ it follows from the closedness of $(I + f)(\partial G)$ that there exists a neighbourhood of zero U in E so that $(a + U) \cap (I + f)(\partial G) = \varnothing$ and let V be a circled neighbourhood of zero in E such that $V + V \subseteq U$.

Then $(a+V+V)\cap(I+f)(\partial G)=\varnothing$. By $A_V \in C(\overline{G}, E_V)$ we denote a mapping of $\overline{G}$ into E_V, which is a finite dimensional subspace of E, such that

$$A_V(x)-f(x)\in V \quad \text{for every } x\in\overline{G}.$$

We can suppose that $a\in E_V$. It is obvious that $a\notin(I+A_V)(\partial G)$ since for every $x\in\overline{G}$

$$(I+A_V)(x)-(I+f)(x)\in V.$$

Let $G_V=G\cap E_V$, $\overline{G}_V=\overline{G}\cap E_V$. Then $a\notin(I+A_V)(\partial G_V)$ since

$$\partial G_V=\overline{G_V}\setminus G_V\subseteq\overline{G}_V\setminus G_V\subseteq\overline{G}\setminus G=\partial G.$$

Then $\deg^n(a,G_V,I+A_V)$ is defined and let, by definition,

$$\operatorname{Deg}(a,G,I+f)=\deg^n(a,G_V,I+A_V). \tag{6.30}$$

Theorem 6.51 *The definition of* $\operatorname{Deg}(a,G,I+f)$, *given by (6.30), is independent of* V, E_V *and* A_V.

Proof. Suppose that V_1 and V_2 are circled neighbourhoods of zero in E such that

$$(a+V_i+V_i)\cap(I+f)(\partial G)=\varnothing,\quad i\in\{1,2\}$$

and let E_i and A_i be as above (for V_i), $i\in\{1,2\}$. Let $V_3\subseteq V_1\cap V_2$ and A_3 and E_3 be defined analogously (for V_3), where we take that $E_1\cup E_2\subseteq E_3$. It is known that if E^ℓ is a subspace of $\mathbb{R}^n (\ell\le n$, where ℓ is the dimension of $E^\ell)$, $G\subseteq\mathbb{R}^n$ is open, $f:\overline{G}\to\mathbb{R}^n$ is continuous, $F=f-I$ is bounded on $\overline{G}$, $F(\overline{G})\subseteq E^\ell$, $a\notin f(\partial G)$, and $G^\ell=G\cap E^\ell$, then

$$\deg^\ell(a,G^\ell,f)=\deg^n(a,G,f).$$

If we apply this result on $G_i=G\cap E_i$ $(i\in\{1,2\})$ then we have

$$\deg^{n_1}(a,G_1,I+A_1)=\deg^{n_3}(a,G_3,I+A_1),$$

where $n_i=\dim E_i$ $(i\in\{1,3\})$. Let $h:[0,1]\times\overline{G}\to E$ be defined in the following way

$$h(t,x)=(1-t)(I+A_1)(x)+t(I+A_3)(x)\quad (t,x)\in[0,1]\times\overline{G}.$$

Now, we can apply the property 4 in order to prove that

$$\deg^{n_3}(a,G_3,I+A_1)=\deg^{n_3}(a,G_3,I+A_3). \tag{6.31}$$

For every $x\in\partial G$, $h(t,x)-(I+f)(x)\in V_1+V_1$, since

$$\begin{aligned} h(t,x)-(I+f)(x) &= (1-t)(A_1(x)-f(x))+t(A_3(x)-f(x))\\ &\in (1-t)V_1+tV_3\\ &\subseteq V_1+V_1. \end{aligned}$$

From $\partial G_3 \subseteq \partial G$ it follows that $a \notin h_t(\partial G_3)$ and by

$$h(t,x) - x = (1-t)A_1(x) + tA_3(x) \quad ((x,t) \in [0,1] \times \overline{G}_3)$$

a bounded mapping is defined, since A_1 and A_3 are compact. Thus for $f_t = h_t|\overline{G}_3$ we can apply 4. and we obtain (6.31).

Further,

$$\deg^{n_3}(a, G_3, I + A_2) = \deg^{n_3}(a, G_3, I + A_3)$$

and therefore

$$\begin{aligned} \deg^{n_1}(a, G_1, I + A_1) &= \deg^{n_3}(a, G_3, I + A_1) \\ &= \deg^{n_3}(a, G_3, I + A_3) \\ &= \deg^{n_3}(a, G_3, I + A_2) \\ &= \deg^{n_2}(a, G_2, I + A_2). \end{aligned}$$ □

Similarly as in the case of locally convex spaces the defined degree $\mathrm{Deg}(a, G, I+f)$ has the following properties:

1. If $a \in G$ then

$$\mathrm{Deg}^n(a, G, I) = 1$$

and if $a \notin \overline{G}$ then

$$\mathrm{Deg}^n(a, G, I) = 0.$$

2. If $\mathrm{Deg}(a, G, I + f) \neq 0$, then there exists $x \in G$ such that

$$(I + f)(x) = a.$$

3. Let $G_1, \ldots, G_k$ be open subsets of E such that

$$\bigcup_{i=1}^{k} G_i \subseteq G, \quad \bigcup_{i=1}^{k} \overline{G}_i = \overline{G}, \quad G_i \cap G_j = \emptyset, \quad \text{for } i \neq j$$

and $a \notin (I + f)(\partial G_i)$ for $i \in \{1, 2, \ldots, k\}$. Then

$$\mathrm{Deg}(a, G, I + f) = \sum_{i=1}^{k} \mathrm{Deg}(a, G_i, I + f).$$

4. Let $f \in CA([0,1]) \times \overline{G}, E)$, $a \in CA([0,1], E)$ and for every $t \in [0,1]$, let $a(t) \notin (I + f_t)(\partial G)$. Then $\mathrm{Deg}(a(t), G, I + f_t)$ is constant on $[0,1]$.

5. Let $X = \{x \mid x \in \overline{G},\ (I + f)(x) = a\}$ and $G_0 \subseteq E$ be an open set such that $X \subseteq G_0 \subseteq G$. Then

$$\mathrm{Deg}(a, G, I + f) = \mathrm{Deg}(a, G_0, I + f).$$

From this equality we obtain the next corollary.

Corollary 6.52 *Let E be a topological vector space, G an open subset of E, $f \in CA(\overline{G}, E)$ and $a \notin (I+f)(\partial G)$. Let V be a circled neighbourhood of zero in E such that*

$$(a+V+V) \cap (I+f)(\partial G) = \varnothing.$$

Then

a) For every $a' \in a+V$ there holds

$$\mathrm{Deg}(a', G, I+f) = \mathrm{Deg}(a, G, I+f).$$

b) If $f' \in CA(\overline{G}, E)$ and $f(x) - f'(x) \in V$ for every $x \in \partial G$, then

$$\mathrm{Deg}(a, G, I+f) = \mathrm{Deg}(a, G, I+f').$$

The following theorem is a generalization of Borsuk's theorem.

Theorem 6.53 *Let E be a real topological vector space, U an open and symmetric neighbourhood of zero in E, $f \in CA(\overline{U}, E)$, $F = I - f$, and for every $s \in [0,1]$ and every $x \in \partial U$ $F(x) \neq sF(-x)$. Then the number $\mathrm{Deg}(0, U, F)$ is odd.*

Proof. Let $h(t,x) = \dfrac{1}{1+t} f(x) - \dfrac{t}{1+t} f(-x)$ for every $(t,x) \in [0,1] \times \overline{U}$. Then $h \in CA([0,1] \times \overline{U}, E)$. Further, since $F(x) \neq sF(-x)$ for every $s \in [0,1]$ and $x \in \partial U$, it follows that

$$\frac{1}{1+t} F(x) - \frac{1}{1+t} F(-x) = x - h(t,x) \neq 0$$

for every $(t,x) \in [0,1] \times \partial U$, and therefore $0 \in (I - h_t)(\partial U)$. Thus we have that for $g(x) = h_1(x) = (f(x) - f(-x))/2$, $G = I - g$

$$\mathrm{Deg}(0, U, F) = \mathrm{Deg}(0, U, G).$$

Let V be a circled neighbourhood of zero in E such that $(V+V) \cap G(\partial U) = \varnothing$. Such a V exists since $0 \notin G(\partial U)$ and $g \in CA(\overline{U}, E)$. Let W be a circled neighbourhood of zero in E such that $W + W \subseteq V$. Since $f \in CA(\overline{U}, E)$ there exists a finite dimensional subspace E_W of E and $A_W \in C(\overline{U}, E_W)$ such that

$$A_W(x) - f(x) \in W, \quad x \in \overline{U}$$

and let $\widetilde{A}_W(x) = (A_W(x) - A_W(-x))/2$, $x \in \overline{U}$. If $A = I - \widetilde{A}_W | \overline{U}_W, \overline{U}_W = \overline{U \cap E_W}$ then

$$\mathrm{Deg}(0, U, G) = \deg^n(0, U_W, A), \qquad n = \dim E_W$$

since $\widetilde{A}_W \in C(\overline{U}, \overline{E}_W)$ and for every $x \in \overline{U}$

$$\widetilde{A}_W(x) - g(x) \in \frac{1}{2} W + \frac{1}{2} W \subseteq V.$$

We prove that $\deg^n(0, U_1, A) = \deg^n(0, U_W, A)$ for some U_1 which is an open bounded and symmetric neighbourhood of zero in E such that $\deg^n(0, U_1, A)$ is odd. First, we prove that for such a neighbourhood U_1 the mapping $A : \overline{U}_1 \to E_W$ is continuous and odd and from Borsuk's theorem it will follow that $\deg^n(0, U_1, A)$ is odd.

The set U_W is open and symmetric, and from the definition of $A : U_W \to E_W$ it follows that A is continuous and odd and $0 \notin A(\partial U_W)$, since $G(x) - A(x) \in V$ for every $x \in \overline{U}_W$. Let

$$M = \{x \mid x \in \overline{U}_W, A(x) = 0\} = \{x \mid x \in U_W,\ x = \widetilde{A}_W(x)\}.$$

Then M is compact and $M \subseteq U_W$. Let U_0 be an open and bounded subset of E_W such that $M \subseteq U_0 \subseteq \overline{U}_0 \subseteq U_W$ and $U_1 = U_0 \cup (-U_0)$. Then U_1 is an open bounded and symmetric set such that $M \subseteq U_1 \subseteq \overline{U}_1 \subseteq U_W$ and $M \subseteq U_1$ implies $0 \notin A(\partial U_1)$. Therefore $A : \overline{U}_1 \to E_W$ is continuous and odd and $\deg^n(0, U_1, A)$ is odd.

From property 5. of $\deg^n$ it follows for $a = 0$, $G_0 = U_1$, $G = U_W$ that $\deg^n(0, U_1, A) = \deg^n(0, U_W, A)$. □

Corollary 6.54 *Suppose that all the conditions of Theorem 6.53 are satisfied. Then there exists $x \in U$ such that $x = f(x)$.*

The following result is a generalization of Landsberg's result from [176].

Theorem 6.55 *Let E be a real topological vector space and U an open neighbourhood of zero in E. If* Fix $(f) = \emptyset$, *where $f \in CA(\overline{U}, E)$, then there exists $t_0 \in (0,1)$ and $x_0 \in \partial U$ such that $x_0 = t_0 f(x_0)$.*

Proof. Let for every $(t, x) \in [0,1] \times \overline{U}, h(t,x) = tf(x)$. Then $h \in CA([0,1] \times \overline{U}, E)$. Suppose that for every $t_0 \in (0,1)$ and $x_0 \in \partial U : x_0 \neq t_0 f(x_0)$. Since for $x_0 \in \partial U$ we have $x_0 \neq 0$ and $x \neq fx$, $x \in \overline{U}$ it follows that $0 \notin (I - h_t)(\partial U)$ for $t \in [0,1]$. Thus

$$\mathrm{Deg}(0, U, I - f) = \mathrm{Deg}(0, U, I) = 1$$

which implies that Fix $(f) \neq \emptyset$. Contradiction and the theorem is proved. □

Further interesting results about the degree theory in topological vector spaces can be found in [3, 214].

Bibliography

[1] **J. Aczél** (1969). *Lectures on Functional Equations and their Applications*, Academic Press, New York.

[2] **R. R. Ahmerov, M. I. Kaminskij, A. S. Potapov** (1986). *Measures of noncompactness and condensing operators*, Novosibirsk, Nauka.

[3] **H. Alex, S. Hahn, L. Kaniok** (1994). The fixed point index for noncompact mappings in non locally convex topological vector spaces, *Math. Univ. Carolinae* **32**, 249–257.

[4] **C. Alsina** (1978). On countable products and algebraic convexifications of probabilistic metric spaces. *Pacific J. Math.* **76**, 291–300.

[5] **C. Alsina** (1983). On convex triangle functions, *Aequat. Math.* **26**, 191–196.

[6] **C. Alsina, B. Schweizer, A. Sklar** (1993). On the definition of a probabilistic normed space, *Aequat. Math.* **46**, 91–98.

[7] **M. A. Altman** (1957). A fixed point theorem in Banach spaces, *Bull. Acad. Polon. Sci., Sér. Sci. Math. Astronom. Phys.* **5**, 89–97.

[8] **H. Amann, S. A. Weiss** (1973). On the uniqueness of the topological degree, *Math. Z.* **130**, 39–54.

[9] **V. G. Angelov** (1987). Fixed point theorems in uniform spaces and applications, *Czechoslovak Math. J.* **37**, 19–33.

[10] **A. Avallone, G. Trombetta** (1991). Measures of noncompactness in the space L_0 and a generalization of the Arzela–Ascoli Theorem, *Bollettino Unione Mat. Ital.* (7), **5-B**, 573–587.

[11] **A. T. Bharucha-Reid** (1976). Fixed point theorems in probabilistic analysis, *Bull. Amer. Math. Soc.* **82**, 641–657.

[12] **Gh. Bocsan** (1973). Some applications of functions of Kuratowski, *Sem. Teor. Funct. si Mat. Apl. Univ. Timisoara,* **5**.

[13] **Gh. Bocsan** (1974a). On the Kuratowski function in random normed spaces, *Sem. Teor. Funct. si Mat. Apl. Univ. Timisoara,* **8**.

[14] **Gh. Bocsan** (1974b). Some remarks on measures of noncompactness in probabilistic metric spaces , *Sem. Teor. Funct. si Mat. Apl. Univ. Timisoara,* **18**.

[15] **Gh. Bocsan** (1976). Masuri aleatore de necompacitate si aplicatii, Ph. D. Thesis, Univ. of Temisoara.

[16] **Gh. Bocsan** (1978). On random operators on separable Banach spaces, *Sem. Teor. Funct. si Mat. Apl. Univ. Timisoara,* **38**.

[17] **Gh. Bocsan, Gh. Constantin** (1973). The Kuratowski function and some applications to the probabilistic metric spaces, *Atti Acad. Naz.Lincei* **55**, 236–240.

[18] **Gh. Bocsan, Gh. Constantin** (1974). On some measures of noncompactness in the probabilistic metric spaces, *Proc. Fifth. Conf. Probab. Theory, Brasov, Romania,* 163–168.

[19] **Gh. Bocsan, Gh. Constantin** (1982). Some properties of random operators and applications to the existence theorems for random equations, *Proc. Seventh. Conf. Probab. Theory, Brasov, Romania,* 403–408.

[20] **H. F. Bohnenblust, S. Karlin** (1950). On a theorem of Ville, *Contribututions to the Theory of Games*, Princeton, 155–160.

[21] **D. Boyd, J. Wong** (1969). On nonlinear contractions, *Proc. Amer. Math. Soc.* **20**, 458–469.

[22] **F. Browder** (1968). Fixed point theory of multivalued mappings in topological vector spaces, *Math. Ann.* **177**, 283–301.

[23] **F. Browder** (1976). On a theorem of Caristi and Kirk, in *Proc. of the Seminar on Fixed Point Theory and its Applications, Dalhouseie University, June 1975,* Academic Press, New York, 23–27.

[24] **L. K. J. Brouwer** (1910). Über Abbilding von Mannigfaltigkeiten, *Math. Ann.* **71**, 97–115.

[25] **G. L. Cain, R. H. Kasriel** (1976). Fixed and periodic points of local contraction mappings on probabilistic metric spaces, *Math. System. Theory* **9**, 4, 289–297.

[26] **J. Caristi** (1976). Fixed point theorem for mappings satisfying inwardness conditions, *Trans. Amer. Math. Soc.* **215**, 241–251.

[27] **Chih-sen Chang** (1983). On some fixed point theorems in probabilistic metric space and its applications, *Z. Wahrsch. Verw. Gebiete* **63**, 463–474.

[28] **Chih-sen Chang** (1984). On the theory of probabilistic metric spaces, *Z. Wahrsch. verw. Gebiete* **67**, 85–94.

[29] **S. S. Chang, Y. J. Cho, S. M. Kang** (1994). *Probabilistic Metric Spaces and Nonlinear Operator Theory*, Sichuan Univ. Press, Chengdu.

[30] **S. S. Chang, Y. J. Cho, S. M. Kang, J. X. Fan** (1994). Common fixed point theorems for multi-valued mappings in Menger PM-spaces, *Math. Japonica* **40**, 2, 289–293.

[31] **S. S. Chang, Y. J. Cho, B.S. Lee, M. Gue** (1997). Fixed point degree and fixed point theorems for fuzzy mappings in probabilistic metric spaces, *Fuzzy Sets and Systems* **87**, 325–334.

[32] **S. S. Chang, Y. J. Cho, B. S. Lee, J. S. Jung, S.M. Kang** (1997). Coincidence point theorems and minimization theorems in fuzzy metric spaces, *Fuzzy Sets and Systems* **88**, 119–127.

[33] **S. S. Chang, Y. J. Cho, X. Wu, B. S. Lee** (1999). Minimax problems and variational inequalities in probabilistic metric spaces, *Nonlinear Anal. Forum* **4**, 15–31.

[34] **S. S. Chang, Non-jing Huan** (1989). On the generalized 2-metric spaces and probabilistic 2-metric spaces with applications to fixed point theory, *Math. Japon.* **34**, 885–900.

[35] **S. S. Chang, B. S. Lee, Y. J. Cho, Y. Q. Chen, S. M. Kang, J. S. Jung** (1996). Generalized contraction mapping principle and differential equations in probabilistic metric spaces, *Proc. Amer. Math. Soc.* **124**, No. 8, 2367–2376.

[36] **A. H. Clifford, G. B. Preston** (1968). *The algebraic theory of semigroups*, Vol. 1, Amer. Math. Soc., Providence, RI.

[37] **A. C. Climescu** (1946). Sur l'équation fonctionelle de l'associativé, *Bull. École Polytechnique Iassy* **1**, 1–16.

[38] **Gh. Constantin** (1985). On some classes of contraction mappings in Menger spaces, *Sem. Teor. Prob. Apl. Univ. Timisoara*, Nr. 76.

[39] **Gh. Constantin, I. Istrăţescu** (1981). *Elemente de analiză probabilistă si aplicaţii*, Bucuresti, Editura Academiei Republicii Socialiste, Romania.

[40] **Gh. Constantin, I. Istrăţescu** (1989). *Elements of Probabilistic Analysis with Applications*, Editura Academiei, Bucuresti, Romania, Kluwer Academic Publishers, Dordrecht, Boston, London.

[41] **L. B. Ćirić** (1974). On fixed points of generalized contraction principle, *Proc. Amer. Math. Soc.* **45**, 267–273.

[42] **L. B. Ćirić** (1975). A generalization of Banach's contractions on probabilistic metric spaces, *Publ. Inst. Math.* **18** (32), 71–78.

[43] **G. Dall'Aglio, S. Kotz, G. Salinetti (Eds.)** (1991). *Advances in probability distributions with given marginals: beyond the copulas*, Kluwer Academic Publishers.

[44] **J. Daneš** (1972). On densifying and related mappings and their applications in nonlinear functional analysis, in: *Theory of Nonlinear Operators, Proceedings of the Summer School 1972*, Neuendorf, GDR.

[45] **G. Darbo** (1955). Punti uniti in transformazioni a condominio non compatto, *Sem. Math. Univ. Padova* **24**, 84–92.

[46] **W. F. Darsow, B. Nguyen, E.T. Olsen** (1992). Copulas and Markov processes, *Ill. J. Math.* **36**, 600–642.

[47] **E. De Pascale, G. Trombetta** (1991). Fixed points and best approximation for convex condensing functions in topological vector spaces, *Rend. Mat.* (7), **11**, 175–186.

[48] **E. De Pascale, G. Trombetta** (1994). Sui sottoinsiemi finitamente compatti di $S(\Omega)$, *Bollettino U.M.I.* (7) **8-A**, 243–249.

[49] **E. De Pascale, G. Trombetta, H. Weber** (1993). Convexly totally bounded and strongly totally bounded sets. Solution of a problem of Idzik, *Ann. Scuola Norm. Sup. Pisa* **20** Fasc. 3.

[50] **A. Deleanu, G. Marinescu** (1963). A fixed point theorem and an implicit function theorem in locally convex spaces, *Rev. Roum. Math. Pures Appl.* **8**, 91-99 (in Russian).

[51] **D. Downing, W. A. Kirk** (1977). A generalization of Caristi's theorem with applications to nonlinear mapping theory, *Pacific J. Math.* **69**, 339–345.

[52] **J. Dugundji, A. Granas** (1978). KKM maps and variational inequalities, *Ann. Scuola Norm. Sup. Pisa* **5**, 679–682.

[53] **J. Dugundji, A. Granas** (1982). *Fixed Point Theory*, PWN–Polish Scientific Publishers, Warsawa.

[54] **C. A. Drossos** (1977). Stochastic Menger spaces and convergence in probability, *Rev. Roum. Math. Pures Appl.* **22**, 1069–1076.

[55] **R. J. Egbert** (1960). Products and quotients of probabilistic metric spaces, *Pacific J. Math.* **24**, 437–455.

[56] **I. Ekeland** (1974). On the variational principle, *J. Math. Anal. Appl.* **47**, 324–352.

[57] **I. Ekeland** (1979). Nonconvex minimization problems, *Bull. Amer. Math. Soc. (New Series)* **1**, 443–474.

[58] **I. Ekeland, S. Terracini** (1990). The ε-variational principle revised, in: A. Cellina (Ed.), *Methods of Nonconvex Analysis*, Lecture Notes in Math., Vol. 1446, Springer, Berlin, 1–15.

[59] **Ky Fan** (1961). A generalization of Tychonoff's fixed point theorem, *Math. Ann.* **142**, 305–310.

[60] **J. X. Fang** (1992a). On fixed point theorems in fuzzy metric spaces, *Fuzzy Sets and Systems* **46**, 107–113.

[61] **J. X. Fang** (1992b). A note on fixed point theorems of Hadžić, *Fuzzy Sets and Systems* **48**, 391–395.

[62] **J. X. Fang** (1996). The variational principle and fixed point theorems in certain topological spaces, *J. Math. Anal. Appl.* **202**, 398–412.

[63] **W. M. Faucett** (1955). Compact semigroups irreducibly connected between two idempotents. *Proc. Amer. Math. Soc.* **6**, 741–747.

[64] **W. Feller** (1971). *Introduction in Probability Theory and its Applications*, Vol. 2. 2nd ed. Willey, New York.

[65] **C. Fenske** (1971). Analitische Theorie des Abbildungsgrades für Abbildungen in Banach-Räumen, *Math. Nachr.* **48**, 279–290.

[66] **H. J. Frank** (1975). Associativity in a class of operations on spaces of distribution functions. *Aequat. Math.* **12**, 121–144.

[67] **H. J. Frank** (1979). On the simultaneous associativity of $F(x,y)$ and $x+y-F(x,y)$. *Aequat. Math.* **19**, 194–226.

[68] **L. Fuchs** (1963). *Partially Ordered Algebraic Systems,* Pergamon Press, Oxford.

[69] **L. Führer** (1971). *Theorie der Abbildungsgrades in endlichdimensionalen Räumen*, Inaugural-Dissertation, FU Berlin.

[70] **L. Führer** (1972). Ein elementarer analytischer Beweis zur Eindeutigkeit des Abbildungsgrades in $\mathbb{R}^n$, *Math. Nachr.* **54**, 259–267.

[71] **S. Gähler** (1963/1964). 2-metrische Räume und ihre topologische Struktur, *Math. Nachr.* **26**, 115–148.

[72] **C. Genest, L.-P. Rivest** (1993). Statistical inference procedures for bivariate Archimedean copulas, *J. Amer. Statist. Assoc.* **88**, 1034–1043.

[73] **I. L. Glicksberg** (1953). A further generalization of the Kakutani fixed point theorem with applications to Nash equilibrium points, *Proc. Amer. Math. Soc.* **3**, 170–174.

[74] **I. Golet** (1999). *Structuri 2-Metrice Probabiliste,* Ed.Polit. Timisoara.

[75] **D. Göhde** (1965). Princip der kontraktiven Abbildung, *Math. Nachr.* **30**, 251–258.

[76] **A. Granas** (1962). *The theory of compact vector fields and some of its applications to the topology of functional spaces*, Rozprawy Matematyczne 30, Warszawa, 1–62.

[77] **L. Guseman** (1970). Fixed point theorems for mappings with a contractive iterate at a point, *Proc. Amer. Math. Soc.* **26**, 615–618.

[78] **O. Hadžić** (1978a). On the (ϵ, λ)-topology of probabilistic locally convex spaces, *Glas. Mat.* **13** (33), 293–297.

[79] **O. Hadžić** (1978b). Fixed point for mappings on probabilistic locally convex spaces, *Bull. Math. Soc. Sci. Math. Rep. Soc. Roum.* **22** (70), No. 3, 287–292.

[80] **O. Hadžić** (1978c). A fixed point theorem in probabilistic locally convex spaces, *Rev. Roum. Math. Pures Appl.* **23**, 735–744.

[81] **O. Hadžić** (1979a). A fixed point theorem in Menger spaces, *Publ. Inst. Math. Beograd* **20**, 107–112.

[82] **O. Hadžić** (1979b). Fixed point theorems in probabilistic metric and random normed spaces, *Math. Sem. Notes Kobe Univ.* **7**, 262–270.

[83] **O. Hadžić** (1979c). Fixed point theorems for multivalued mappings in probabilistic metric spaces, *Mat. Vesnik* **3** (16)(31), 125–133.

[84] **O. Hadžić** (1979d). A fixed point theorem for multivalued mappings in random normed spaces, *L'Anal. Numer. Teor. Approx.* **81**, 49–52.

[85] **O. Hadžić** (1980a). A generalization of the contraction principle in PM-spaces, *Univ. u Novom Sadu, Zb. Rad. Prirod.-Mat. Fak. Ser. Mat.* **10**, 13–21.

[86] **O. Hadžić** (1980b). On the topological structure of random normed spaces, *Univ. u Novom Sadu, Zb. Rad. Prirod.-Mat. Fak. Ser. Mat.* **10**, 31–35.

[87] **O. Hadžić** (1981a). Some fixed point and almost fixed point theorems for multivalued mappings in topological vector spaces, *Nonlinear Anal. Theory, Methods, Appl.* **5**, No. 9, 1009–1019.

[88] **O. Hadžić** (1981b). On multivalued mappings in paranormed spaces, *Comm. Math. Univ. Carol.* **22**,1, 129–136.

[89] **O. Hadžić** (1981c). On a generalization of Kakutani's fixed point theorem in paranormed spaces, *Univ. u Novom Sadu, Zb. Rad. Prirod.-Mat. Fak. Novi Sad Ser. Mat.* **11**, 19–28.

[90] **O. Hadžić** (1982a). On Kakutani's fixed point theorem in topological vector spaces, *Bull. Acad. Pol. Sci. Sér. Sci. Math.* **30**, 141–144.

[91] **O. Hadžić** (1982b). On common fixed points in probabilistic metric spaces, *Math. Sem. Notes Kobe Univ.* **10**, 31–39.

[92] **O. Hadžić** (1982c). Some theorems on the fixed point in probabilistic metric and random normed spaces, *Boll. Unione Mat. Ital.* **1-B** (6), 381–391.

[93] **O. Hadžić** (1982d). On equilibrium point in topological vector spaces, *Comm. Math. Univ. Carol.* **23**, 727–738.

[94] **O. Hadžić** (1982e). Fixed point theorem for sum of two mappings, *Proc. Amer. Math. Soc.* **85**, 37–41.

[95] **O. Hadžić** (1983). Fixed point theorems for multivalued mappings in uniform spaces and its applications to PM-spaces, *Analele Univ. Timisoara* **21**, Fasc. 1-2, 45–57.

[96] **O. Hadžić** (1984). *Fixed Point Theory in Topological Vector Spaces*, University of Novi Sad, Institute of Mathematics, Novi Sad.

[97] **O. Hadžić** (1985a). Some fixed point theorems in PM-spaces, *Univ. u Novom Sadu, Zb. Rad. Prirod.-Mat. Fak. Ser. Mat.* **15**, 23–35.

[98] **O. Hadžić** (1985b). Fixed point theorems in random paranormed spaces, *Univ. u Novom Sadu, Zb. Rad. Prirod. - Mat. Fak. Ser. Mat.* **15**,2, 15–30.

[99] **O. Hadžić** (1988a). Common fixed point theorems in probabilistic metric spaces with a convex structure, *Univ. u Novom Sadu, Zb. Rad. Prirod. - Mat. Fak. Ser. Mat.* **18**,2, 165–178.

[100] **O. Hadžić** (1988b). Some properties of measures of noncompactness in paranormed spaces, *Proc. Amer. Math. Soc.* **102**, 843–849.

[101] **O. Hadžić** (1989). Fixed point theorems for multivalued mappings in some classes of fuzzy metric spaces, *Fuzzy Sets and Systems* **29**, 115–125.

[102] **O. Hadžić** (1990). On multivalued contractions in probabilistic metric spaces, *Univ. u Novom Sadu, Zb. Rad. Prirod.-Mat. Fak. Ser. Mat.* **20**, 2, 161–171.

[103] **O. Hadžić** (1991a). Continuous dependence of the fixed point on parameters in random normed spaces, *Univ. u Novom Sadu, Zb. Rad. Prirod.-Mat. Fak. Ser. Mat.* **21**, 1, 203–215.

[104] **O. Hadžić** (1991b). *Some classes of random operator equations*, Institute of Mathematics, Novi Sad, (in Serbian).

[105] **O. Hadžić** (1991c). Fixed point theorems for multivalued mappings in probabilistic metric spaces with a convex structure, *Proc. IWAA, Kupari, 1990, University of Novi Sad*, 237–262.

[106] **O. Hadžić** (1992). On (n, f, g)-locally contractions in probabilistic metric spaces, *Univ. u Novom Sadu, Zb. Rad. Prirod.-Mat. Fak. Ser. Mat.* **22**, 1, 1–10.

[107] **O. Hadžić** (1994). Fixed point theorems for multivalued probabilistic (Ψ)-contractions, *Indian J. Pure Appl. Math.* **25** (8), 825–835.

[108] **O. Hadžić** (1995a). *Fixed point theory in probabilistic metric spaces*, Serbian Academy of Sciences and Arts, Branch in Novi Sad, University of Novi Sad, Institute of Mathematics, Novi Sad.

[109] **O. Hadžić** (1995b). Continuous dependence of the fixed points on parameters in probabilistic metric spaces, *Univ. u Novom Sadu, Zb. Rad. Prirod.-Math. Fak. Ser. Math.* **25**,2, 81–91.

[110] **O. Hadžić** (1997). Fixed point theorems for multivalued mappings in probabilistic metric spaces, *Fuzzy Sets and Systems* **88**, 219–226.

[111] **O. Hadžić, M. Budinčević** (1978). A fixed point theorem in probabilistic metric spaces, *Coll. Math. Soc. Janos Bolyai, 23 Topology, Budapest*, 579–584.

[112] **O. Hadžić O., Z. Ovcin** (1996). A variational principle in fuzzy metric spaces, *Bull. Acad. Serbe Sci. Art. Sci. Math.* **21**, 73–84.

[113] **O. Hadžić, E. Pap** (2000). On some classes of t-norms important in the fixed point theory, *Bull. Acad. Serbe Sci. Art. Sci. Math.* **25**, 15–28.

[114] **O. Hadžić, E. Pap** (to appear). A fixed point theorem for multivalued mappings in probabilistic metric spaces and an application in fuzzy metric spaces, *Fuzzy Sets and Systems.*

[115] **O. Hadžić, E. Pap, M. Budinčević** (to appear). Countable extension of triangular norms and their applications to fixed point theory in probabilistic metric spaces.

[116] **O. Hadžić, E. Pap, V. Radu** (to appear). Generalized contraction mapping principles in probabilistic metric spaces.

[117] **O. Hadžić, B. Stanković** (1970). Some theorems on the fixed points in locally convex spaces, *Publ. Inst. Math.* **10** (24), 9–19.

[118] **O. Hadžić, M. Stojaković** (1979). On the existence of a solution of the system $x = H(x,y), y = K(x,y)$ in random normed spaces, *Univ. u Novom Sadu, Zb. Rad. Prirod. Mat. Fak. Ser. Mat.* **9**, 43–48.

[119] **S. Hahn** (1976). Fixpunktsätze für mengenwertige Abbildungen in lokalkonvexen Räumen, *Math. Nachr.* **73**, 269–283.

[120] **S. Hahn** (1977). A remark on a fixed point theorem for condensing set-valued mappings, TU Dresden, Informationen, 07-5-77.

[121] **S. Hahn** (1978). Zur Theorie kompakter Vektorfelder in topologischen Vektorräumen, *Math. Nachr.* **85**, 273–282.

[122] **S. Hahn, F. Pötter** (1970). Eine verallgemeinerung eines Satzes von H.H. Schaefer, *Wiss. Z. TU Dresden* **19**, 1383–1385.

[123] **S. Hahn, F. Pötter** (1974). Über Fixpunkte kompakter Abbildungen in topologischen Vektorräumen, *Studia Math.* **50**, 1–16.

[124] **S. Hahn, T. Riedrich** (1973). Der Abbildungsgrad kompakter Vektorfelder in nicht notwendig lokalkonvexen topologischen Vektorräumen, *Wiss. Z. TU Dresden* **22**, 37–42.

[125] **P. J. He** (1992). The variational principle in fuzzy metric space and its applications, *Fuzzy Sets and Systems* **45**, 389–394.

[126] **M. Hegedüs , S. Kasahara** (1979). A contraction principle in metric spaces, *Math. Sem. Notes Kobe Univ.* **7**, no. 3, 597–603.

[127] **S. Heikkila, S. Seikkalla** (1983). On generalized contractions in probabilistic metric spaces, *Rev. Roum. Math. Pures. Appl.* **28**, 297–306.

[128] **T. L. Hicks** (1983). Fixed point theory in probabilistic metric spaces, *Univ. u Novom Sadu, Zb. Rad. Prirod.-Mat. Fak. Ser. Mat.* **13**, 63–72.

[129] **T. L. Hicks** (1989). Some fixed point theorems, *Radovi Mat.* **5**, 115–119.

[130] **T. L. Hicks** (1995). Fixed point theorems in F-complete topological spaces, *Far East J. Math. Sci.* **3** (2), 205–213.

[131] **T. L. Hicks** (1996a). Fixed point theory in probabilistic metric space II, *Math. Japonica* **44**, No. 3, 487–493.

[132] **T. L. Hicks** (1996b). Random normed linear structures, *Math. Japonica* **44**, No. 3, 483–496.

[133] **T. L. Hicks, B. E. Rhoades** (1992). Fixed point theorems for d-complete topological spaces II, *Math. Japonica* **27**, 847–853.

[134] **T. L. Hicks, P. L. Sharma** (1984). Probabilistic metric structures: Topological classification, *Univ. u Novom Sadu, Zb. Rad. Prirod.-Mat. Fak. Ser. Mat.* **14**,1, 43–50.

[135] **C. J. Himmelberg** (1972). Fixed points of compact multifunctions, *J. Math. Anal. Appl.* **38**, 205–207.

[136] **C. J. Himmelberg** (1975). Measurable relations, *Fund. Math.* **87**, 53–72.

[137] **C. J. Himmelberg, J. R. Porter, F.C. Van Vleck** (1969). Fixed points theorems for condensing multifunctions, *Proc. Amer. Math. Soc.* **23**, 635–641.

[138] **A. Idzik** (1978). Remarks on Himmelberg's fixed point theorems, *Bull. Acad. Polon. Sci. Sér. Math. Astronom. Phys.* **26**, 909–912.

[139] **A. Idzik** (1988). Almost fixed point theorems, *Proc. Amer. Math. Soc.* **104**, 779–784.

[140] **M. Iosifescu, G. Mihoc, R. Theodorescu** (1966). *Teoria Probabilităţilor şi Statistică Matematică*, Ed. Tehnică, Bucureşti.

[141] **S. Itoh, W. Takahashi** (1977). Singlevalued mappings, multivalued mappings and fixed point theorems, *J. Math. Anal. Appl.* **59**, 514–521.

[142] **J. Ishii** (1965). On the admissibility of function spaces, *J. Fac. Sci. Hokkaido Univ. Series I* **19**, 49-55.

[143] **V.I.Istrăţescu** (1974). *Introducere in Teoria Spatiilor Metrice Probabiliste cu Aplicaţii*, Editura Technica, Bucuresti.

[144] **I.I.Istrăţescu** (1976). On some fixed point theorems in generalized Menger spaces, *Boll. Unione Mat. Ital.* (5), **13-A**, 96–100.

[145] **I.I.Istrăţescu** (1981). A fixed point theorem for mappings with a probabilistic contractive iterate, *Rev. Roum. Math. Pures Appl.* **26**, 431–435.

[146] **Sh. Itoh** (1977). A random fixed point theorem for a multivalued contraction mapping, *Pacific J. Math.* **68**, 1 , 85–90.

[147] **Sh. Itoh** (1979). Some fixed point theorems in metric spaces, *Fund. Math.* **102**, 109–117.

[148] **J. R. Jachymski** (1998). Caristi's fixed point theorem and selections of set-valued contractions, *J. Math. Anal. Appl.* **277**, 55–67.

[149] **S. Jenei** (1999). Fibred triangular norms, *Fuzzy Sets and Systems* **103**, 67–82.

[150] **S. Jenei, E. Pap** (1999). Smoothly generated Archimedean approximation of continuous triangular norms, *Fuzzy Sets and Systems* **104** (1999), 19-25.

[151] **T. Jerofski** (1983). *Zur Fixpunkttheorie mengenwertiger Abbildungen*, Dissertation A, TU Dresden.

[152] **J. S. Jung, Y. J. Cho, S. M. Kang, S. S. Chang** (1996). Coincidence theorems for set-valued mappings and Ekeleland's variational principle in fuzzy metric spaces, *Fuzzy Sets and Systems* **79**, 239–250.

[153] **W. Kaballo** (1973). Zum Abbildungsgrad in Hausdorffschen topologischen Vektorräumen, *Manuscripta Math.* **8**, 209–216.

[154] **S. Kakutani** (1941). A generalization of Brouwer's fixed point theorems, *Duke Math. J.* **8**, 457–459.

[155] **O. Kaleva, S. Seikalla** (1984). On fuzzy metric spaces, *Fuzzy Sets and Systems* **12**, 215–229.

[156] **B. G. Kang, S. Park** (1990). On generalized ordering principle in nonlinear analysis, *Nonlinear Anal. Theory, Methods, Appl.* **14**, 159–165.

[157] **A. W. Kaplan** (1972). On F. Browder's theorem in topological vector spaces, *Trudy mat. fakulteta, VGU* **6**, 38–42 (in Russian).

[158] **G. Kayser** (1975). A fixed point theorem for noncompact set-valued mapping, *VII International Conference on Nonlinear Oscillation, Abh. Akad. Wiss. Abt. Natur. W. Techn., DDR, Berlin*, 8–13.

[159] **J. F. C. Kingman** (1964). Metrics for Wald spaces. *J. London Math. Soc.* **39**, 129–130.

[160] **V. Klee** (1960). Leray–Schauder theory without local convexity, *Math. Ann.* **141**, 286–296.

[161] **E. P. Klement, R. Mesiar, E. Pap** (1996). On the relationship of associative compensatory operators to triangular norms and conorms, *Internat. J. Uncertain. Fuzziness Knowledge-Based Systems* **4**, 129–144.

[162] **E. P. Klement, R. Mesiar, E. Pap** (1997). A characterization of the ordering of continuous t-norms, *Fuzzy Sets and Systems* **86**, 189–195.

[163] **E. P. Klement, R. Mesiar, E. Pap** (1999). Quasi- and pseudo-inverses of monotone functions, and the construction of t-norms, *Fuzzy Sets and Systems* **104**, 3–13.

[164] **E. P. Klement, R. Mesiar, E. Pap** (2000a). *Triangular Norms*, Kluwer Academic Publishers, Trends in Logic 8, Dordrecht.

[165] **E. P. Klement, R. Mesiar, E. Pap** (2000b). Integration with respect to decomposable measures, based on a conditionally distributive semiring on the unit interval, *Internat. J. Uncertain. Fuzziness Knowledge-Based Systems* **8**, 701–717.

[166] **E. P. Klement, R. Mesiar, E. Pap** (to appear). Uniform approximation of associative copulas by strict and non-strict copulas, *Illinois J. Math.*.

[167] **E. P. Klement, R. Mesiar, E. Pap** (to appear). Triangular norms as ordinal sums in the sense of A. H. Clifford, *Semigroup Forum.*

[168] **E. P. Klement, S. Weber** (1991). Generalized measures, *Fuzzy Sets and Systems* **40**, 375–394.

[169] **K. Knopp** (1944). *Theory and Applications of Infinite Series*, Black, London.

[170] **A. Kolesárová** (1999). A note on Archimedean triangular norms, *BUSEFAL* **80**, 57–60.

[171] **V. N. Kolokoltsov, V. P. Maslov** (1997). *Idempotent Analysis and Its Applications,* Kluwer Academic Publishers, Dordrecht.

[172] **I. Kramosil, J. Michalek** (1975). Fuzzy metric and statistical metric spaces, *Kybernetika* **11**, 336–314.

[173] **M.A. Krasnoselski, P. P. Zabreiko** (1975). *Geometricheskie metody nelineinogo analiza*, Nauka (in Russian).

[174] **C. Krauthausen** (1976). *Der Fixpunktsatz von Schauder in nicht notwendig konvexen Räumen sowie Anwendungen auf Hammersteinsche Gleichungen*, Doktors Dissertation, Aachen.

[175] **K. Kuratowski** (1930). Sur les espaces complets, *Fund. Math.* **15**, 301–309.

[176] **M. Landsberg** (1964). Über die Fixpunkte kompakter Abbildungen, *Math. Ann.* **154**, 427–431.

[177] **M. Lassonde** (1983). On the use of KKM multifunctions in fixed point theory and related topics, *J. Math. Anal. Appl.* **97**, 151–201.

[178] **C. H. Ling** (1965). Representation of associative functions, *Publ. Math. Debrecen* **12**, 189–212.

[179] **T.W. Ma** (1972). *Topological degree of set valued compact fields in locally convex spaces*, Diss. Math. 92, Warszawa.

[180] **J. Mach** (1972). Die Zulässigkeit und gewisse Eigenschaften der Funktionenräume $L_{\Phi,k}$ und L_{Φ}, *Ber. Ges. f. Math. u. Datenverarb. Bonn* Nr **61**.

[181] **K. Margolis** (1968). On some fixed point theorems in generalized complete metric spaces *Bull. Amer. Math. Soc.* **74**, 275–282.

[182] **B. Margolis, J. B. Diaz** (1968). A fixed point theorem of the alternative for contractions on a generalized complete metric space, *Bull. Amer. Math. Soc.* **74**, 305–309.

[183] **V. P. Maslov, S. N. Samborskij** (eds.) (1992). *Idempotent Analysis,* Advances in Soviet Mathematics 13, Providence, Rhode Island, Amer. Math. Soc..

[184] **R. D. Mauldin**, ed. (1981). *The Scottish Book: Mathematics from the Scottish Café*, Birkhäuser, Boston, MA.

[185] **A. Meir, E. Keeler** (1969). A theorem on contractive mappings, *J. Math. Anal. Appl.* **28**, 328–329.

[186] **K. Menger** (1942). Statistical metric, *Proc. Nat. Acad. USA* **28**, 535–537.

[187] **K. Menger, B. Schweizer and A. Sklar** (1959). On probabilistic metrics and numerical metrics with probability 1, *Czech. Math. Journal* **9**, 459–465.

[188] **R. Mesiar, M. Navara** (1999). Diagonals of continuous triangular norms, *Fuzzy Sets ann Systems* **104**, 35–41.

[189] **R. Mesiar, H. Thiele** (2000). On T-quantifiers and S-quantifiers, in: V.Novak, I.Perfilieva, eds., *Discovering the World with Fuzzy Logic*, Studies in Fuzziness and Soft Computing vol. 57, Physica-Verlag, Heidelberg, 310–326.

[190] **D. Miheţ** (1993). A fixed point theorem for mappings with contractive iterate in H-spaces, *Analele Univ. Timişoara* **31**, 2.

[191] **D. Miheţ** (1994a). B-contractions in σ- Menger spaces, *West Univ.of Timisoara* **109**.

[192] **D. Miheţ** (1994b). A type of contraction mappings in PM-spaces, *West Univ.of Timisoara* **114**.

[193] **D. Miheţ** (2001). *Inegalitatea triunghiului si puncte fixe in PM-spatii*, Doctoral Thesis, West Univ. of Timisoara,1997, in English.

[194] **D. Miheţ** (in print). On a theorem of O. Hadžić, *Univ. u Novom Sadu, Zb. Rad. Prirod.-Mat. Fak. Ser. Mat.*.

[195] **D. Miheţ, V. Radu** (1997). A fixed point theorem for mappings with contractive iterate, *Analele St. univ. Al. I. Cuza Iasi.*

[196] **Morrel, J. Nagata** (1978). Statistical metric spaces as related to topological spaces,*Gen. Top. Appl.* **9**, 233–237.

[197] **P. Mostert, A. Shields** (1957). On the structure of semigroups on a compact manifold with boundary, *Annals of Math.* **65**, 117–143.

[198] **R. Moynihan** (1978a). Infinite τ_T products of probability distribution functions, *J. Austral. Math Soc. Ser. A* **26**, 227–240.

[199] **R. Moynihan** (1978b). On τ_T semigroups of probability distribution functions II, *Aequationes Math.* **17**, 19–40.

[200] **R. Moynihan, B. Schweizer** (1979). Betweeness relations in PM-spaces, *Pacific J. Math.* **81**, 175–196.

[201] **T. Murofushi, M. Sugeno** (1991). Fuzzy t-conorm integrals with respect to fuzzy measures: generalization of Sugeno integral and Choquet integral, *Fuzzy Sets and Systems* **42**,57–71.

[202] **D. H. Mushtari, A.N. Serstnev** (1966). On methods of introducing a topology in random metric spaces, *Izv. vysh. Uch. Zav. Math.* **6**(55), 99–106.

[203] **M. Nagumo** (1951a). A theory of degree of mappings based on infinitesimal analysis, *Amer. J. Math.* **73**, 485–496.

[204] **M. Nagumo** (1951b). Degree of mappings in convex linear topological spaces, *Amer. J. Math.* **73**, 497–511.

[205] **S. B. Nadler** (1969). Multivalued contraction mappings, *Pacific J. Math.* **30**, 475–478.

[206] **S. A. Naimpally, K. L. Singh and J. H. M. Whitfield**(1983). Common fixed points for nonexpansive and asymptotically nonexpansive mappings, *Comm. Math. Univ. Carolinae* **24**, 2, 287–300.

[207] **R. B. Nelsen** (1999). *An Introduction to Copulas*, Lecture Notes in Statistics 139, Springer, New York.

[208] **J. V. Neumann** (1937). Über ein ökonomisches Gleichungssystem und eine Verallgemeinerung des Brouwerschen Fixpunktsatzes, *Ergebn. eines Math. Kolloq.* **8**, 73–83.

[209] **T. N. Nguyen, H. T. Le** (1994). No Roberts space is a counter-example to Schauder's conjecture, *Topology* **33**, 371–378.

[210] **T. N. Nguyen** (1996). The fixed point property for weakly admissible compact convex sets: searching for a solution to Schauder's conjecture, *Top. and its Appl.* **68**, 1–12.

[211] **R. D. Nussbaum** (1972). Degree theory for local condensing map, *J. Math. Anal. Appl.* **37**, 741–766.

[212] **V. V. Obukhovskij** (1971). On some fixed point principles for multivalued densifying operators, *Trudy mat. fak. VGU* **4**, 70–79 (in Russian).

[213] **W. Oettli, M. Théra** (1993). Equivalents of Ekeland's principle, *Bull. Austral. Math. Soc.* **48**, 385–392.

[214] **T. Okan** (1995). A generalized topological degree in admissible linear spaces, *Zeit. für Anal. Anwend.* **14**, No 3, 469–496.

[215] **A. B. Paalman-de Miranda** (1964). *Topological Semigroups*, Mathematical Centre Tracts 11,Matematisch Centrum Amsterdam.

[216] **D. V. Pai, P. Veeramani** (1980). Fixed point theorems for multimappings, *Yokohama Math. J.* **28**, 7–14.

[217] **E. Pap** (1995). *Null-Additive Set Functions*, Kluwer Academic Publishers, Dordrecht and Ister Science, Bratislava.

[218] **E. Pap** (1997a). Decomposable measures and nonlinear equations, *Fuzzy Sets and Systems* **92**, 205–222.

[219] **E. Pap** (1997b). Solving Nonlinear Equations by Non-Additive Measures, *Nonlinear Anal. Theory, Methods, Appl.* **30**, 31–40.

[220] **E. Pap** (1999). Applications of decomposable measures, in Handbook Mathematics of Fuzzy Sets-Logic, *Topology and Measure Theory* (Eds. U. Höhle, S.R. Rodabaugh), Kluwer Academic Publishers, 675–700.

[221] **E. Pap** (2000). Pseudo-convolution and Its Applications, (Eds. M. Grabisch, T. Murofushi, M. Sugeno) *Fuzzy Measures and Integrals, Theory and Applications*, Physica-Verlag (Springer-Verlag Company), 171–204.

[222] **E. Pap** (to appear). Pseudo-analysis and nonlinear equations, *Soft Computing.*

[223] **E. Pap** (to appear). Pseudo-additive measures and their applications, *Handbook of Measure Theory* (Editor E. Pap), North-Holland.

[224] **E. Pap, O. Hadžić, R. Mesiar** (1996). A fixed point theorem in probabilistic metric spaces and applications in fuzzy set theory, *J. Math. Anal. Appl.* **202**, 433–449.

[225] **E. Pap, I. Štajner** (1999). Generalized pseudo-convolution in the theory of probabilistic metric spaces, information, fuzzy numbers, optimization, system theory, *Fuzzy Sets and Systems* **102**, 393–415.

[226] **E. Pap, Dj. Takači, A. Takači** (1997). *Partial Differential Equations through Examples and Exercises*, Kluwer Academic Publishers, Dordrecht.

[227] **E. Părău, V. Radu** (1997). Some remarks on Tardiff's fixed point theorem on Menger spaces, *Portugaliae Mathematica* **54** (4), 431–440.

[228] **E. Părău, V. Radu** (to appear). On the triangle inequality and probabilistic contractions in PM-spaces, *Analele Univ.Bucuresti.*

[229] **S. Park** (1985). Some applications of Ekeland's variational principle to the fixed point theory, in *"Approximation Theory and Applications"* (S.P. Singh, Ed.,) Pitman, 159–172.

[230] **S. Park** (1998). A unified fixed point theory of multimaps on topological vector spaces, *J. Korean Math. Soc.* **35**, 4, 803–829.

[231] **S. Park** (2000). Fixed points, intersection theorems, variational inequalities, and equilibrium theorems, *Inter. J. Math. Math. Sci.* **24**,2, 73–93.

[232] **S. Park, B. G. Kang** (1993). Generalizations of the Ekeland type variational principles, *Chinese J. Math.* (Taiwan, R.O.C.) **21**, No. 4, 313–325.

[233] **W. T. Park, K. S. Park, Y. J. Cho** *Coincidence point theorems in probabilistic metric spaces*, Department of Mathematics, Gyeongsang National University, Jinju, Korea.

[234] **C. M. Pearcy, A. L. Shields** (1974). *A survey of the Lomonosov technique in the theory of invariant subspaces, Topics in operator theory*, Amer. Math. Soc. Surveys 13, Providence RI.

[235] **He Pei-jun** (1992). The variational principle in fuzzy metric spaces and its applications, *Fuzzy Sets and Systems* **45**, 389–394.

[236] **J. P. Penot** (1986). The drop theorem, the petal theorem and Ekeland's variational principle, *Nonlinear Anal.* **10**, 813–822.

[237] **W. V. Petryshyn, P. M. Fitzparick** (1973). Degree theory for noncompact multivalued vector fields, *Bull. Amer. Math. Soc.* **79**, 609–613.

[238] **V. Radu** (1982). On some metrics for the (ϵ, λ)-topology in PM-spaces, *Sem. Teor. Prob. Apl. Univ. Timisoara* **61**.

[239] **V. Radu** (1983a). On the t-norms of Hadžić type and fixed points in PM - spaces, *Univ. u Novom Sadu, Zb. Rad. Prirod.-Mat. Fak. Ser. Mat.* **13**, 81–85.

[240] **V. Radu** (1983b). A remark on contractions in Menger spaces, *Sem. Teor. Prob. Apl. Univ. Timisoara* **64**.

[241] **V. Radu** (1983c). On the t-norms of Hadžić type and fixed points in probabilistic metric spaces, *Sem. Teor. Prob. Apl. Univ. Timisoara, Apl.* **66**.

[242] **V. Radu** (1984a). On the t-norms of Hadžić type and locally convex random normed spaces, *Sem. Teor. Prob. Apl. Univ. Timisoara* **70**.

[243] **V. Radu** (1984b). On the t-norms with the fixed point property, *Sem. Teor. Prob. Apl. Univ. Timisoara* **72**.

[244] **V. Radu** (1984c). On the contraction principle in Menger spaces, *Analele Univ. Timisoara* **22**, fasc. 1-2, 83–88.

[245] **V. Radu** (1985a). On some contraction type mappings in Menger spaces, *Analele Univ. Timisoara* **23**, 61–65.

[246] **V. Radu** (1985b). A family of deterministic metrics on Menger spaces, *Sem. Teor. Prob. Apl. Univ. Timisoara* **78**.

[247] **V. Radu** (1987). On some fixed point theorems in PM - spaces, Lecture Notes in Mathematics, Springer Verlag, 1233, 125–133.

[248] **V. Radu** (1988). Deterministic metrics on Menger spaces and applications to fixed point theorems, *Babeş-Bolyai University, Fac. of Math. and Ph. Research Seminar, Preprint nr.2*, 163–166.

[249] **V. Radu** (1992). On a theorem of Hadžić and Budinčević, *Sem. Teor. Prob. Apl. Univ. Timisoara* **104**.

[250] **V. Radu** (1994).*Lectures on probabilistic analysis*, Surveys, Lectures Notes and Monographs Series on Probability, Statistics & Applied Mathematics, No 2, Universitatea de Vest din Timişoara.

[251] **V. Radu** (to appear). A fixed point principle in probabilistic metric spaces under Archimedean triangular norms, *Analele Univ. Timişoara.*

[252] **V. Radu, D. Miheţ** (1994). A fixed point principle in σ-Menger spaces, *Analele Univ. Timişoara* **32**, fasc. 2, 99–110.

[253] **E. Rakotch** (1962). A note on contractive mappings, *Proc. Amer. Math. Soc.* **13**, 459–465.

[254] **B.E. Rhoades, K.L. Singh and J.H.M. Whitfield** (1982). Fixed points for generalized nonexpansive mappings, *Comm. Math. Univ. Carolinae* **23**, 3, 443–451.

[255] **T. Riedrich** (1963). Die Räume $L^p(0,1)(0 < p < 1)$ sind zulässig, *Wiss. Z.TU, Dresden* **12**, 1149–1152.

[256] **T. Riedrich** (1964). Der Raum $S(0,1)$ ist zulässig, *Wiss. Z.TU Dresden* **13**, 1–6.

[257] **T. Riedrich** (1976). *Vorlesungen über nichtlineare Operatorengleichungen*, Teubner-Texte zur Mathematik, Leipzig.

[258] **E. Rothe** (1937). Zur theorie der topologischen Ordnung und der Vektorfelder in Banachschen Räumen, *Comp. Math.* **5**, 177–197.

[259] **L. Rüschendorf, B. Schweizer, M.D. Taylor (Eds.)** (1996). *Distributions with fixed marginals and related topics*, Lecture Notes Monograph Ser., Vol. 28, Inst. Math. Stat..

[260] **B. Sadovskij** (1972). Ultimately compact and condensing mappings, *Uspekhi Mat. Nauk.* **27**, 81–146.

[261] **J. Schauder** (1930). Der Fixpunktsatz in Funtionalräumen, *Studia Math.* **2**, 171–180.

[262] **B. Schweizer, H. Sherwood, R. M. Tardiff** (1988). Contractions on probabilistic metric spaces: Examples and counterexamples, *Stochastica* **12**-1, 5–17.

[263] **B. Schweizer, A. Sklar** (1958). Espaces Métriques Aléatoires, *Comptes Rendus Acad. Sci. Paris* **247**, 2092–2094.

[264] **B. Schweizer, A. Sklar** (1960). Statistical metric spaces, *Pacific J. Math.* **10**, 313–334.

[265] **B. Schweizer, A. Sklar** (1961). Associative functions and statistical triangle inequalities, *Publ. Math. Debrecen* **8**, 169–186.

[266] **B. Schweizer, A. Sklar** (1963). Associative functions and abstract semigroups, *Publ. Math. Debrecen* **10**, 69–81.

[267] **B. Schweizer, A. Sklar** (1969). Mesures aléatoires de l'information, *Comptes Rendus Acad. Sci. Paris* **269 A**, 721–723.

[268] **B. Schweizer, A. Sklar**(1983). *Probabilistic Metric Spaces*, Elsevier North-Holland, New York.

[269] **B. Schweizer, J. Smital** (1994). Measures of chaos and a spectral decomposition of dynamical systems on the interval, *Trans. Amer. Math. Soc.* **344**, 737–754.

[270] **B. Schweizer, A. Sklar, E. Thorp** (1960). The metrization of SM-spaces, *Pacific J. Math.* **10**,673–675.

[271] **V. M. Sehgal** (1966). *Some fixed point theorems in functional analysis and probability*, Ph. D. dissertation, Wayne State Univ..

[272] **V. M. Sehgal, A. T. Baharucha-Reid** (1972). Fixed points of contraction mappings on probabilistic metric spaces, *Math. Syst. Theory* **6**, 97–102.

[273] **A. N. Sherstnev** (1962). Random normed spaces: problems of completness, *Kazan. Gos. Univ. Učen. Zap.* **122**, 3–20.

[274] **A. N. Sherstnev** (1963). Sluchainie normirovanije prostranstva,*DAN* **149**, 280–283.

[275] **A.N. Sherstnev** (1964). On the probabilistic generalization of metric spaces, *Kazan Gos. Univ. Uch. Zap.* **124**, 3–11.

[276] **H. Sherwood** (1971). Complete probabilistic metric spaces, *Z. Wahr. verw. Geb.* **20**, 117–128.

[277] **H. Sherwood** (1976). A note on PM spaces determined by measure preserving transformations, *Z. Wahrsch. Verw. Geb.* **33**, 353–354.

[278] **H. Sherwood** (1984). Characterizing dominates on a family of triangular norms, *Aequationes Math.* **27**, 255–273.

[279] **H. Sherwood, and M.D. Taylor** (1974). Some PM structures on the set of distribution functions, *Rev. Roum. Math. Pures et Appl.* **19**, 1251-1260.

[280] **Zhang Shisheng** (1985). On the theory of probabilistic metric spaces with applications, *Acta Math. Sinica, New Series* **1**, No. 4, 366–377.

[281] **Zhang Shisheng, Chen Yuqing, Guo Jinli** (1991). Ekeland's variational principle and Caristi's fixed point theorem in probabilistic metric space, *Acta Mathematicae Applicatae Sinica* **7**, 2, 217–228.

[282] **R. Sine** (Editor) (1983). *Fixed Points and Nonexpansive Mappings*, American Mathematical Society, Contemporary Mathematics, Vol. 18.

[283] **S. L. Singh and R. Talwar** (1994). Coincidences and fixed points in probabilistic analysis, *Bull. Malaysian Math. Soc. (Second Series)* **17**, 29–43.

[284] **S. L. Singh, B. M. L. Tivari and V. K. Gupta** (1980). Common fixed point of commuting mappings in 2-metric spaces and applications, *Math. Nachr.* **95**, 293–297.

[285] **S. P. Singh, S. Thomeier and B. Watson** (Editors) (1983). *Topological Methods in Nonlinear Functional Analysis*, American Mathematical Society, Contemporary Mathematics, Vol. 21.

[286] **S. Singh, B. Watson and P. Srivastava** (1997). *Fixed Point Theory and Best Approximation: The KKM-map Principle*, Kluwer Academic Publishers, Dordrecht/Boston/London.

[287] **D. Smutná** (1998). On a peculiar t-norm, *BUSEFAL* **75**, 60–67.

[288] **F. Stenger** (1975). Computing the topological degree of a mapping in n-space, *Bull. Amer. Math. Soc.* **81**, 179–182.

[289] **R.R. Stevens** (1968). Metrically generated probabilistic metric spaces, *Fund. Math.* **61**, 259–269.

[290] **M. Stojaković** (1985). Fixed point theorems in probabilistic metric spaces, *Kobe J. Math.* **2**, 1–9.

[291] **M. Stojaković** (1986). A common fixed point theorem for the commuting mappings, *Indian J. Pure Appl. Math.* **17**, 4, 466–475.

[292] **M. Stojaković** (1987). Coincidence point theorems for multivalued mappings in Menger spaces, *J. Natur. Phys. Sci.* **1**, 53–62.

[293] **M. Stojaković** (1988a). Common fixed point theorems in complete metric and probabilistic metric spaces, *Bull. Austral. Math. Soc.* **36**, 73–88.

[294] **M. Stojaković** (1988b). On some classes of contraction mappings, *Math. Japonica* **33** (2), 311–318.

[295] **M. Stojaković, Z. Ovcin** (1994). Fixed point theorems and variational principle in fuzzy metric space, *Fuzzy Sets and Systems* **66**, 353–356.

[296] **M. Sugeno, T. Murofushi** (1987). Pseudo-additive measures and integrals. *J. Math. Anal. Appl.* **122**, 197–222.

[297] **W. Takahashi** (1970). A convexity in metric spaces and nonexpansive mappings I, *Kodai Math. Rep.* **22**, 142–149.

[298] **L. A. Tallman** (1977). Fixed points for condensing multifunctions in metric spaces with convex structures, *Kodai Math. Sem. Rep.* **29**, 62–70.

[299] **D. H. Tan** (1980). On the contraction principle in uniformizable spaces, *Acta Math. Vietnamica,* **5** , 88–99.

[300] **D. H. Tan** (1981). On probabilistic condensing mappings, *Rev. Roum. Math. Pures Appl.* **26**,10, 1305–1317.

[301] **D. H. Tan** (1982). A fixed point theorem for multivalued quasi-contractions in probabilistic metric spaces, *Univ. u Novom Sadu, Zb. Rad. Prirod.-Mat. Fak. Ser. Mat.* **12**, 43–54.

[302] **D. H. Tan** (1983). A note on probabilistic measure of noncompactness, *Rev. Roum. Math. Pures Appl.* **28**, 283–288.

[303] **D. H. Tan** (1985). Some remarks on probabilistic measures of noncompactness, *Rev. Roum. Math. Pures Appl.* **30**, 1, 43–47.

[304] **D. H. Tan** (1998). A classification of contractive mappings in probabilistic metric spaces, *Acta Mathematica Vietnamica,* **23** No 2, 295–302.

[305] **E. Tarafdar** (1982). On minimax principle and sets with convex sections, *Publ. Math. Debrecen* **29**, 219–226.

[306] **E. Tarafdar, T. Husain** (1978). Duality in fixed point theory of multivalued mappings with applications, *J. Math. Anal. Appl.* **63** (2), 371–376.

[307] **R. M. Tardiff** (1976). Topologies for probabilistic metric spaces, *Pacific J. Math.* **65**, 233–251.

[308] **R. M. Tardiff** (1980). On a functional inequality arising in the construction of the product of several metric spaces, *Aequat. Math.* **20**, 51–58.

[309] **R. M. Tardiff** (1984). On a generalized Minkowski inequality and its relation to dominates for t-norms, *Aequat. Math.* **27**, 308–316.

[310] **R. M. Tardiff** (1992). Contraction maps on probabilistic metric spaces, *J. Math. Anal. Appl.* **165**, 517–523.

[311] **W. W. Taylor** (1972). Fixed point theorems for nonexpansive mappings in locally convex spaces, *J. Math. Anal. Appl.* **49**, 162–173.

[312] **E. Thorp** (1960). The metrization of statistical metric spaces. *Pacific J. Math.* **10**, 673–675.

[313] **G. Trombetta** (2000). The measure of nonconvex total boundedness and of nonstrongly convex total boundedness for subsets of L_0, *Ann. Soc. Math. Pol. Ser. I Comm. Math.* **40**, 191–207.

[314] **G. Trombetta** (2001). A compact convex set non convexly totaly bounded, *Bull. Pol. Acad. Sci.* **49**, No 3 (in print).

[315] **G. Trombetta, H. Weber** (1986). The Hausdorff measure of noncompactness for balls of F-normed linear spaces and for subsets of L_0, *Bollettino U.M.I. Analisi Funzionale e Applicazioni, Serie VII,* **5-C** N1, 213–232.

[316] **A. Tychonoff** (1935). Ein Fixpunktsatz, *Math. Ann.* **111**, 767–776.

[317] **J. van der Bijl, K. P. Hart, J. van der Mill** (1992). Admissibility, homeomorphism extension and the AR-property in topological linear spaces, *Top. and its Appl.* **48**, 63–80.

[318] **P. Viceník** (1999a). Generated t-norms and the Archimedean property, *Proceedings EUFIT '99, Aachen.* (CD-Rom).

[319] **P. Viceník** (1999b). A note to a construction of t-norms based on pseudo-inverses of monotone functions, *Fuzzy Sets and Systems* **104**, 15–18.

[320] **A. Wald** (1943). On Statistical Generalization of Metric Spaces. *Proc. Nat. Acad. Sci. U.S.A.* **29**, 196–197.

[321] **S. Weber** (1984). $\perp$-decomposable measures and integrals for Archimedean t-conorm $\perp$, *J. Math. Anal. Appl.* **101**, 114–138.

[322] **R. Wegrzyk** (1982). *Fixed point theorems for multivalued functions and their applications to functional equations*, Dissert. Math. **201**.

[323] **Zeng Wenzhi** (1987). Probabilistic 2-metric spaces, *J. Math. Research Expo.* **2**, 241–245.

[324] **Ding Xie Ping** (1986). Common fixed points for nonexpansive type mappings in convex and probabilistic convex metric spaces, *Univ. u Novom Sadu, Zb. Rad. Prirod.-Mat. Fak. Ser. Mat.* **16**, 1, 73–84.

[325] **H. K. Xu** (2000). *Metric fixed point theory for multivalued mappings*, Diss. Math. 389, Warszawa.

[326] **L. A. Zadeh** (1965). Fuzzy sets, *Inform. and Control* **8**, 338–353.

[327] **G. X. Z. Yuan** (1998). *The study of minimax inequalities and applications to economies and variational Inequalities*, AMS Memoires 625.

[328] **S. S. Zhong** (1987). The equivalence between Ekeland's variational principle and Caristi's fixed point theorem, *Advan. Math.* **16**, 203–206.

[329] **C. K. Zhong** (1997). On generalized Ekeland's principle and minimization theorems, *J. Math. Anal. Appl.* **205**, 239–250.

[330] **J. Zhu, C. K. Zhong, G. P. Wang** (2001). Vector-valued variational principle in fuzzy metric space and its applications, *Fuzzy Sets and Systems* **119**, 343–354.

[331] **Liu Zuoshi, Chen Shaozhong** (1983). On fixed point theorems of random set-valued maps, *Kexue Tong bao* **28**, 4, 433–435.

List of Symbols

$\mathbb{N}$	set of natural numbers
$\mathbb{Z}$	set of integers
$\mathbb{Z}_+$	set of non-negative integers
$\mathbb{R}$	set of real numbers
$\mathbb{R}^n$	n-dimensional real Euclidean space
$\mathbb{C}$	set of complex numbers
A^c	complement of the set A
χ_A	characteristic function of the set A
$\cap$	intersection of sets
$\cup$	union of sets
$\setminus$	set difference
∂A	the boundary of the set A
$\mathcal{A}$	σ-algebra of subsets of a set Ω
2^S	powers set of a non-empty set S
2^S_{co}	the family of all non-empty and convex subsets of S
$\overline{A}$	closure of the set A
$\liminf$	limes inferior
$\limsup$	limes superior
$\mathbf{m}_0$	Lebesgue measure
$(\Omega, \mathcal{A}, P)$	the probability measure space
$\mathbf{S}$	triangular conorm
T	triangular norm
δ_T	the diagonal of the t-norm T
$\mathbf{s}$	an additive generator of the t-conorm S
$\mathbf{t}$	an additive generator of the t-norm T
θ	a multiplicative generator of the t-norm T
$T_{\mathbf{M}}$	the minimum t-norm
$T_{\mathbf{L}}$	the Lukasiewicz t-norm
$T_{\mathbf{P}}$	the product t-norm
$T_{\mathbf{D}}$	the drastic product t-norm
Fix (f)	the set of fixed points of the mapping f
Δ	the family of all distribution functions on $[-\infty, \infty]$
Δ^+	the family of all distance distribution functions

$\mathcal{D}^+$	the subfamily of Δ^+ with $\lim_{x\to\infty} F(x) = 1$
τ	triangle function
H_0	the Dirac distribution function
$\dim A$	dimension of a set A
Lin A	the linear hull of a set A
co	the convex hull of a set A
$\overline{\text{co}}$	the closed convex hull of a set A
$\mathcal{S}(Z)$	a family of non-empty subsets of the set Z
$\mathcal{R}(Z)$	the family of all non-empty, convex and closed subsets of TVS Z
diam A	the diameter of the set A
$CB(S)$	the family of all non-empty closed (in the (ε, λ)-topology) probabilistic bounded subsets of S
$C(S)$	the family of all non-empty closed subsets of S (in the (ε, λ)-topology)
$\deg^n(a, G, f)$	topological degree

Index